U.S. Marine Corps Wilderness Medicine Survival Course

Plus

Marine Scouting and Patrolling

Department of Defense

Pentagon
Publishing

UNITED STATES MARINE CORPS
Mountain Warfare Training Center
Bridgeport, California 93517-5001

WILDERNESS MEDICINE COURSE STUDENT HANDBOOK

TABLE OF CONTENTS

TOC

TABLE OF CONTENTS (CONT.)

UNITED STATES MARINE CORPS
Mountain Warfare Training Center
Bridgeport, California 93517-5001

FMST.07.18
04/02/02

STUDENT HANDOUT

MOUNTAIN SAFETY (SUMMER)

TERMINAL LEARNING OBJECTIVE: Given a unit in a wilderness environment and necessary equipment and supplies, apply the principles of mountain safety to prevent death or injury per the reference. (FMST.07.18)

ENABLING LEARNING OBJECTIVE: Without the aid of references and given the acronym "BE SAFE MARINE", list in writing the 12 principles of mountain safety, in accordance with the references. (FMST.07.18a)

OUTLINE

1. **PLANNING AND PREPARATION:** (FMST.07.18a) As in any military operation, planning and preparation constitute the keys to success. The following principles will help the leader conduct a safe and efficient operation in any type of mountainous environment. We find this principle in the acronym "BE SAFE MARINE". Remember the key, think about what each letter means and apply this in any type of environment.

B - Be aware of the group's ability.

E - Evaluate terrain and weather constantly.

S - Stay as a group.

A - Appreciate time requirements.

F - Find shelter before storms if required.

E - Eat plenty and drink lots of liquids.

M - Maintain proper clothing and equipment.

1

A - Ask locals about conditions.

R - Remember to keep calm and think.

I - Insist on emergency rations and kits.

N - Never forget accident procedures.

E - Energy is saved when warm and dry.

a. Be Aware of the Group's Ability. It is essential that the leader evaluates the individual abilities of his men and uses this as the basis for his planning. In his evaluation, the leader must include the group's overall physical conditioning, and the consideration of change in climate and how long the unit has had to acclimatize.

(1) Mental attitude of your group. Is morale high? How much tactical training has the group had in a particular type of terrain?

(2) Technical aspect of your group. Have they been on skis, snowshoes, etc.?

(3) Individual skills. At this point, you must choose who is most proficient at the individual skills that will be required for your mission, navigation techniques, security, call for fire, rope installations, track plans, bivouac site selection, skijoring, etc.

b. Evaluate Terrain and Weather Constantly.

(1) Terrain. During the planning stages of your mission, the leader must absorb as much information as possible on the surrounding terrain and key terrain features involved in your area of operation. Considerations to any obstacles must be clearly planned for. Will you need such things as fixed ropes, rope bridges, climbing gear, etc?

(a) Stress careful movement in particularly dangerous areas, such as loose rock and steep terrain.

(b) Always know your position. Knowing where you are on your planned route is important.

(2) Weather. Mountain weather can be severe and variable. Drastic weather changes can occur in the space of a few hours with the onset of violent storms, reduced visibility, and extreme changes. In addition to obtaining current weather data, the leader must plan for the unexpected "worst case". During an operation he must diagnose weather signs continually to be able to foresee possible weather changes.

(a) Constantly evaluate the conditions. Under certain conditions it may be advisable to reevaluate your capabilities. Pushing ahead with a closed mind could spell disaster for the mission and the unit.

c. Stay as a Group. Individuals acting on their own are at a great disadvantage in this environment.

(1) Give the unit adequate rest halts based upon the terrain and elevation, physical abilities of the unit, combat load and mission requirements.

(2) Remember to use the buddy system in your group.

(3) Maintain a steady pace so that it will allow accomplishment of the mission when all members of the unit reach the objective area.

d. Appreciate Time Requirements. Efficient use of available time is vital. The leader must make an accurate estimate of the time required for his operation based on terrain, weather, unit size, abilities, and on the enemy situation. This estimate must take into account the possibility of unexpected emergencies and allow sufficient leeway to make unplanned bivouacs in severe conditions.

(1) Time-Distance Formula (TDF). This formula is designed to be a guideline and should not be considered as the exact amount of time required for your movement. Furthermore, this formula is for use in ideal conditions:

3 km/mph + 1 hour for every 300 meters ascent; and/or + 1 hour for every 800 meters descent.

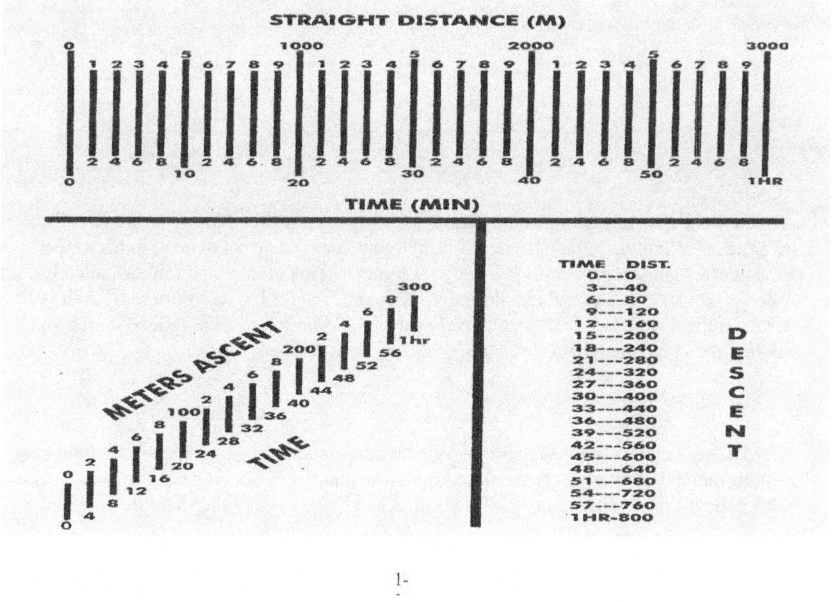

(2) Route Planning. Route cards are not to be used in place of an overlay, but as a tool to be used in route planning. Overlays/Route cards should contain the following information at the minimum:

Unit Designation:

Unit Commander.
Number of personnel.
Inclusive dates and times of movement.
Grid coordinates of each checkpoint and bivouac.
Map references.
Azimuth and distances for each leg.
Elevation gain/loss per leg.
Description of the ground.
ETA and ETD.

ROUTE CARD

UNIT I.D.		UNIT COMMANDER		NUMBER OF PERSONNEL		DATE AND TIME	MAP REFERENCE
LEG	AZM	DIST	GRID	ETA	ETD	ELEVATION	DESCRIPTION

(3) As in any military operation, route planning and execution are of vital importance. Prior to departure, the unit commander must submit a route card or patrol overlay to his higher headquarters and keeps a duplicate copy for himself. This preplanned route should be followed as closely as possible, taking into account changes based on the tactical situation. In non-tactical situations, the preplanned route should be followed to reduce search and rescue time in an emergency situation.

e. Find Shelter before Storms if required. Under certain conditions, inclement weather can provide tactical advantages to the thinking unit commander, but by the same token it can reduce the efficiency of a unit to nil if an incorrect evaluation of the situation is made.

(1) Bivouac. If the group decision is to bivouac, then it's vital that we know the principles for an unplanned bivouac.

(a) Unplanned bivouac. The principles and techniques discussed here apply both to unplanned and tolerated bivouacs. In any survival situation, especially in a mountainous environment, the most immediate danger is from exposure to the elements. Being lost will not directly kill an individual. Starvation takes time, but hypothermia can manifest itself in a matter of hours resulting in death. Adhering to the following principles will give an individual the best chance to spend a relatively safe bivouac with the prospect of continued effort toward mission accomplishment.

1. Make shelter. The requirements for expedient shelters and the building procedures will be covered in another section. The basic requirement for protection from the elements is essential.

2. Keep warm. The retention of body heat is of vital importance; any action in which body heat is lost should be avoided. The following points should be considered:

a. Adequate shelter.

b. Insulation from the ground using branches, a rucksack, etc.

c. Wear extra clothing.

d. Use extra equipment for insulation.

e. Produce external heat while trying to conserve fuel for future use.

3. Keep dry. Being wet causes the loss of body heat 24 times faster than when dry. Adequate protection from the elements is of prime importance to prevent the onset of hypothermia.

f. Eat Properly and Drink Plenty of Fluids

(1) Food. The human body can be compared to a furnace, which runs on food to produce energy (warmth). By planning the consumption of food to suit the specific situation, adequate nutrition and extra warmth can be supplied.

(2) Water. The intake of adequate amounts of water will maintain the body in proper working order. Danger from dehydration is as high in mountain regions as in hot dry areas. Loss of liquids is easily seen and felt in hot climates; whereas in the

mountains, the loss of body fluids is much less noticeable. High water intake, at least 6 quarts per day when in bivouac, 8 quarts per day when active, will help to prevent dehydration.

g. Maintain Proper Clothing and Equipment

(1) Clothing

(a) Our clothing has to perform an important function in our mission; therefore, when choosing our clothing we have to take into account some essential requirements.

1. Protection against wind and rain.

2. Layered and easily adjustable.

3. Lightweight and durable.

(b) To help us remember how to maintain and wear our clothing, we use the acronym, "COLD".

C - Keep clothing Clean.

O - Avoid Overheating.

L - Wear clothing loose and in Layers.

D - Keep clothing Dry.

NOTE: In the mountains a man should never be separated from his gear. Here are some basic and essential items that should be considered during your planning stage.

(2) Required equipment:

(a) A daypack should always be carried per squad and one Marine should carry a combat load.

(b) Map and compass. Every individual in a leadership position and his assistant should carry a map and compass. The maps should be weatherproofed and extra maps should be distributed throughout the unit.

(c) Repair kit. This kit should include those items necessary to do emergency repairs on your equipment.

(d) Survival Kit. Always carried on your person. The contents of a survival kit will be covered in another period of instruction dealing specifically with survival kits.

h. Ask Locals About Conditions. An often-overlooked source of information is the indigenous population of an area. Local weather patterns, rockslide areas, watering points, and normal routes can all be obtained by careful questioning. The leader must obtain current information of the actual conditions along his intended route. Of particular importance are recent precipitation and enemy sightings.

i. Remember to Keep Calm and Think.

(1) Emergency situation. Having recognized that you are lost and that an emergency situation exists, the following principles should be followed:

(a) Keep calm and do not panic. At this point you must make every effort to conserve body heat and energy.

(b) Think. When an individual is cold, tired, hungry or frightened he must force himself to organize his thoughts into a logical sequence.

(c) The group must try to help itself by either finding the way back to safety or by preparing shelters and procuring food.

(d) Above all else, the group must act as a tight-knit unit. In emergency situations, individual dissension can cause a total loss of control and unit strength.

(2) If the decision is reached that the group should seek its way back to safety, several possibilities exist. In most situations, the safest approach will be to retrace the route to the last known point and continue from there. The other course of action is to get a group consensus on the present location and send out a small search party to locate a known point. This party must ensure that they mark their trail adequately to return to the group. If all attempts at finding a way back to known terrain fails, a definite survival situation exists and actions discussed later in this section must be instituted.

j. Insist on Emergency Rations and Kits. Just like what was covered in the SUMMER MOUNTAIN WARFIGHTING LOAD REQUIREMENTS class, survival rations and a survival kit should always be carried.

k. Never Forget Accident/Emergency Procedures.

(1) Causes of accidents. The general procedures used to handle accidents differ little in this environment, but several distinct points should be kept in mind. The most frequent causes of accidents are as follows:

(a) Overestimation of physical and technical abilities.

(b) Carelessness.

(c) General lack of observation of one's surroundings.

(d) Lack of knowledge and experience by leaders.

(e) The failure to act as a group.

(f) Underestimation of time requirements to move through mountainous terrain and underestimation of the terrain itself.

(2) Preventive measures. The only truly effective preventive measures for the above lie in the education and experience of leaders at all levels. Too often, leaders sit by watching during training and as a result have no concept of the requirements involved in the mountainous environment. Only by active involvement can a leader gain the knowledge and experience needed to effectively lead in this environment.

(3) General procedures for handling an accident. These require only a good dose of common sense as outlined below.

(a) Perform basic first aid.

(b) Protect the patient from the elements to include insulation on top and bottom.

(c) Evacuate if necessary.

(d) Send for help if required. If possible, never send a man for help alone.

(e) Send the following information regarding the accident:

1. Time of accident.
2. Nature and location of accident.
3. Number injured.
4. Best approach route to accident scene.

(4) If one man of a two-man team is injured, the injured man must be given all available aid prior to going for help. If the injured man is unconscious, he should be placed in all available clothing and sleeping gear and anchored if on steep terrain. A note explaining the circumstances, and reassuring him, should be left in a conspicuous spot. This note must also contain the following information:

(a) When you expect to return.

(b) Where you went.

(c) What you did before you left (medication, etc.).

(5) International distress signal (whistle):

(a) Six short blasts in 1 minute from person requesting help.

(b) The return signal is three blasts in 1 minute from the respondent.

(6) Other methods if help is required:

(a) Red pyrotechnics.

(b) SOS, (... --- . . .).

(c) “Mayday" by voice communications.

l. Energy is saved when warm and Dry. With the previous 11 principles in mind this one should fall right into place. Save your heat and energy by following these steps:

(1) Dress properly.

(2) Eat properly.

(3) Drink properly.

(4) Ensure shelter meets criteria.

(5) Produce external heat (fires, stove, extra clothing, etc.) to save body heat and energy for future use.

(6) Don't lose body heat by getting wet.

UNITED STATES MARINE CORPS
Mountain Warfare Training Center
Bridgeport, California 93517-5001

FMST.07.01
03/08/02

STUDENT HANDOUT

NUTRITION

TERMINAL LEARNING OBJECTIVE: Given a unit in a mountainous environment, and the necessary equipment and supplies, apply the principles of nutrition in a mountainous environment to prevent death or injury per the reference. (FMST.07.01)

ENABLING LEARNING OBJECTIVES:

(1) Without the aid of references and from a given list, choose the correct caloric intake needed per day in a mountainous environment, in accordance with the references. (FMST.07.01a)

(2) Without the aid of references and from a given list, choose which type of diet increases your tolerance to cold, in accordance with the references. (FMST.07.01b)

(3) Without the aid of references and from a given list, choose one of the preventive measures of excessive water loss in a mountainous terrain, in accordance with the references. (FMST.07.01c)

OUTLINE.

1. **STARVATION FACTORS.** Factors that combine to result in "Near Starvation" during military operations in any mountainous environment:

a. Difficulty in carrying sufficient food.

b. Lack of time for preparing meals and beverages.

c. Loss of appetite and low food palatability resulting from hypoxia/hypothermia.

d. Fat intolerance resulting from altered metabolism.

2. **PHYSIOLOGICAL ASPECT TO COLD STRESS.**

a. Man's adaptation to cold is essentially behavioral. Shelters and clothing are utilized to provide a micro-environment, which will allow survival. Body composition adjustments such as increases in body weight and fat are also examples of the body's adaptation to the cold. The winter months lend themselves to greater periods of inactivity, leading to these changes.

b. In the Falkland War, constipation was a major concern. Dehydration was found to be the culprit in the majority of cases, but diet had a significant role as well. MRE's are low in fiber, containing less than 10 grams. Man needs at least 31 grams of fiber per day to ease passage of stool. Constipation can be a debilitating condition, affecting overall unit readiness.

(1) Low fiber diets can lead to diverticulosis, diverticulitis and Irritable Bowel Syndrome (IBS). IBS flare-ups tend to occur during upsetting situations and emotional stress. Educate your troops on the importance of fiber, especially in a cold weather environment.

(2) Fiber is supplied by foods derived from plants. The best sources are:

(a) Whole grain cereals.

(b) Whole wheat breads.

(c) Fresh and dried fruits.

NOTE: Vegetables also help prevent constipation, but the fiber from whole grain cereals and whole wheat bread is more effective than vegetables.

3. **CALORIC REQUIREMENTS.** (FMST.07.01a) Recommended caloric intake (Naval Medical Command Instruction 10110.1) for troops operating in a cold weather environment is 4,500 calories per day.

a. Energy Expenditure. Military operations in a cold weather environment usually involve strenuous activity. Walking on snow expends roughly twice the energy that walking on hard ground does. If troops are unfamiliar with personal over-snow mobility equipment, i.e. snowshoes and skis, awkward movement and thus wasted energy will result.

(1) The heavy clothing necessary for functioning in the cold increases energy expenditures by 5-15%. You will have the opportunity to experience this effect in the upcoming field evolution.

(2) Properly clothed, experienced individuals should have negligible heat loss from radiation, conduction, and convection. However, losses associated with respiration and evaporation can be significant.

(3) In intense cold, particularly with high winds, heat loss to the environment may be inevitable.

4. **NUTRITION IN A HIGH ALTITUDE ENVIRONMENT.** A diet high in carbohydrates and high in fat is the most beneficial, while a high protein diet is generally the least desirable.

a. The marked weight loss seen in individuals at high altitude is largely due to increased metabolic demands and anorexia secondary to hypoxia. Total caloric intake commonly does not meet energy demands.

b. Altered metabolic processes that have been reported at high altitude include delayed gastric emptying, intestinal malabsorption and decreased gastric enzyme secretion.

5. **BEST DIET FOR COLD WEATHER**. (FMST.07.01b)

a. High carbohydrate and high fat foods should be included rather than high protein foods.

b. Small meals (snacks) spaced at time intervals of two hours or less should be consumed rather than the usual three meals-a-day.

c. During a typical eight-hour day of cold exposure, three high-fat (unsaturated fat) meals providing 60% of the day's calories, coupled with two normally scheduled meals (breakfast and dinner) should provide good cold tolerance.

6. **BASIC CONSTITUENTS OF FOOD**. The three basic constituents of food are carbohydrates, fats and proteins.

a. Carbohydrates. Also known as the quick energy foods that provide energy to produce heat with the byproducts of carbon dioxide and water when oxidized by the body.

(1) Carbohydrates are present in food mainly as sugars and starches and are broken down during digestion into simple sugars that are converted into glucose.

(2) Carbohydrates are stored as glycogen in the liver and muscles and can be broken down quickly into glucose to provide a rapid source of energy. However, these stores are not large and are markedly depleted by fasting for as little as 24 hours.

(3) Eating a high carbohydrate diet for several days will double one's glycogen stores and can increase endurance by up to three times compared to the ordinary balanced diet.

(4) The main carbohydrate food sources are: CARBOHYDRATE SOURCES

(a) Fruits.

(b) Vegetables.

(c) Cereals.

(d) Sugar.

b. Fats: In the body, fat serves as the main storage form of energy. One gram of fat has the equivalent of 9.3 Cal/gm as opposed to 4.1 Cal/gm for carbohydrates and proteins. They also produce energy, heat, CO_2 and H_2O.

(1) In cases of starvation, body fat tends to be broken down into acidic compounds. If these compounds accumulate in the blood faster than they can be burned, they cause the body tissues and blood to become excessively acidic (acidosis).

(2) During brief physical activity, energy is derived equally from fat and carbohydrates, but as the duration of the activity lengthens, the percent of energy supplied by fat increases as carbohydrate energy is quickly depleted.

(a) Fat sources:

1. Butter.
2. Lard.
3. Cooking oil.
4. Mayonnaise.
5. Ice cream.

FAT SOURCES

(4) Fat is also found in lesser amounts in:

(a) Dairy products. (d) Nuts.

(b) Meat. (e) Vegetables.

(c) Eggs. (f) Cereals.

c. Proteins: Protein is a reparative food of complicated molecules composed of chains of amino acids.

(1) The protein structure is made up of 25 different amino acids. Eleven of these are important for growth while eight are essential for normal tissue maintenance.

(2) Common protein food sources are:

PROTEIN SOURCES

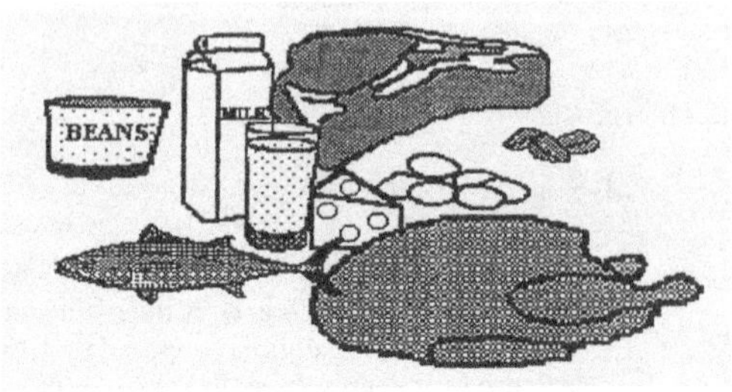

(a) Eggs.

(b) Dairy products.

(c) Meat.

(d) Poultry.

(e) Fish.

(f) Legumes (peas and beans, nuts and cereals with animal products, especially eggs, being better sources of complete protein).

(3) A pure protein diet may cause fatalities in 3-8 weeks from "rabbit starvation". This refers to those who have tried to live on relatively fat-free rabbit meat only. Fat must be a part of a normal diet.

7. **<u>PERCENTAGE OF FATS, PROTEINS AND CARBOHYDRATES IN A SUMMER MOUNTAINOUS ENVIRONMENT</u>**.

a. Requirements:

(1) 20% Protein.

(2) 35% Fats.

(3) 45% Carbohydrates.

b. When troops move to higher altitudes and/or are doing strenuous activity, carbohydrates should definitely be increased to at least 70 - 80% of energy intake. (Experienced mountaineers find carbohydrates more palatable at altitude).

8. **<u>VITAMINS</u>.** Vitamins are essential to the metabolic functioning of the body and are even more important in a cold weather environment because of an increase in metabolism and stress. Because the body cannot make vitamins, they are supplied in the food we consume. A daily multivitamin tablet supplement is recommended in a cold weather environment, but mega doses (more than 10 times the RDA) may be harmful.

a. Water-soluble vitamins are not stored in the body and need to be consumed daily. They are not stable to cooking and oxidation.

(1) B complex, Niacin, Pantothenic acid, Biotin, Choline, Folic Acid and C. (Note: Vitamin C facilitates absorption of iron from food.)

b. Fat-soluble vitamins can be stored in the body and are more stable to cooking and oxidation.

(1) A, D, E and K.

9. **<u>FOODS THAT MAKE YOU FEEL WARM</u>.**

a. Hot Beverages. Increases peripheral vasodilatation.

b. Adequate Dietary Iron. Mild iron deficiency increases chilling, thus decreasing cold tolerance.

c. Fat in the Diet. Fat slows gastric emptying and anecdotal evidence suggests less chilling and more restful sleep at night.

d. Carbohydrates. Keeps energy levels up. "Tired" is associated with "Cold".

e. Spices. Red pepper (Tabasco sauce) and monosodium glutamate increase peripheral vasodilatation, while capsicum and capsaicin, which are found in Cayenne, produces the same result when applied externally to the feet.

f. Water. Dehydration leads to lethargy, which tends to reduce food intake resulting in less heat production.

10. **<u>INCREASING COLD TOLERANCE</u>.**

a. Frequent Feedings. A high carbohydrate diet is superior to a high protein diet in increasing tolerance to cold.

(1) Diets deficient in both calories and protein reduce physical work capacity in the cold. The provision of both components is required to maintain work capacity.

11. **<u>HIGH FAT DIETS AND SUSCEPTIBILITY TO FROSTBITE</u>.**

a. High fat diets consumed at frequent (2 hour) intervals, maintain body temperatures better than high carbohydrate diets, but not if normal (4 hour) intervals are observed between meals.

b. Fat enriched diets lead to a propensity toward increased arterial thrombosis and increased platelet aggregation, which translates into increased potential for thrombosis.

c. Frostbite injuries are associated with disruption of the microcirculation, progressive micro vascular thrombosis, alteration of platelet function, and fibrin deposition in affected areas.

d. Therefore, diets high in (saturated) fat may predispose to frostbite.

CAUTION! A high-fat diet is important in a high-altitude, cold-weather environment. The benefits associated with the high-energy source called Fat far outweigh the risks of arteriosclerosis from the diet itself. A high-fat diet is appropriate in this specific environment during the time frame of cold, high-altitude exposure. In garrison, at sea level, a high fat diet should be avoided in order to maintain a healthy lifestyle.

12. **IRON DEFICIENCY AND THERMOREGULATION**.

a. Iron deficiency results in 9% less heat production.

b. Consuming only 1/3 of the RDA for iron results in a 29% greater heat loss during cold exposure. "The furnace is going, but the windows are open" is a phrase that helps describe this effect.

c. Iron absorption is increased by consuming iron rich foods:

(1) Lean meat.

(2) Poultry.

(3) Beets.

(4) Beans.

(5) Green leafy vegetables.

(6) Fish.

(7) Vitamin C.

d. Bottom line, IRON improves the body's response to cold.

13. **WATER REQUIREMENTS**.

a. On average, water makes up about 60% of the one's entire body weight. The average sedentary person excretes about 2.7 quarts of water each day. Of this total, 1.3 quarts are lost in the urine, 1.1 quarts through the skin and lungs and 10 ounces in the stool.

b. At high altitude, and during strenuous exercise, the amount of water lost through the skin and lungs due to the cold, dry air increases greatly.

c. Cold weather decreases the sense of thirst, which may lead to a state of chronic, mild dehydration. At temperatures below freezing and at elevations above the snow line, the lack of liquid water and the time and effort required to melt snow compound the problem.

d. Prevention of excessive water loss: (FMST.07.01c)

(1) Force fluids. Drink 6-8 quarts of water daily.

(2) Observe for dark urine and other symptoms of dehydration. (Headache, nausea/vomiting, dizziness, constipation, etc.)

(3) Limit caffeine to 1-2 cups per day.

(4) Do not rely on thirst as an indicator of hydration status.

NOTE: Eating snow or ice will cause extreme inflammation of the mucous membranes of the mouth after two days. Warm canteens of water placed in your sleeping bag will give you water to drink in the morning. They will initially act as hot water bottles. (Place them upright, as they tend to leak ...keep bag dry.)

e. Health Problems Associated with Excessive Water Loss:

(1) Dehydration.

(2) Constipation.

(3) Hypothermia.

(4) Frostbite.

(5) Heat Exhaustion.

f. The importance of hydration is essential while in a cold weather environment. If these guidelines are followed then complications arising from poor hydration should be prevented.

14. **SPECIAL PURPOSE MILITARY RATIONS**.

RATION	APPLICATION
Ration, Cold Weather, RCW.	Cold weather operations where weight and freezing are a problem.
MRE Supplement Packet.	Cold weather calorie supplements for the MRE.
Ration, Light Weight, RLW.	Long range patrols, reconnaissance, and assault.
Nutrition Sustainment Module, NSM.	Calorie dense food, bar ration for mission lasting 3-5 days where soldier must carry his own food.

Family of Operational Rations, FOR.	Modular rations that allow an optimum mix of meals for tactical operations under all climatic conditions.

15. **FIELD STRIPPING OF MRE'S.**

a. Tips on how to conserve space and weight:

(1) Remove the entire contents from the plastic wrapper.

(2) Remove the cardboard, and all materials that you will not use.

(3) Be sure to place the toilet paper and other materials from the accessory pack either back in your pocket or into the brown packaging.

(4) Carefully repack the stripped meal back into the wrapper and seal with tape.

(5) Your meal will be half its original size and two-thirds its original weight.

UNITED STATES MARINE CORPS
Mountain Warfare Training Center
Bridgeport, California 93517-5001

FMST.07.09
03/31/02

WILDERNESS PATIENT ASSESSMENT

TERMINAL LEARNING OBJECTIVE: Given a casualty in a cold weather environment and the necessary equipment and supplies, perform a patient assessment in a mountainous environment to determine the nature and extent of injury per the reference. (FMST.07.09)

ENABLING LEARNING OBJECTIVES:

(1) Without the aid of references, from a given list, choose when CPR may be legally discontinued, in accordance with the references. (FMST.07.09a)

(2) Without the aid of references and from a given list, choose which situations CPR may not be started in a remote wilderness environment, in accordance with the references. (FMST.07.09b)

(3) Without the aid of references, in the proper sequence, correctly perform a patient assessment in the proper sequence in accordance with the references. (FMST.07.09c)

OUTLINE

1. **INTRODUCTION.**

    a. An organized prioritized assessment of a wilderness trauma victim is essential to proper care. A modified Advanced Trauma Life Support (ATLS) assessment system will be introduced in this chapter.

2. **TRIAGE.**

    a. Two situations of triage may exist in a wilderness environment.

(1) The number of patients and the severity of their injuries do not exceed the ability of the care-giver and his/her equipment to render care. In this situation, treat life-threatening problems first.

3

(2) The number of patients and the severity of their injuries exceed the capacity of the care-giver and his/her equipment to render care. In this situation, those with the greatest chance of survival are treated first

3. **CPR: URBAN vs. WILDERNESS SPECIFICS.**

a. In a remote wilderness situation where advanced life support assistance may be hours or days away, CPR probably has no chance of success. In addition, administering CPR under wilderness conditions may put group members in serious danger, due to physical hazards and the effects of exhaustion.

b. Based on the above, several suggestions regarding CPR in remote wilderness areas can be made. Even though the following suggestions seem reasonable, they should be viewed as recommendations, without legal force, and not doctrine. At the present time, a rescuer who begins CPR is legally obligated to continue unless or until one of the following conditions is fulfilled:

(1) CPR may be discontinued legally when: (FMST.07.09a)

(a) The patient revives.

(b) ACLS is started by higher authority.

(c) Relieved by another rescuer.

(d) Too exhausted to continue.

(e) Pronounced dead by proper authority.

(2) In a remote wilderness environment, CPR may not be started under one of the following situations: (FMST.07.09b)

(a) The patient is in cardiac arrest caused by trauma.

(b) The patient is a drowning victim who has been immersed for over an hour.

(c) The patient is in cardiac arrest and advanced life support is more than an hour away, especially if the patient must be carried out.

(d) The patient's cardiac arrest was unwitnessed and the time of onset is unknown.

(e) The patient appears to be dead, based on rigor mortis, lethal injuries, or a body temperature below 60°F.

(f) Giving CPR would be hazardous to rescuers.

(3) After 30 minutes of CPR with no signs of life, further CPR is probably useless and may reasonably be discontinued. Administering CPR to a victim who is being evacuated by litter or sled is very difficult, if not impossible. Unless an ambulance or helicopter can be brought in rapidly, chances of survival are slim.

(4) Exceptions include patients in cardiac arrest caused by hypothermia; patients with another illness or injury complicated by hypothermia (avalanche burial and near drowning in cold water), and patients in cardiac or respiratory arrest caused by lightning injury. In these cases, the outlook is probably more favorable and CPR should be aggressively administered.

4. **PATIENT ASSESSMENT** (FMST.07.09c). There are different considerations to think about in a wilderness setting. The environment plays a major role in treatment, care and even the extraction of the patient. Not only do we have to be mindful of the enemy situation, by keeping tactical. We also have to take into consideration the elements. Snow, rain ice, and the terrain can and will pose a great danger to the patient and the rescuers. We must protect the patient from the elements. We can do this by placing an insulating barrier between the patient and the ground. Also exposing one body part at a time and recovering as you move on will help to fight hypothermia.

A) SCENE SIZE-UP.

1) Determine scene safety:

a. Rescuers: Your safety is priority. You are no good to the team as a casualty.

b. Patient: Protect the patient from further injuries, enemy, and the elements.

c. Bystanders: They can cause many accidents. Limit the personnel on scene.

d. Tactical: Have the Marines maintain security and find cover for the rescuer and the patient.

e. Fire: Forest fires can pose a formidable threat.

f. Electrical: Make sure to shut off electricity first. Rushing in injures many rescuers.

g. Special Rescues: Swift water and high angle rescues are very dangerous and only trained professionals should attempt the rescue.

2) Determine the mechanism of injury / nature of illness:

a. Obtain SAMPLE and Mechanism of Injury (MOI) history from patient, witnesses, or other caregivers.

S = Signs and/or Symptoms.
A = Allergies.
M = Medications.
P = Past Illnesses.
L = Last Meal.
E = Events of Injury.

MOI = What, when, where, and how injury occurred.

b. Environment: Survey the area to get clues. (i.e. falling rocks, next to a cliff, broken branches, etc…)

c. Bystanders: Ask them what they saw.

3) Determine number of patients:

a. Survey area for number of patients.

b. Look for the hidden patients. Patients could be buried in rock, sand or snow. They also could be spread out in bushes and trees. One of the most over looked places is in the bystanders. Be sure to have people search and identify all patients.

4) Request additional help / Organize help if needed:

a. Medical assistance: Due to the lack of medical personnel and equipment, you might need to request additional help.

b. Technical assistance: When the scope of care is over your ability or higher than the casevac corpsman can provide, you should request higher trained personnel on the casevac bird.

c. Organize help at the scene: Use the people around you to help out.

5) Consider stabilization of the spine:

a. If indicated from: Falls of 10 feet or greater, high speed collisions, unwitnessed loss of consciousness, etc…

b. Designate an assistant: Use the bystanders / marines. Tell them exactly what you want them to do and show them how to do it.

c. Improvise control: If you are alone you can use the patient's boots, sand bags, rocks, etc as forms of temporary control.

B) INITIAL ASSESSMENT.

1) Make general impressions of the patient: What can you obtain from the patient's condition as you walk up? Is the patient talking, or unconscious?

2) Determine responsiveness / level of consciousness.

a) Use AVPU.

b) Assess pupils for size, equality, and reactivity.

3) Determine chief complaint / Apparent life threats.

4) Assess Airway and breathing.

a) Assessment of airway.

- Perform chin lift or jaw thrust maneuver, with C-Spine precautions.
- Clear the airway of foreign bodies.
- Insert an oropharyngeal or nasopharyngeal airway.
- Use of other airways as needed, and as equipment and training allow.

b) Assessment of breathing.

- Expose the neck and chest, assure immobilization of the head and neck.
- Determine the rate and depth of respirations.
- Auscultate the chest bilaterally. (Ear to chest wall, if necessary)
- Check for symmetry. (Rise and fall of the chest)

c) Management.

- Ventilate. (Face-mask, bag-valve-mask, if available)
- Alleviate tension pneumothorax.
- Seal open pneumothorax.
- Administer high concentration oxygen, if available.

5) Assess Circulation.

a) Assess and control major bleeding.

- Identify and control major bleeding.
- Pulse: Quality, rate, and regularity.
- Pulse Pressure: Check radial, then carotid.
- Skin color: cyanosis, pallor and temperature.

b) Management.

- Insert two large bore IV's with saline locks. IV fluid replacement, if indicated.
- Place in shock position.
- Prevent hypothermia.

6) Identify priority patients and make transportation decisions.

a) Triage patients at this point.

b) Make transportation decisions. In this type of environment it is possible that a casevac vehicle might not gain access to your position. At this point you will have to organize how the patient will be transported out. Also, you might have to set up relay points and rewarming stations.

C) DETAILED PHYSICAL EXAM: This is the head to toe assessment. We will find and treat all secondary injuries at this time. When performing this exam in the wilderness, you will need to expose one body part at a time and cover it back up as you go. When assessing the patient, you are looking for deformities, contusions, abrasions, penetrations/punctures, bruising, tenderness, lacerations, and swelling (DCAP-BTLS).

1) Head.

a. Assessment.

- Inspect and palpate entire head and face for lacerations, contusions, and fractures.
- Re-evaluate pupils.
- Re-evaluate level of consciousness.
- Assess eyes for retinal hemorrhage, optic disc bulging, visual acuity disturbances, and contact lenses.
- Evaluate cranial nerve function.
- Inspect ears and nose for CSF leakage, hemorrhage
- Inspect mouth, tongue, and teeth.

b. Management.

- Maintain airway, continue ventilation, and oxygenate as indicated.
- Control hemorrhage.
- Prevent secondary brain injury.
- Remove Contact Lenses.

2) Assess the neck.

a. Assessment.

- Inspect for signs of blunt trauma, penetrating injuries, tracheal deviation, and use of accessory muscles during respiration, and jugular vein distention.
- Palpate for tenderness, deformity, swelling, or subcutaneous emphysema.

b. Management.

- Maintain open airway.
- Maintain immobilization and protection of cervical spine.
- Mark trachea.

3) Assess the chest.

3

a) Assessment.

- Inspect chest wall for signs of blunt trauma, penetrating injuries, and use of accessory muscles during respiration.

- Auscultate the chest wall and posterior bases for breath sounds.
- Auscultate for heart sounds.
- Palpate the entire chest wall for evidence of blunt trauma, penetrating injuries, flail chest, tenderness, crepitus, or subcutaneous emphysema.
- Percuss for hyperresonance or dullness.

c. Management.

- Needle thoracenthesis, if tension pneumothorax is suspected and proper training and equipment are present.
- Three-sided dressing for open chest wounds.
- Pericardiocentesis, if cardiac tamponade is suspected and proper training and equipment are present.

4) Abdomen.

a. Assessment.

- Inspect abdomen for signs of blunt trauma, penetrating injuries, and internal bleeding.
- Auscultate for the presence / absence of bowel sounds.
- Palpate the abdomen to elicit subtle tenderness, rebound, or guarding.

- Check pelvis for stability.

b b. Management.

c

- Correctly dress wounds.
- Keep patient NPO until definitive care is delivered.

5) Perineum / Rectum / Penis / Vagina.

d a) Assessment.

e

(1) Perineum.

- Contusions / hematomas.
- Lacerations.

(2) Rectum.

- Blood. Sphincter tone.
- High-riding prostate.

(3) Penis.

- Urethral bleeding.

(4) Vagina.

- Bleeding.
- Lacerations.

6) Musculoskeletal.

a) Assessment.

- Inspect extremities for evidence of blunt / penetrating injury. (deformities, altered range of motion).
- Palpate extremities for tenderness, crepitus, and abnormal movements.
- Palpate all pulses.
- Assess pelvis for stability. (Fractures) - Palpate spine for tenderness, crepitus, step-off's, depressions, and abnormal movements.

b) Management.

- Splint all suspected fractures in the position of function or comfort.

- Reduce dislocations as soon as possible, depending on training.
- Maintain immobilization and alignment of cervical, thoracic, and lumbar spine.

7) Reassess the vitals signs.

a) Assessment.

- Assess the vitals, pulse rate, breathing rate, pulse pressure.
- Re-evaluate the pupils and level of consciousness (GCS).
- Evaluate extremities for motor and sensory deficits.

b) Management.

- Maintain adequate immobilization of entire patient, as indicated.

5. **PATIENT PACKAGING AND EVACUATION.**

a. See the Mountain Casualty Evacuation chapter.

UNITED STATES MARINE CORPS
Mountain Warfare Training Center
Bridgeport, California 93517-5001

FMST.07.13
03/30/02

STUDENT HANDOUT

HIGH-ALTITUDE ILLNESSES

TERMINAL LEARNING OBJECTIVE: Given a simulated casualty, in a high altitude wilderness environment, treat high altitude health problems, in accordance with the references. (FMST.07.13)

ENABLING LEARNING OBJECTIVES:

(1) Without the aid of references, select from a given list the definition of acute mountain sickness, in accordance with the references. (FMST.07.13a)

(2) Without the aid of references, select from a given list the four main mechanisms of acute mountain sickness, in accordance with the references. (FMST.07.13b)

(3) Without the aid of references, select from a given list the most common and prominent symptom of acute mountain sickness, in accordance with the references. (FMST.07.13c)

(4) Without the aid of references, select from a given list the field management of acute mountain sickness, in accordance with the references. (FMST.07.13d)

(5) Without the aid of references, select from a given list the seven high altitude health preventive measures, in accordance with the references. (FMST.07.13e)

(6) Without the aid of references, select from a given list the correct definition of high altitude cerebral edema, in accordance with the references. (FMST.07.13f)

(7) Without the aid of references, select from a given list the pathophysiology of high altitude cerebral edema, in accordance with the references. (WMC.6.4g)

4

(8) Without the aid of references, select from a given list five of the symptoms of high altitude cerebral edema, in accordance with the references. (FMST.07.13h)

(9) Without the aid of references, select from a given list the field management of high altitude cerebral edema, in accordance with the references. (FMST.07.13i)

(10) Without the aid of references, select from a given list the correct definition of high altitude pulmonary edema, in accordance with the references. (FMST.07.13j)

(11) Without the aid of references, select from a given list three of the nine symptoms of high altitude pulmonary edema, in accordance with the references. (FMST.07.13k)

(12) Without the aid of references, select from a given list the field management of high altitude pulmonary edema, in accordance with the references. (FMST.07.13l)

(13) Without the aid of references, select from a given list choose the most important treatment for both high altitude cerebral edema and high altitude pulmonary edema, in accordance with the references. (FMST.07.13m)

OUTLINE

1. **ACCLIMATIZATION**.

a. Successful acclimatization depends on three factors:

(1) Degree of hypoxic stress (i.e. altitude)

(2) Rate of onset of hypoxic stress (i.e. ascent rate)

(3) Individual Physiology (i.e. genetic differences between individuals)

b. Physiologic Changes:

(1) Respiratory - Hypoxic Ventilatory Response (HVR)

- Hypobaric hypoxia
- Triggers carotid body
- Central respiratory center stimulated
- Ventilatory rate increases

HVR can be measured:

- "High" leads to successful acclimatization and superior performance.
- "Low" tends to be found in victims of altitude illnesses.
- HVR can be influenced by depressants (ETOH) and stimulants. (caffeine), and progesterone.
- **May** be negatively influenced by physical conditioning.

(2) Respiratory – Fast response: minutes to hours

- HVR leads to increased Ventilatory rate.
- Increased ventilations "blows off' C02."
- Low C02 levels yield a respiratory alkalosis and rising pH.
- Elevation of pH feeds back to depress the central respiratory center.
- No further immediate rise in ventilatory rate (a brake is placed on any further increase in respiration.)

(3) Renal – Slower response: kidneys respond over the next 24 - 48 hours:

- Kidneys respond to alkalosis by inducing a bicarbonate diuresis.
- PH slowly declines - brake on central respiratory center eases.
- Respiratory rate gradually rises over 4 - 7 days to new baseline.
- Final respiratory rate is altitude dependent.

(4) Cardiovascular - changes are a result of increased sympathetic tone (due to hypoxia).

- Increased: Blood Pressure, venous tone and Cardiac output (increased heart rate and contractility.)
- Stroke Volume declines in first 1 - 3 days due to bicarbonate diuresis.
- Left ventricular pressures are normal, usually; therefore, wedge pressures are usually normal.

(5) Pulmonary Circulation - critical in understanding the pathophysiology of HAPE.

- Ascent results in global Hypoxic Pulmonary Vasoconstriction.
- Elevated Pulmonary Artery Pressure (PAP) occurs.
- Exertion at altitude worsens this Pulmonary Hypertension.
- Damage to pulmonary vascular endothelium leading to leaky capillaries.

(6) Cerebral Circulation - critical in understanding pathophysiology of HACE

- Hypoxia -----------Vasodilatation
- Hypocapnia ----------Vasoconstriction
- Hypoxia is the dominant influence and vasodilatation results.
- On average, Cerebral Blood Flow (CBF) increases 20-40% on ascent to 12.000ft.

(7) Hemopoietic Response

- Erythropoietin levels increase within hours of ascent.

- New immature Red Blood Cells (RBC) seen within days.
- New mature RBC's present within 4-10 days
- RBC mass increase seen over weeks to months (altitude dependent).
- Early changes due to hemoconcentration.
- Polycythemia is a potential danger – this can lead to a condition known as CMS, or Chronic Mountain sickness.

(8) Muscle Tissue

- Intramuscular capillary and mitochondrial density increases after three weeks exposure to altitude.

(9) Sleep Pattern

- Stage 1 increase, Stage 2 unaffected
- Stage 3 & 4 decreases
- Rapid Eye Movement (REM) decreases
- Arousals increase
- Periodic Breathing (Cheyne-Stokes).
  Increased respiratory rate leads to respiratory alkalosis
  Elevated pH depresses central respiratory center
  Apnea
  02 saturation declines, C02 levels rise
  Carotid body stimulated
  Ventilatory rate rises

(10)Summary: Physiologic changes seen at altitude result from the body's responses to hypo baric hypoxia. It has been speculated that man would be limited to an 8000-foot elevation limit without supplemental oxygen if not for the acclimatization process. As Messner and Haebler demonstrated on Mt. Everest, the process of acclimatization enables man to explore the entire planet.

2. **ACUTE MOUNTAIN SICKNESS (AMS).**

a. Definition - AMS is an acute, self-limiting illness, which results when an unacclimatized individual ascends rapidly to high altitude. (FMST.07.13a)

(1) It is rare below 8,000 feet but will occur in approximately 42% of persons rapidly exposed to altitudes greater than 10,000 feet in elevation.

(2) Aerobic fitness is no predictor for risk of AMS with exposure to altitude. There is some evidence that prior aerobic fitness leads to increased AMS incidence, most likely due to excessively rapid ascent or over-exertion at altitude. Elite athletes may also be at increased risk due to depressed cardiovascular and Ventilatory responses to exertion.

(3) Prior ascents to altitude without symptoms of AMS are no guarantee against having symptoms of AMS with future ascents. However, those with past AMS are at an increased risk of AMS with repeat exposure to high altitude. Generally, there is no correlation between the severity of the illness and increasing altitudes. The incidence is slightly increased in younger persons.

(4) There is still much we do not know about altitude-induced physiologic changes. Current explanations in the literature center around genetic differences in hypoxic ventilatory response and vascular endothelium release of inflammatory mediators and Nitric Oxide.

b. Pathophysiology of AMS: The physiologic changes associated with the symptom complex known as AMS are linked to: (FMST.07.13b).

(1) Poor ventilatory response: An individual's Hypoxic Ventilatory Response (HVR) dictates how well he/she will respond to the hypobaric hypoxia of altitude. Those with a brisk HVR will do well, whereas those with a blunted response tend to suffer the symptoms of AMS. A poor ventilatory response leads to hypoxia and hypercapnia (elevated C02 levels). These two combine to augment cerebral blood flow by inducing cerebral vasodilatation.

(2) Fluid retention: The increased sympathetic tone associated with hypobaric hypoxia explains many of the changes seen in the renal system at altitude. Low Renal Blood Flow, increased anti-diuretic hormones, low Glomerular Filtration Rates, and decreased Urine Output all contribute to a Net Fluid Gain. This retained fluid redistributes within the body's fluid spaces. As a result, intravacular and extravascular fluid shifts occur, especially in the brain, lung, and peripheral interstitial tissues.

(3) Increased Intra-cranial Pressure: Increased cerebral blood flow coupled with CNS inflammation resulting in increased vascular permeability leads to increased intra-cranial pressure and associated symptoms.

(4) Altered Pulmonary Mechanics. On ascent to altitude, there is an increase in extravascular lung water. This is thought to be mediated, in part, by a phenomenon known as hypoxic pulmonary vasoconstriction. Blood vessels in the lungs constrict when they are exposed to low oxygen levels, this global vasoconstriction leads to increased right-sided heart pressures. The result of increased right-sided heart pressures causes fluid to leak from the vascular space into the alveoli, which results in impaired gas exchange. This has been demonstrated via Pulmonary Function Testing bronchoalveolar lavage, right sided-heart catheterization and arterial blood gas analysis showing decreased Vital Capacity, decreased Peak Expiratory Flows, pulmonary exudates rich in proteins and Increased Alveolar-Arterial oxygen Gradients. One study showed a 54% increase in estimated lung mass. It is not uncommon to hear crackles in all lung fields when auscultating asymptomatic individuals. The presence of intra-alveolar fluid inhibits the diffusion of oxygen

across the alveolar membrane, thereby reducing oxygenation across the pulmonary capillary membrane.

Signs and Symptoms of AMS. Symptoms tend to occur within the first 1- 3 days of exposure to altitude. Symptoms should resolve spontaneously within 15 - 24 hours, if ascent is arrested. The three most prominent symptoms are Headache, Nausea and Vomiting. (FMST.07.13c) Although nonspecific, headache must be present to make a diagnosis of AMS. 95% of AMS victims will complain of one or more of these three. Other common complaints include dizziness, anorexia, drowsiness, malaise, weakness, and insomnia. It is possible that findings of dyspnea on exertion and mental status changes along with ataxic gait may accompany the classic AMS signs and symptoms. These are clear indications for descent, as these are the hallmark signs of progression to HACE and the development of HAPE. Again, headache is the most common and prominent symptom of AMS. The Lake Louise Consensus criteria was created to give clinicians formal guidance in diagnosing AMS. The criteria require the presence of headache plus one or more symptoms from the above list. (i.e. nausea, vomiting, anorexia, insomnia, dizziness, fatigue, and dyspnea on exertion.)

c. Field Management (FMST.07.13d)

(1) Stop any further ascent. Light duty. No tobacco.

(2) Symptomatic Treatment. Mild analgesics such as Aspirin or Tylenol for the headache, Compazine or Phenergan for nausea/vomiting.

(3) Other Meds: Acetazolamide (Diamox), Dexamethasone (Decadron), and oxygen. The effects of Diamox and Decadron are additive when used together and are a very effective means for treating AMS.

(4) Hyperventilate: It is unrealistic to try to consciously control one's respiratory rate over any appreciable period of time. It is nearly impossible to maintain conscious hyperventilation while functioning in the field.

(5) Descend. If none of these methods help, then individuals should descend 1,000-3,000 feet. This will usually result in marked relief of symptoms. Most cases, however, improve in 1-2 days with symptomatic treatment.

(6) ALL PATIENTS WITH AMS MUST BE EVALUATED FOR HAPE AND HACE.

HOW?

- Field sobriety test.
- Mini-mental status exam.
- Auscultation of lung fields.

d. Preventive Measures: (FMST.07.13e)

(1) The best approach to high-altitude travel is staged, gradual ascent, when moving at altitudes greater than 3000 meters. Gradual ascent means no faster than 3000 ft/day up to 14,000 and no greater than 1,000 ft/day over 14,000 ft.

(2) Avoid any alcohol and/or sedatives during the first two nights.

(3) Maintain adequate fluid intake.

(4) Work high - Sleep low.

(5) High Carbohydrate Diet (approximately 70% of diet).

(6) Avoid overexertion, but remember that mild exercise assists acclimatization.

(7) No smoking.

(8) Pharmacological options: Prevention of AMS is possible with Acetazolamide (Diamox). It dramatically lowers the incidence of AMS when taken prophylactically. The usual dose is 125 mg PO BID starting 24-48 hours before ascent and should be continued for 3-4 days after the start of the mission. It is also effective when used as a treatment especially when combined with Decadron. In individuals with contraindications or allergies to Diamox use Decadron (dosed PO/IM/IV 4 mg Q6).

3. **HIGH ALTITUDE CEREBRAL EDEMA (HACE).**

a. Definition. HACE is a high-altitude illness characterized by swelling of the brain. (FMST.07.13f)

(1) HACE can occur as low as 8,000 feet, but typically occurs at more than 12,000 feet.

(2) HACE, considered by most, to be a progression of AMS, occurs as a failure to heed the warnings of AMS signs and symptoms and continue ascent.

b. Pathophysiology. Increased cerebral blood flow along with the fluid shifts mentioned above lead to vasogenic edema (leaky blood vessels) inside the confined space of the skull. It is also thought that inflammation plays a key role in the development of HACE and occurs as follows. Hypoxia triggers the release of inflammatory mediators (likely Transforming Growth Factor Beta, TGF-β and Vascular Endothelial Growth Factor, VEGF.) These mediators cause increased vascular permeability that result in fluid accumulation within the brain – edema and increased intracranial pressure. Once this pressure nears arterial pressure, cerebral blood flow will be impaired. Eventually, intra-cranial pressure will rise further with a concomitant increase in the risk of brain herniation leading to marked alterations in mental status, ataxia and even cranial nerve palsies. Increased CSF production further compounds the problem. (FMST.07.13g)

c. Differential Diagnosis: Other illnesses are possible in individuals ascending to altitude. Your differential should include the following when neurologic deficits are involved:

   (1) Cerebrovascular spasm and Transient Ischemic Attacks.

   (2) Cerebral vascular thromboembolic events.

   (3) Intra-cranial hemorrhage (aneurysm, AVM).

   (3) Hypothermia.

   (4) Meningitis

   (5) Encephalitis

   (6) "Hangover"

d. Signs and Symptoms:

   (1) Early signs and symptoms are those found in AMS. (i.e. H/A, NV)

   (2) When the above warning signs go unheeded the true signs and symptoms of HACE can quickly follow: (FMST.07.13h)

      (a) Ataxia (loss of muscle coordination leading to difficulty maintaining balance), especially prominent in the victim's gait.

      (b) Mental Status Changes, poor judgement, personality change which may progress to stupor, coma and death relatively rapidly without proper treatment.

      (c) Lassitude, confusion, hallucinations, convulsions, behavioral aberrations.

      (d) Paralysis of one or more extremities (see Differential Diagnosis).

   (3) The most important impediment to early recognition of HACE is its insidious onset. Early signs and symptoms frequently go unrecognized or are ignored by patients, as well as their companions, who also may be suffering to some degree from the effects of altitude.

e. Field Treatment (FMST.07.13i)

(1) Early recognition is the KEY. Look for Ataxia and Mental Status Changes.

(2) Treatment should be immediate since fatalities can occur within a few hours in severe cases.

(3) Once diagnosed, the patient should be placed in the most comfortable position possible, descent should be directed immediately and administration of high concentration O2 should be given, in route, if available.

(4) Decadron (early), 10mg PO/IM/IV followed by 4 mg Q6. Diamox can also be used 125-250 mg BID.

(5) Consider diuretics – use carefully.

(6) If loss of consciousness, think airway management and ABC's.

(7) Casevac to an advanced medical facility ASAP!

(8) A Gamow Bag can be a lifesaver if decent is not immediately possible. Perform 4-6 compressions per day each treatment should last 40-60 minutes.

NOTE: The Gamow Bag is a one-man portable hyperbaric chamber. It weighs 14.5 lbs., is operated by a foot-pump, and is made of nylon with an airtight zipper. With the HAPE or HACE victim inside, the bag is pressurized with the foot-pump to an internal pressure of 2 psi. This is the equivalent to a 5000 - 6000-foot descent in regards to the partial pressure of oxygen inside the bag. Air is circulated within the bag via the foot-pump. This can be a life-saving resource in the event a victim is caught at altitude and cannot descend due to weather or other uncontrollable variables.

f. Prevention

(1) Preventive measures are the same as those discussed for AMS.

(2) Decadron can be used as a prophylactic medication along with Diamox, in cases of unavoidable rapid ascent to extreme altitude. However one must continue this regimen, while at altitude, or face rapid onset of symptoms once the medication is stopped.

4. **HIGH ALTITUDE PULMONARY EDEMA (HAPE)**.

a. Definition. HAPE is a high-altitude illness that is characterized by filling of the lungs with edema fluid. (FMST.07.13j)

(1) HAPE rarely occurs below 12,000 feet.

(2) HAPE victims in the past were commonly described, after autopsy, as victims of "pneumonia" and "Inflammation of the lungs". Since 1960, medical science has had the technology to investigate the complexities of the physiological changes associated with hypobaric hypoxia.

(3) Incidence: Occurs in 1-2 % of people brought rapidly to 12,000ft.

(4) Commonly occurs within 2-4 days of ascent.

(5) Risk for developing HAPE remain the same as above:

(a) Rate of ascent
(b) Sleeping altitude
(c) Level of exertion
(d) Gender (M>F)
(f) Individual physiology

b. Pathophysiology of HAPE:

(1) Hypoxic Pulmonary Vasoconstriction

(a) Normal part of the acclimatization process (elevated PAP).
(b) Adaptive at sea level (i.e. lobar pneumonia).
(c) Maladaptive at altitude (global hypoxia).
(d) Tends to be uneven, leading to high pressure/high flow area.

(2) Overperfusion

(a) Localized areas exposed to high pressure and high flow.
(b) Shearing forces and stress failure at the microvascular level.
(c) Endothelial damage to capillaries leads to inflammatory response and increased vascular permeability.
(d) Elevated hydrostatic pressure combined with increased vascular permeability leads to leakage of protein rich fluids into the alveoli and decreased oxygenation

(3) Nitric Oxide deficiency

Individuals prone to HAPE may have a deficiency in the amount of Nitric Oxide produced in the vascular endothelium resulting in more profound hypoxic pulmonary vasoconstriction.

c. Pathologic Findings of HAPE: Postmortem studies of victims of HAPE have yielded some interesting results:

(1) Grossly, the lungs are congested and swollen, as seen with other causes of pulmonary edema. Average lung weight on autopsy is 2-4 times normal. Histologically, one sees pulmonary edema with a protein rich exudate filling the alveoli. It is also common to find the alveolar spaces filled with hyaline membranes.

(2) The left heart is completely normal in appearance. However, the right ventricle, right atrium, and pulmonary arteries are distended and dilated.

(3) Greater than 50% of HAPE victims are found to have evidence of HACE.

d. Signs and Symptoms. These tend to occur within 2-4 days of arrival at altitude. Usually the symptoms of AMS are present before or occur with the symptoms of HAPE. (FMST.07.13k)

(1) Early signs:

(a) Dry cough, frequently occurring at night.

(b) Dyspnea on Exertion (DOE), especially with ambulation uphill.

(c) Mild chest pain - usually perceived as an ache beneath the sternum.

(d) Decreased work performance and increased recovery time between events.

(e) Peripheral or Central Cyanosis.

(2) Later signs:

(a) Dyspnea at Rest.

(b) Productive cough which yields large amounts of pink, frothy sputum.

(c) Rapid pulse and respiratory rates.

(d) Audible crackles on auscultation.

(e) **Mental status changes, Ataxia**, Loss of Consciousness.

(f) CXR findings:

- Prominent pulmonary arteries.
- Normal heart size.
- Patchy infiltrates.

<u>ECHO findings:</u>

-Non-cardiogenic pulmonary edema -increased PAP
-Tricuspid regurgitation.
-Normal left ventricle
-Prominent pulmonary arteries -enlarged right atrium
-Enlarged right ventricle.

NOTE: Crackles heard on lung exam, as an isolated finding is NOT an indication for descent or grounds for the diagnosis of HAPE. It is common (up to 15% of climbers on Mt. Rainier) to find asymptomatic rales in climbers at altitudes of 10,000-12,000ft. Concern is heightened when crackles occur in constellation with other signs and symptoms.

e. Field Treatment (FMST.07.131)

(1) The most important emergency care measure is immediate descent to a lower altitude, since fatalities can occur within 6-12 hours in severe cases.

(a) A descent of at least 2,000-3,000 feet can be a definitive treatment if the condition is caught early enough.

(2) The patient should be placed in the most comfortable position (usually sitting) and given **high-flow O2 if available**. It is important to minimize activity, as exertion increases pulmonary artery pressures. Remember oxygen is Lasix for the lungs.

(3) Medication regimens include:

(a) Nifedipine (10mg PO initially then 30mg SR Q12) – calcium channel blocker - pulmonary artery/arteriole dilator. SR=sustained relief.

(b) Diamox and Decadron may be appropriate regimens, depending on your medication availability.

(c) Morphine (5mg IV initially decreased PAP) as a last resort.

(4) Oxygen is always appropriate, however, it is not always available.

(5) Gamow Bag can be a lifesaver if decent is not immediately possible.

f. Prevention Measures. Preventive measures are the same as those discussed for AMS.

(1) Acetazolamide (Diamox) has been proven to be successful in the prevention of HAPE. 125 - 250 mg PO bid helps prevent HAPE in individuals with recurrent episodes.

(2) Nifedipine (30 mg SRq12h) - also effective in individuals with recurrent HAPE.

(3) It should be noted that a history of HAPE is NOT a contraindication to the individual participating in high-altitude activities. Each case should be evaluated individually.

NOTE: Remember decent is the most important treatment for both HACE and HAPE. (FMST.07.13m)

5. **SYSTEMIC EDEMA AT HIGH ALTITUDES**.

Occasionally, certain individuals, especially women, will develop edema of the face, hands, and feet at altitude. This edema usually tends to occur in the absence of other symptoms. In susceptible persons, repeat episodes are common. The edema goes away upon return to lower altitude. There is no need for descent in these cases because the patient is usually able to tolerate the edema and continues to function well. However, they should be examined carefully to rule out the presence of HAPE or HACE. Diuretics are usually not warranted. It is also probably reasonable to restrict salt intake. The only indication for descent is if the edema is causing functional disability, i.e. inability to don one's boots, or facial edema severe enough to restrict vision.

6. **SICKLE CRISIS**.

Persons with sickle cell trait are at risk for developing sickling attacks when exposed to hypoxic stress, as might occur at high altitudes. However, sickling can occur at lower elevations. Individuals with Mediterranean family ties should be considered at risk for having sickle cell trait. The trait is more likely in black individuals as well. Any patient complaining of Left Upper Quadrant pain, SOB, and/or arthralgia at altitude should have sickling crisis included in the differential diagnosis regardless of race. It has been known to occur in white females as well. It has been recommended by some that individuals with sickle cell trait should be restricted to an altitude of 8,000 feet and below. However, more liberal recommendations exist, suggesting that the individual and their symptoms dictate ultimate altitude.

7. **ALTITUDE THROAT**.

At higher altitudes, mouth breathing is often increased during exertion. Inhalation of this cold, dry air can result in drying of the mucous membranes in the throat and upper airway. This condition responds to humidified air, salt-water gargles or lozenges. If exudate or edema is present, throat culture and/or antibiotics may be warranted.

8. **HIGH ALTITUDE FLATUS EXPULSION HAFE).**

Ascent to altitude can result in rapid expansion of intestinal gas, due to the decreased pressure of altitude. This produces abdominal discomfort and the passage of colonic gas. This condition is not associated with serious consequences; however, it may create tension amongst tent-team members. Treatment consists of oral administration of anti-flatulence such as simethicone, Mylanta II, or Gaviscon. Know and avoid your gas producing foods!

9. **HIGH ALTITUDE RETINAL HEMORRHAGE.**

Up to 60% of individuals ascending to altitudes greater than 18,000 feet will develop asymptomatic retinal hemorrhages. On fundoscopic exam, there will be hyperemia of the disc and dilation and tortuosity of the vessels. "Flame" hemorrhages and "cotton wool spots" can be seen on exam. The only indication for descent in an individual with retinal hemorrhages is loss of visual acuity. This usually results when a large hemorrhage involves the macular region of the retina. Remember that Carbon Monoxide poisoning can present with retinal hemorrhage as well.

10. **THROMBOEMBOLIC EVENTS.**

Thromboembolic disease occurs with increased frequency at higher altitudes. It is believed that the common circumstances of volume depletion due to dehydration and Polycythemia due to altitude exposure predispose individuals to coagulopathic events. In addition, extended periods of inactivity due to "going to ground" during storms contribute to this predisposition. There does not seem to be any direct influence of hypobaric hypoxia on the body's normal coagulation system.

11. **ACETAZOLAMIDE.**

Acetazolamide (Diamox) is a Carbonic Anhydrate Inhibitor, which is especially active in the brain, the lung, and the kidney.

a. Kidney:

(1) Uptake of bicarbonate ion is inhibited.

(2) Bicarbonate diuresis is induced.

(3) Metabolic acidosis ensues within one hour of medication.

(4) Sequence mimics and accelerates the natural acclimatization process.

(5) Enhanced HVR is the end result.

(6) Diuresis addresses problems associated with fluid retention.

b. Brain:

(1) Direct effects at the level of the Central Respiratory Center.

(2) Diminishes periodic breathing during sleep.

(3) Fewer episodes of apnea and extreme hypoxemia.

(4) CSF production is inhibited and CSF absorption is enhanced.

c. Side Effects:

(1) Polyuria.

(2) Flattens taste of carbonated beverages.

(3) Peripheral paraesthesias, myopia, impotence, nausea, and drowsiness.

(4) Contraindicated for Sulfa Allergic and G6PD-deficient individuals (reversible when medication is discontinued.)

UNITED STATES MARINE CORPS
Mountain Warfare Training Center
Bridgeport, California 93517-5001

FMST.07.35
03/13/02

STUDENT HANDOUT

HEAT ILLNESSES

TERMINAL LEARNING OBJECTIVE: Given a casualty in a wilderness environment, and necessary equipment and supplies, manage common heat related injuries to prevent death or further injury per the reference (FMST.07.35)

ENABLING LEARNING OBJECTIVES:

(1) Without the aid of references, given a list of heat illnesses match the definition to the correct condition, in accordance with the references. (FMST.07.35a)

(2) Without the aid of references, select from a given list the correct field treatment for heat cramps, in accordance with the references. (FMST.07.35b)

(3) Without the aid of references, select from a given list the correct field treatment for heat exhaustion, in accordance with the references. (FMST.07.35c)

(4) Without the aid of references, select from a given list the correct field treatment for heat stroke, in accordance with the references. (FMST.07.35d)

1. **CAUSES OF HEAT ILLNESSES**. Heat illness is caused by one of or a combination of the following factors:

a. Excessive environmental heat and humidity.

b. Faulty heat dissipation mechanisms within the body.

c. Stressed heat dissipation mechanisms.

d. Prolonged strenuous physical activity with inadequate fluid and electrolyte replacement.

2. **THERMOREGULATION**. The body is constantly adjusting itself in order to maintain its optimal temperature. This is called thermoregulation. A basic physics principle applies, in that heat is transferred from a higher to a lower temperature. What this means is that when the environmental temperature is higher than the skin temperature, the body gains heat, but when the environmental temperature is lower, the body loses heat. As a result, elevated environmental temperature adds to the heat burden of the body and can interfere with heat dissipation.

a. Heat Transfer Mechanisms. Keeping in mind the above principle we will now discuss the four methods of heat transfer.

1) Conduction: This is heat exchange between two surfaces in direct contact. An example would be lying on a hot or cold surface.

2) Convection: This is heat exchange from a surface to a gas or liquid, usually air or water. The amount of heat transferred depends on many variables, such as density, flow rate and surface area exposed.

3) Radiation: This is heat transfer between the body and its surroundings. All matter either absorbs heat or emits thermal radiation. The sun emits heat and the body absorbs it. A interesting fact is that highly pigmented skin while protective against UV radiation absorbs about 20% more heat than lightly pigmented skin.

4) Evaporation: This is the heat flow from the body surface to the environment by sweating.

b. If the body is unable to maintain thermal equilibrium by radiation, convection, and conduction and core temperature rises, sweating must occur to permit heat loss by vaporization of water. Since gastric emptying of water is about 1L/hr and sweat rates can exceed 1L/hr, a healthy diet along with constant replenishment of water lost is imperative to prevent illness.

3. **PREDISPOSING FACTORS**.

a. An underlying illness such as infection.

b. The temperature and humidity of previous days effect the likelihood of becoming a heat casualty.

c. Most heat illnesses occur during the morning hours.

d. Thirst is not an accurate indication of hydration; therefore you must drink even when you don't feel thirsty.

e. People usually do not voluntarily drink as much water as they lose.

f. All persons in the heat should be considered dehydrated, unless they have been forced to drink more than they desired.

4. **PREVENTIVE MEASURES**.

a. Drink 6-8 qts every day to replace fluids lost by sweating or from other bodily functions.

b. Consumption of a solute rich meal or a fluid replacement beverage about 2 hours before exercise can potentially counteract the sweat/salt losses.

5. **EARLY SIGNS OF HEAT RELATED ILLENSSES**. Changes in cognitive functions appear to occur frequently prior to the physical symptoms of heat illnesses. It is therefore imperative that medical personnel and leaders watch for confusion and impaired ability to function or work in the members of their group.

6. **SPECIFIC HEAT RELATED ILLNESSES**.

a. Heat Induced Syncope. (FMST.07.35a) This event is caused by the gravitational displacement of blood into the legs of an upright and stationary person combined with hemodynamic displacement of blood into peripheral circulation to support heat transfer at the body surface.

1) Prodromal symptoms.

(A) Restlessness

(D) Nausea

(B) Sighing

(E) Yawning

(C) Dysphoria

2) Treatment.

(A) The fainting is usually brief and responds to horizontal positioning.

(B) Rest in a cool, shaded area.

(C) Oral fluids.

(D) Persons should be cautioned against standing in hot environments and advised to flex leg muscles repeatedly while standing still to enhance venous return and warned to sit or lie down if warning signs reappear.

b. Heat Cramps. These are painful spasms of muscles, caused by sweating resulting in excessive loss of body salt. Generally, the spasms involve voluntary skeletal muscle subjected to prior exertion. This can occur even if a person is replacing water loss, but not salt losses.

1) Precipitating Factors: Cramps are preceded by several hours of sustained effort, combining heavy sweating in hot surroundings, with the ingestion of large volumes of water.

2) Symptoms: Painful muscle cramps in the arms, legs, and abdomen.

3) Treatment: (FMST.07.35b)

(A) Rest in a cool, shaded area.

(B) Gentle massage of affected muscles (may relieve spasms) or place the affected muscle in a stretch.

(C) Generally, heat cramps respond quickly to salt solutions.

1.) 1/4 to 1/2 teaspoon of salt in a quart of water (sip).

2.) In severe cases IV isotonic saline 0.9% NaCL.

NOTE: Salt tablets are not recommended due to the gastric irritation they cause.

c. Heat Exhaustion. This occurs when body salt losses and dehydration from sweating are so severe that a person can no longer maintain adequate blood pressure. It tends to develop over several days, that present ample opportunity exists for electrolyte and water imbalances to occur. Some 20% of cases are found to also have some form of gastroenteritis. Additionally, 60% of heat exhaustion casualties will also have symptoms of heat cramps.

1) Problems in recognition.

(A) Early signs may go unrecognized or be mistaken for malingering.

(B) Headache and visual disturbances may mimic prodromal of migraine headache.

(C) Dizziness and syncope may lead to an evaluation for arrythmia.

(D) Chills, ^ temperature, and sweating may lead to an evaluation for infection.

(E) Altered mental status or ataxia may lead to an evaluation for Meningitis or high altitude medical problem.

2. Symptoms:

(A) Headache

(B) Fatigue

(C) Piloerection

(D) Nausea/Vomiting

(E) Profuse sweating

(F) Dizziness

(G) Hyperirritability, Anxiety

(H) Chills

(I) Heat cramps

3. Clinical Findings:

(A) Tachycardia

(B) Hypotension

(D) Hyperventilation

(E) Syncope

(C) Core temperature may be normal or elevated. Spontaneous cooling may occur, usually with profuse sweating. This will not happen with Heat stroke.

(4) Treatment is based on symptoms and focuses on reduction of body temperature and rehydration. (FMST.07.35c)

(A) Lay down in a cool, shaded area.

(B) If body temperature is elevated, lower by any means available.

(C) Sprinkling with water and fanning.

(D) Ice bags over superficial great vessels and fanning.

(E) Periodic toweling of the skin renews the evaporative process.

(F) If patient is alert, orally rehydrate with cool water or commercially available rehydration mixture. Goal is 2L over 1st 2 hours.

(G) If patient is not alert, rehydrate via IV with LR.

(H) Victim must rest and continue to be rehydrated over next 24 to 36 hours.

d. Heatstroke (Hyperthermia). " A true medical emergency". This is a failure of the body's cooling mechanisms that rid the body of excessive heat build-up. Prior to 1950 the mortality rate was 40 to 75%. Long term survival of victims is directly related to the speed of effective treatment measures. Traditionally, the diagnosis of Heatstroke required the following 3 signs: a core temperature > 104°F(41°C), altered mental status, and cessation of sweating. Although, this symptom complex does represent full-blown Heatstroke adherence to this strict criteria may delay critical interventions. Unless an alternative etiology is obvious, the previously healthy person who collapses after physical exertion in hot weather should be considered to have exertional Heatstroke. Treat for Heatstroke whenever you suspect it and always suspect it in the heat.

1. Symptoms:

(A) Headache	(H) Dry mouth
(B) Shortness of breath	(I) Core temp > 104°F (41°C)
(C) Nausea/ vomiting	(J) Weakness/ dizziness
(D) Low BP	(K) Hot, red skin
(E) Skin may be wet or dry	(L) Tachycardia
(F) Constricted pupils	(M) Confusion, Aggression, Combative
(G) Seizures	(N) Coma

2. Clinical Findings.

   (A) CNS: altered mental status, agitation, ataxia, delirium, hallucinations, convulsions, seizures, coma, pupillary constriction.

   (B) Cardiovascular: hypotension (a late and ominous finding), shock, sinus tachycardia with ST segment & T wave abnormalities.

   (C) Pulmonary: hyperventilation, pulmonary edema.

   (D) Renal: acute renal failure, hematuria, pyuria, proteinuria, UA casts.

   (E) Gastrointestinal: nausea, vomiting, +/- disseminated intravascular coagulation, hematemesis, melena.

   (F) Hematologic: WBC 20-30 thousand, < platelets, < clotting factors.

   (G) Hepatic: increased liver function tests.

3. Emergency Treatment. (FMST.07.35d)

(A) Rapid reduction of body core temperature by any means available until core temperature is 102°F.

1) Ice water bath (controversial).
2) Cool water bath.
3) Pouring or hosing cool water over victim and fanning.
4) Placing ice bags at neck, armpits, groin, forehead, chest etc..

(B) Maintain ABC's.

(C) Monitor core temperature.

(D) IV access infuses RL or normal saline 500cc/hr x 4hrs. (watch for pulmonary edema).

(E) Control seizures with valium.

(F) Transport STAT. Cool enroute.

UNITED STATES MARINE CORPS
Mountain Warfare Training Center
Bridgeport, California 93517-5001

FMST.07.15
04/01/02

STUDENT HANDOUT

COMBAT CASUALTY CARE

TERMINAL LEARNING OBJECTIVE: Given a simulated casualty in a simulated combat environment, treat the casualty in accordance with the references. (FMST.07.15)

ENABLING LEARNING OBJECTIVES:

(1) Without the aid of references, choose from a given list the three correct phases of care of tactical casualty management, in accordance with the references. (FMST.07.15a)

(2) Without the aid of references, given a simulated casualty and the necessary equipment, conduct combat casualty care, in accordance with the references. (FMST.07.15b)

OUTLINE

1. Summary of Findings and Recommendations.

a. Pre-hospital care of combat casualties is divided into three phases: (FMST.07.15a)
   1. Care under Fire.
   2. Tactical Field Care
   3. Combat Casualty Evacuation (CASEVAC) Care.

b. Suppression of hostile fire may take temporary precedence over the rendering of care in the Care under Fire phase.

c. Cervical spine immobilization is not required for victims of penetrating head or neck trauma.

d. CPR should not be attempted on the battlefield for victims of blast or penetrating trauma who have no pulse, respirations, or other signs of life.

e. The use of tourniquets is encouraged in the Care under Fire phase. The tourniquet is removed and bleeding controlled with direct pressure in subsequent phases when feasible.

f. The nasopharyngeal airway is the airway of first choice for unconscious patients until the CASEVAC phase.

g. For patients with airway obstruction in the Tactical Field Care phase unrelieved by a nasopharyneal airway, the next airway maneuver is cricothyroidotomy.

h. Endotracheal intubation is not the standard of care until the CASEVAC phase.

i. Evaluation of the laryngeal mask airway and the Combitube for use by SOF corpsmen and medics is recommended.

j. Progressive, severe respiratory distress in the setting of unilateral blunt or penetrating chest trauma on the battlefield should result in a presumptive diagnosis of tension pneumothorax and decompression of that hemithorax with needle thoracostomy.

k. Chest tube insertion is not indicated until the CASEVAC phase of care.

l. Oxygen is not expected to be carried onto the battlefield, but CASEVAC assets should have it.

m. No IV fluids are to be given until the CASEVAC phase unless the patient is in hypovolemic shock from a bleeding site which has now been controlled. Fluid resuscitation is not necessary for patients who are not in shock and not appropriate for patients in hypovolemic shock from uncontrolled hemorrhage (penetrating wounds of the chest and abdomen).

n. Saline locks may be used instead of IV's if fluid resuscitation is not required.

o. 18 gauge catheters may be used instead of 14 or 16 gauge.

p. Hespan (1000cc) is the fluid of choice for initial resuscitation of patients with controlled hemorrhagic shock.

q. Morphine is to be used IV (5mg) instead of IM.

r. IV cefoxitin is to be used as soon as possible for patients with penetrating abdominal trauma, grossly contaminated wounds, massive soft tissue trauma, open fractures, or any patient in whom a long delay until definitive treatment is expected.

s. Casualties should not be completely undressed for a secondary survey in the field. Removal of clothing should be limited to that necessary to expose known or suspected wounds.

t. More specific combat casualty care planning should be carried out and based on specific mission-oriented scenarios.

u. The establishment of Combat Casualty Transport Teams and their use on CASEVAC assets is recommended.

v. Electronic monitoring should be routinely used on CASEVAC assets unless conditions prohibit.

w. MAST trousers are not recommended as a standard of care in any phase.

2. Phases of Care:

a. Phase One: Care Under Fire

1. Return fire as directed or required.

2. Keep yourself from getting shot.

3. Try to keep the casualty from sustaining additional wounds.

4. Stop any massive external hemorrhage with a tourniquet.

5. Take the casualty with you when you leave.

b. Phase Two: Tactical Field Care

1. Airway management

- Unconscious casualty without airway obstruction: > Nasopharyngeal airway.

- Unconscious casualty with airway obstruction: > Cricothyroidotomy.

- Cervical spine immobilization is not necessary for casualties with penetrating head or neck trauma.

2. Breathing

- Consider tension pneumothorax and decompress with needle thoracostomy if a casualty has unilateral penetrating chest trauma and progressive respiratory disstress.

3. Bleeding

- Control any remaining bleeding with a tourniquet or direct pressure.

4. IV

- Start an 18 gauge IV or saline lock.

5. Fluid Resuscitation

- Controlled hemorrhage without shock: > No fluids necessary.

- Controlled hemorrhage with shock: > Hespan 1000cc

- Uncontrolled (infra-abdominal or thoracic) hemorrhage: > No IV fluid resuscitation.

6. Inspect and dress wound.

7. Check for additional wounds.

8. Analgesia as necessary: > Morphine (5mg) IV, wait 5minutes, repeat as necessary.

9. Splint fractures and recheck pulses.

10. Antibiotics

Cefoxitin 2gm slow IV push (over 3-5 minutes) for penetrating abdominal trauma, grossly contaminated wounds, massive soft tissue trauma, open fractures, or any patient in whom a long delay until CASEVAC or definitive treatment is expected.

11. Cardiopulmonary Resuscitation

CPR on the battlefield for victims of blast or penetrating trauma who have no pulse, no respirations, and no other signs of life will not be successful if attempted.

c. Phase Three: Combat Casualty Evacuation (CASEVAC) Care

1. Airway management

Unconscious casualty without airway obstruction: > Nasopharyngeal airway, endotracheal intubation, Combitube or laryngeal mask airway.

Unconscious casualty with airway obstruction: > Cricothyroidotomy if endotracheal intubation and/or other airway devices are unsuccessful.

2. Breathing

Consider tension pneumothorax or hemothorax and decompress with needle thoracostomy if a casualty has unilateral penetrating chest trauma and progressive respiratory distress.

Consider chest tube insertion if a suspected tension pneumothorax is not relieved by needle thoracostomy

Oxygen.

3. Bleeding

Consider removing tourniquets and using direct pressure to control bleeding if possible.

4. IV

Start an 18 gauge IV or saline lock if not already done.

5. Fluid resuscitation

No hemorrhage or controlled hemorrhage without shock: > Lactated Ringer's at 250cc/hr.

Controlled hemorrhage with shock: > Hespan 1000cc initially.

Uncontrolled (infra-abdominal or thoracic) hemorrhage: > No IV fluid resuscitation.

Head Wound patient: > Hespan at KVO unless there is concurrent controlled hemorrhagic shock.

UNITED STATES MARINE CORPS
Mountain Warfare Training Center
Bridgeport, California 93517-5001

FMST.07.37
04/01/02

STUDENT HANDOUT

BURN MANAGEMENT

TERMINAL LEARNING OBJECTIVE. Given a casualty in a wilderness environment, and necessary equipment and supplies, treat burn injuries to prevent death or further injury per the reference. (FMST.07.37)

ENABLING LEARNING OBJECTIVES

1. Without the aid of references, given a list of the depths and appearances of burns, match them to the type of burn, in accordance with the references. (FMST.07.37a)

2. Without the aid of references, given a list choose the major burns that require rapid casualty evacuation, in accordance with the references. (FMST.07.37b)

3. Without the aid of references, given a list choose the first step in the treatment of burn injuries, in accordance with the references. (FMST.07.37c)

OUTLINE

1. **GENERAL:** Each year there are about 2 million individuals burned severely enough to seek medical attention. Around 70,000 require hospitalization. Deaths are on the decrease largely due to the increase in the use of smoke detectors. It has been estimated that 90% of all burn injuries are preventable. Children and young adults are frequently affected by burn injuries.

2. **PHYSIOLOGY:** For non-chemical burns, the primary events of injury occur during the time of heat contact. Coagulation necrosis takes place within cells and denaturation of collagen in the dermis. Blood vessels are completely destroyed or endothelium damage is severe enough to cause clotting which leads to ischemic necrosis of remaining viable cells. Burn wounds are not static. Surrounding the "coagulation zone" is a zone of capillary and small vessel stasis. Circulation becomes stagnant from a strip formed by red blood cells, and aggregation of platelets and white blood cells. The fate of the burn wound depends on the progression of this zone of stasis.

7

3. In patients with burns less than 10% of the total body surface area (TBSA), the inflammatory process is generally limited to the burn site itself. As burns approach 20% TBSA, the local

response becomes systemic. Burns of 30% TBSA or greater, often lead to multiple organ system failure, which in a patient with a severe burn almost invariably leads to a fatal outcome, often due to acute renal failure.

Cardiovascular changes begin immediately after a burn. The degree of these changes depends initially on the size of the burn and to a lesser extent the depth of the burn. Burns less than 15% TBSA can be fluid resuscitated with crystalloid fluids. As burns pass 20% TBSA, massive shifts of fluid and electrolytes occurs. This results in massive fluid shift from the intravascular to the extravascular space. Unless intravascular volume is repleted, classic hypovolemic shock occurs.

3. **TYPES OF BURNS**

a. Scald Burns: Scalding is the most common cause of burns, usually resulting from hot water. Water at 140°F creates a deep partial-thickness or full-thickness burn in three seconds. At 156°F the same burn occurs in one second. Freshly brewed coffee is generally about 180°F. Boiling water always causes a deep burn. Soups and sauces are thicker, remaining in contact with the skin longer. Exposed areas tend to burn less deeply than areas covered with a thin layer of clothes. Clothing retains heat and keeps liquid in contact with the skin longer. Immersion scalds are deep and cause severe burns resulting from longer contact time. These burns are more common in children and elderly patients because of their thin skin. Burns from hot oils are generally deep partial-thickness or full-thickness. Temperatures for cooking oils reach around 400°F, and tar can be up to 500°F.

b. Flame Burns: The next most common burn injuries are flame burns. Although household injuries have gone down with the advent of smoke detectors, a high number of burn injuries still occur from careless smoking, improper use of flammable liquids, automobile accidents, and clothing ignited from stoves or space heaters.

c. Flash Burns: Flash burns result from intense heat for a brief time caused by explosions of natural gas, propane, gasoline, and other flammable liquids.

d. Contact Burns: These burns usually result from direct contact of hot metals, plastics, glass, or coals. Contact burns are usually limited in size, but are deep wounds. With the increase of wood stove usage, this type of burn injury is on the rise in children. Contact burns, dealing with an unconscious patient or involving molten materials, are usually fourth degree. The most common contact burn in the wilderness is from hot coals.

e. Electrical Burns: Electrical burns are actually thermal burns from very high-intensity heat. When electricity encounters resistance from the body it turns into heat. Severity of the burn depends on the portion of the body affected relative to the amperage. The smaller the body part, the more intense the heat is, and the part is less able to dissipate the heat. Consequently the fingers and toes are almost always destroyed with severe damage to the foreams. Larger areas can dissipate the current enough to prevent extensive damage. Arc burns take the most direct route, not the path of least resistance. These are deep and destuctive burns which mostly occur at the joints.

Electrical burns may have other associated injuries that should not be overlooked on the initial assessment. Fall related injuries are common. The intense muscle contraction resulting from electrocution can cause fractures of the lumbar vertebrae, humerus, or femur, and also may cause dislocation of the hips and shoulders. The casualty may have cardiac symptoms of a myocardial contusion or infarction. There can be a conduction system failure or an actual rupture of the heart wall. If there are no signs present of cardiac problems intially after a shock of 110 or 220 volts there is a minute chance they will appear later. Nervous system damage is possible whenever a current passes from one side of the body to the other more than likely affecting the spinal cord.

f. Chemical Burns: Usually caused by strong acids or alkalis and are most often the result of industrial accidents, home use of drain cleaners, assault, and improper use of harsh solvents. Chemical burns continue to burn until they are inactivated by reaction with tissue or dilution by flushing with water. Acids tend to tan the skin, creating an impermeable barrier that limits futher penetration. Alkalis combine with cutaneous lipids and saponify[1] the skin until they are neutralized. The appearance of the wound can be misleading. Unless the observer can be 100% sure, all chemical burns should be considered full thickness.

g. NBC Burns: Nuclear weapons cause burns in two ways: by direct absorption of thermal radiation through exposed surfaces (flash burns); or by the indirect action of fires caused by the weapon (flash burns). The first step in treatment is to prevent contamination to the rescuers and other patients.Vesicants such as mustard gas cause skin burns with edema and blister formation. Burns resulting from vesicants should be flushed with copious amounts of water. Absorbent powder (flour, talcum powder, Fuller's earth) can be used if water is scarce. The powders should be wiped away with a moist towel. The military uses M258A1 kits for skin decontamination. White Phosphorus ignites spontaneously when exposed to air. Remove the air with an air-tight seal; mud is optional. A brief rinse with 1% copper sulfate[2] will impead further oxidation.

4. **CLINICAL PRESENTATION:** The severity of the burn injury is related to the size of the burn, the depth of the burn, and the part of the body that is burned.

a. Burn Size: Burns are the only quantifiable form of trauma. The burn size is the single most important factor in predicting mortality, need for specialized care, and the complications expected from the burn. We can measure burn size with the "rule of nines". The "rule of nines" is where areas of the body represent approximately 9% of a person's total body surface area (TBSA), each leg 18% (the front of the leg 9%, and the back of the leg 9%), the front of the torso represents 18%, the back of the trunk is 18%, the head represents 9%, and the perineum 1%. For infants and small children under four years of age, the head represents a larger percentage 18%, and the legs a smaller percentage 13%, each arm 9%, front troso 13%, back troso 13%, and the perineum 1%. The percentage of the burn area and that of the unburned area need to add up to equal 100%. For smaller burn areas, use the Rule of Palmar Surface: the patient's palmar surface equals about 1% TBSA.

---

[1] Saponify: to convert fat into soap by treating it with an alkali.
[2] Navy uses sodium bicarbonate instead of copper sulfate

b. Burn Depth: The depth of the burn is either described in degree: first, second, third or fourth, or by the depth of the injury. The following describes burns in terms of partial-thickness or full-thickness. While these descriptions appear to separate burns into defined categories, many burns have a mixture of characteristics making a precise diagnosis difficult.

(1) First-Degree: These burns involve the epidermis only and do not blister. They are most commonly caused by ultraviolet light. An example would be a sunburn. The burned skin is painful and red. It should heal in 7 days without scarring.

(2) Superficial Second-Degree, or Superficial Partial-Thickness: These burns include the epidermis and upper layers of the dermis. They characteristically form blisters, under which the skin is red and moist, these burns are painful to the touch. Blisters can take up to two days to appear. These burns are usually caused by hot liquids. Wounds should heal in 14 to 21 days; they may scar depending on the extent of the burn.

(3) Deep Second-Degree, or Deep Partial-Thickness: Burns that involve deeper layers of the dermis. There is damage to hair follicles and sweat glands. They are characterized with a molted pink and white color with blisters forming immediately. These wounds may be less sensitive to touch than the surrounding normal skin, or maybe tender to touch. The patient often complains of discomfort rather than pain. Capillary refill may be slow or absent, when pressure is applied to the wound. These burns are caused by hot liquids, oil, steam, or flame. They may be difficult to distinguish from Third-Degree/Full-Thickness burns. Healing takes 3-9 weeks. Scarring is probable, the degree is related to the amount and the depth of dermal injury. Surgical grafting may be required.

(4) Third-Degree, or Full-Thickness: Involves the entire thickness of the skin, epidermis through the dermis, down to subcutaneous fat. All structures of the epidermis and dermis are destroyed. The wound is classically described as leathery, firm, and depressed when compared to normal skin. The tissue is charred, pale, and insensitive to light and touch. It is often misdiagnosed as a Deep Second-Degree/Deep Partial-Thickness due to the similar clinical findings. This burn will not spontaneously heal. Surgical repair and/or skin grafts are necessary. There is significant scarring.

(5) Fourth-Degree: These burns involve all layers of the skin, subcutaneous fat, muscle, and bone. These are devastating, life threatening injuries. They almost always have a charred appearance, and often only the cause of the burn gives a clue to the amount of underlying tissue destruction. (FMST.07.37a)

c. Burn Classification: The American Burn Association has devised a classification of burns, dividing them into major, moderate, and minor burns. Patients are placed in groups related to their risk. Low-risk patients are between the ages of 10 and 50 years old. High-risk patients are less than 10 and greater than 50 years old. Poor-risk are patients with underlying medical illnesses such as heart disease, diabetes, or chronic pulmonary problems.

(1) Minor Burns: Imply outpatient treatment. Minor burns involve TBSA of less than 15% in the low-risk group, or less than 10% TBSA in the high-risk group; full-thickness burns of less than 2% TBSA in anyone; and no other injuries.

(2) Moderate Burns: Are partial-thickness burns of 15-25% TBSA in the low-risk group, 10-20% TBSA in the high-risk group; full-thickness burns of equal to or less than 10% TBSA in anyone; burns not involving the hands, face, feet, perineum, or circumferential limbs.

(3) Major Burns: Defined as partial-thickness burns greater than 25% TBSA in the low-risk group, or greater than 20% TBSA in the high risk group; full-thickness burns of greater than 10% TBSA in anyone; burns involving the hands, face, feet, or perineum; burns crossing major joints; circumferential limb burns; burns complicated by inhalation injury; electrical burns; burns complicated by fractures or other forms of trauma; and burns in poor-risk patients. These burns require rapid casualty evacuation. (FMST.07.37b)

5. **TREATMENT:**

a. Initial Care:

(1) The first step in the treatment of burn injuries is to stop the burning process. (FMST.07.37c) Remove patients from heat, remove all their clothes that are affected by hot liquids, shut off electricity (very important), remove chemicals by flushing the affected area for 10 to 15 minutes, and decontamination of NBC. Heat can continue to injure tissue for a surprisingly long time. No first aid will be effective until the burning process has stopped.

(2) Manage the ABC's. Once an airway has been established give oxygen. If qualified, consider intubation earlier rather than later.

(3) Assess for associated injuries such as fractures or lacerations and inhalation injury.

(4) Remove clothing and jewelry from the burn area. This will help prevent constriction from swelling of the burn tissue. Do not try to remove anything that has adhered to the wound.

(5) Evaluate the Burn (depth, extent, pain).

b. General Treatment for the Patient:

(1) Gently wash the burn with tepid water and mild soap to remove any debris and to clean the skin surface around the burn site. Pat dry. Remove the skin from blisters that have popped open (do not open blisters unless necessary for function of hands or feet).

(2) Dress burn with a thin layer of antibiotic ointment.

(3) Cover the burn with Second Skin (if the burn is small enough) or cover with a thin layer of gauze, or clean, dry clothing. Burns on the face, neck and hands may be left open to the air after applying silvadene ointment. Covering the wound reduces pain and evaporative losses, but do not use an occlusive dressing.

(4) When evacuation is imminent, do not redress or reexamine the injury. If evacuation is prolonged, redress once a day. Remove old dressings, reclean (removing the old ointment), and apply fresh ointment and a clean dry covering. (Note: if stuck, soak off old dressings with clean, tepid water.)

(5) Do not pack wounds, which are larger than 20% of the body surface area, in ice. Do not leave wet coverings on burns for more than two hours at a time to reduce the risk of hypothermia.

(6) Stabilize the body temperature. When skin is lost, so is the patient's ability to thermoregulate.

(7) Have the patient drink as much fluid as he or she can tolerate without vomiting. Include some salt in the oral fluids, but do not make these solutions stronger than . 9%. This is the equivalent of a pinch of salt per 8 ounce glass.

(8) IV therapy as follows, Baxters (Parkland) Formula: In the first 24 hours give Ringers Lactate, 4ml times the weight in kilograms times the percent TBSA burned. Half of the solution is given in the first eight hours. The second half is given over the next 16 hours. The second 24 hours should be given fluids to maintain blood pressure, colloids (usually in the form of albumin) are the preferred choice.

4 x wt (kg) x TBSA= amount LR given in the first 24 hours

Example a 70kg patient with 50% TBSA this is the formula for IV therapy for the first 24 hours.

4 x 70kg x 50% = 14,000ml in 1ST 24 hours
7,000ml hours 1-8
3,500ml hours 9-16
3,500ml hours 17-24

c. General Management of the Burn: Caring for the wound itself is often the least important aspect of burn care. All burn wounds are sterile for the first 24 to 48 hours. Burn management is aimed at keeping the wound clean and reducing the pain.

(1) Elevate burned extremities to minimize swelling. Swelling retards healing and encourages infection. Have the patient gently and regularly move burned areas as much as possible.

(2) Burns are tetanus-prone wounds. Check on the last known tetanus shot. The patient may need a renewal.

(3) Ibuprofen is probably the best over the counter analgesic for burn pain (including sunburn).

(4) If you have no ointment or dressings, leave the burn alone. The burn's surface will dry into a scab like covering that provides a significant amount of protection.

6. **INHALATION INJURIES:** Half of all fire-related deaths are from smoke inhalation. Smoke inhalation doubles the mortality rate for any burn. Smoke inhalation is a general term for carbon monoxide poisoning, thermal airway injury, and smoke poisoning.

a. Carbon Monoxide Poisoning: Carbon monoxide (CO) is a colorless, odorless, tasteless gas that has an affinity for hemoglobin 200 times greater than that of oxygen. Carboxyhemoglobin (COHb) levels are measured as the percentage of hemoglobin bound to carbon monoxide. The following COHb levels correlate with clinical symptoms:

<10%	No symptoms
20%	Headache, nausea, vomiting, loss of dexterity
30%	Confusion, lethargy, ECG ST depression
40%	Coma
50%	Death

The half-life of COHb is 4 to 5 hours at room air, 90 minutes with 100% oxygen, and 20 to 25 minutes at 3 ata in a hyperbaric chamber. Treatment for less than 40% COHb is a 100% oxygen. At 40% or greater the hyperbaric chamber should be used. If patient is still in a coma after COHb levels are normal, prognosis is poor, and it has been experience that they rarely awaken.

b. Thermal Airway Injury: Injury to the respiratory tract from inhalation of hot gas or steam. The lower respiratory tract is usually not damaged, unless there was an inhalation of flame or steam. Any patients who were in an explosion, with burns of the hands, face, and upper torso are at risk. Signs of carbon such as soot in the pharnyx should alert the caregiver to airway injury. Maintaining the airway is the important issue here.

c. Smoke Poisoning: Is the inhalation of noxious gases that are the products of combustion. Cyanide poisoning is known to occur in victims of smoke inhalation. Other toxic gases released are sulfur dioxide, hydrogen chloride, phosgene, and ammonia. These gases cause damage to the respiratory tract similar to chemical burns. Physical signs include facial burns, intraoral or pharyngeal burns, singed nasal hairs, soot in mouth or nose, hoarseness, carbonaceous sputum, and expiratory wheezing.

UNITED STATES MARINE CORPS
Mountain Warfare Training Center
Bridgeport, California 93517-5001

WMC
04/09/02

LESSON PLAN

NOMENCLATURE AND CARE OF MOUNTAINEERING EQUIPMENT

LESSON PURPOSE. The purpose of this period of instruction is to introduce you to the types of equipment used here and how to care for it so as to prevent its untimely failure. This lesson relates to all climbing and installation work that you perform here.

1. **ROPES.** All ropes used in the military must meet UIAA standards or U. S. Federal Test Standard 191A. Most ropes have a 5-year shelf life and maximum 2-year service life.

   a. Static. Black in color.

      (1) Construction. Kernmantle

      (2) Minimum tensile strength. 7500 lbs.

      (3) Maximum elongation. 1.5%

      (4) Diameter. 11mm

      (5) Sizes.

         (a) 165 ft $\pm$ 5 ft

         (b) 300 ft $\pm$ 10 ft

      (6) Usage. Rescue operations and bridging where a low amount of elongation is desirable under a working load.

   b. Dynamic. Olive Drab in color.

      (1) Construction. Water-resistant treated Kernmantle to reduce friction.

      (2) Minimum tensile strength. 6500 lbs.

      (3) Maximum elongation. 6%

(4) Diameter. 10.5mm and 11mm

(5) Size. 165 ft ± 5 ft.

(6) Usage. For lead climbing/party climbing.

c. Maxim Dry Rope. Olive Drab or Multi-Colored.

(1) Construction. Water-repellent treated Kernmantle.

(2) Minimum tensile strength. 3472 lbs.

(3) Maximum elongation. 6%

(4) Diameter. 9mm

(5) Size. 150 ft ± 5 ft

(6) Usage. For glacier travel/ice climbing

d. Gold Line II.

(1) Construction. Eight strand braided nylon plymor.

(2) Minimum tensile strength. 4500 lbs.

(3) Maximum elongation. 20%

(4) Diameter. 11 mm

(5) Size. 300 ft or 600 ft spools

(6) Usage. Sling Ropes and litters only.

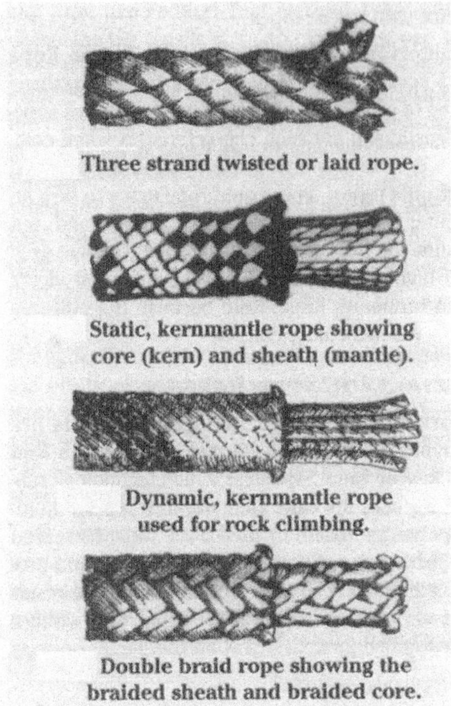

Three strand twisted or laid rope.

Static, kernmantle rope showing core (kern) and sheath (mantle).

Dynamic, kernmantle rope used for rock climbing.

Double braid rope showing the braided sheath and braided core.

INSIDES AND OUTSIDE SHEATHS OF VARIOUS TYPES OF ROPES

NOTE: Sling ropes are made from 15 foot lengths of plymor or dynamic rope ONLY. Twenty-five foot practice coils should be constructed with static rope, but dynamic rope can be used.

2. (5 Min) **ADVANTAGES/DISADVANTAGES**

a. Advantages of Nylon Rope

(1) High strength to weight ratio.

(2) Good energy absorption in dynamic ropes.

(3) Flexible.

(4) Rot resistant, not affected by frost.

b. Disadvantages of Nylon Rope

(1) Low melting point. Nylon fuses at 400°F and melts at 480°F.

(2) Susceptible to abrasions and cuts.

(3) Affected by chemicals and light.

c. <u>Advantages of Manila Rope</u>

(1) Easily gripped.

(2) Hard wearing.

(3) Does not deteriorate in heat.

d. <u>Disadvantages of Manila Rode</u>

(1) Heavy, kinks, especially when wet. Absorbs water and swells.

(2) Burns at +300°F.

(3) Edible by rodents.

3. (3 Min) **GENERAL INFORMATION**

a. Nylon rope stretches under tension and will rupture at between 30% and 70% elongation, depending on construction.

b. Nylon rope loses as much as 30% strength when wet.

c. Temperatures as low as 250°F will damage a nylon rope.

4. (5 Min) **NYLON WEBBING**

a. The type of nylon webbing available is tubular. Tubular nylon webbing is very strong and flexible. All rules that apply to nylon rope apply to tubular nylon webbing. The size of nylon webbing used is:

(1) 1 inch tubular nylon. Tensile strength approximately 4,000 - 4,500 lbs., depending on the manufacturer.

b. Pre-sewn Spectra Runners. Tensile strength approximately 5,500 lbs.

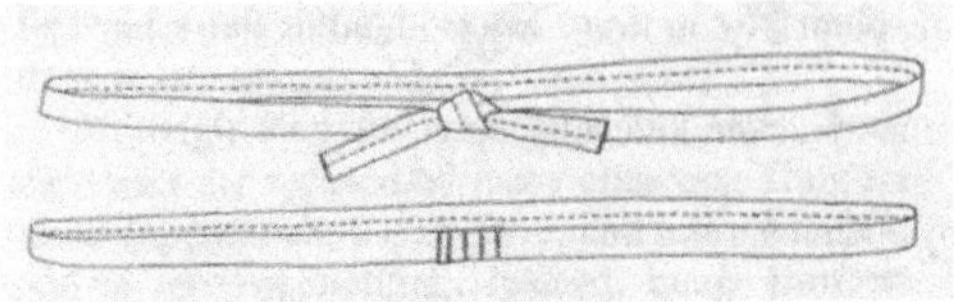

NYLON WEBBING

NOTE: These are minimum strengths. Some manufactures make even stronger webbing.

TRANSITION: We have just discussed general information in nylon webbing, are there any questions? Not only do we use ropes, but we also use carabiners in our installations, we will discuss the types of carabiners used:

5. (5 Min) **CARABINERS**. Also commonly known as snaplinks. Both locking and non-locking are used.

   a. Purpose. Carabiners are used for the following purposes:

      (1) To attach ropes or runners to pieces of protection.

      (2) To attach the rappel rope to the rappel seat for seat-hip rappels or for crossing rope bridges.

      (3) To attach the individuals safety rope to a safety line on a rope installation.

      (4) To form field expedient pulley systems.

   b. Nomenclature of a non-locking carabiner

(1) Gate

(2) Gate pivot pin.

(3) Locking pin.

(4) Body.

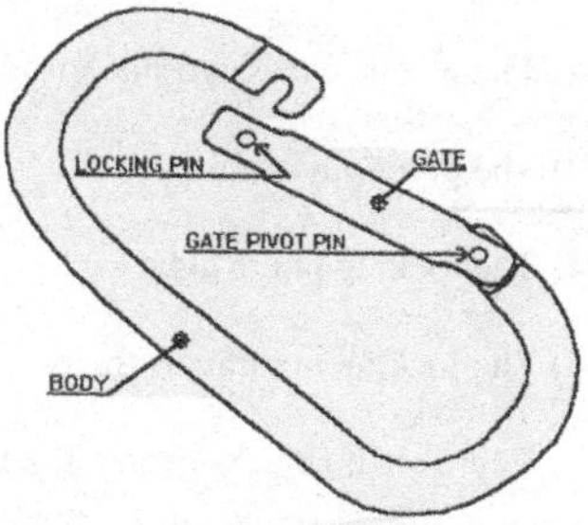

c. Nomenclature of a locking carabiner

NON-LOCKING ALUMINUM CARABINER

(1) Gate.

(2) Gate pivot pin.

(3) Locking notch.

(4) Locking nut.

(5) Body.

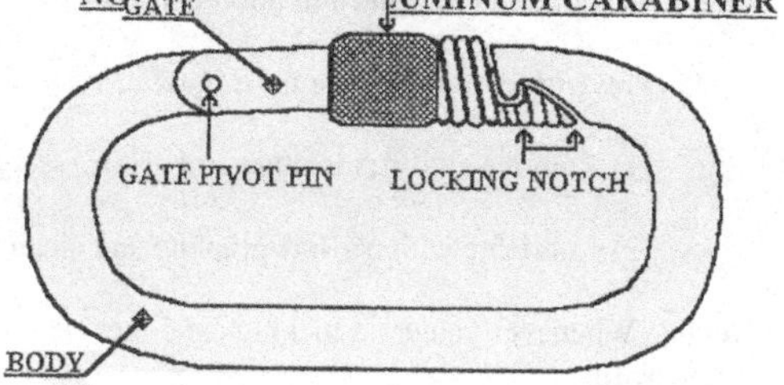

STEEL LOCKING CARABINER

d. There are two types of carabiners used. The two types and their characteristics are:

(1) Steel locking carabiners

(a) Large steel locking "D" (various manufacturers): Minimum tensile strength of 5,500 lbs.

(b) Steel-locking oval Stubai 82 is not in the MAC Kit and obsolete. However, it is being used at MWTC to save money, even though they are beyond the service life. Tensile strength of only 3,300 lbs.

(2) Aluminum non-locking carabiners

(a) Aluminum non-locking oval (various manufacturers): minimum tensile strength of 4,200 lbs.

e. Serviceability Check for a Carabiner. The following steps are used for you to check a carabiner for serviceability:

(1) The gate snaps shut with no friction and with no gap between the locking pin and locking notch.

(2) There is no excessive side to side movement of the gate.

(3) The pivot pin is tight.

(4) The locking pin is tight.

(5) The locking nut travels freely and locks securely.

(6) There are no cracks or flaws in the metal.

NOTE: The weakest part of a carabiner is the gate. If an engraver is used to mark a carabiner, it should be applied to the gate and not the load bearing side.

f. Preventive Maintenance for a Carabiner.

(1) Remove all dirt, moisture and grime.

(2) Lubricate with tri-flow graphite and clean off thoroughly.

NOTE: Whenever you use a locking carabiner ensure that the locking nut is always locked down (tightened).

6. (2 Min) **CARE OF THE CARABINER**. Do not drop the carabiner as this may result in either actual damage to the carabiner or in dirt getting into the workings of the carabiner and damaging it.

UNITED STATES MARINE CORPS
Mountain Warfare Training Center
Bridgeport, California 93517-5001

FMST.07.11
05/01/01

STUDENT HANDOUT

HYPOTHERMIA/REWARMING

TERMINAL LEARNING OBJECTIVE: Given a hypothermia casualty in a cold weather environment and necessary equipment and supplies, manage a hypothermia casualty to prevent death or further injury per the reference. (FMST.07.11)

ENABLING LEARNING OBJECTIVES:

(1) Without the aid of references, select from a given list the definition of hypothermia, in accordance with the references. (FMST.07.11a)

(2) Without the aid of references, select from a given list three of the five mechanisms by which the body loses heat, in accordance with the references. (FMST.07.11b)

(3) Without the aid of references, select from a given list the reason why gentle handling of a hypothermia victim is important, in accordance with the references. (FMST.07.11c)

(4) Without the aid of references, select from a given list the four rewarming techniques used to rewarm a hypothermia victim, in accordance with the references. (FMST.07.11d)

(5) Without the aid of references, select from a given list the proper steps of the sleeping bag rewarming method of rewarming a hypothermia victim, in accordance with the references. (FMST.07.11e)

(6) Without the aid of references, select from a given list the temperature of the water when rewarming a hypothermia casualty in a warm water bath, in accordance with the references. (FMST.07.11f)

8

1. **HYPOTHERMIA**:

a. Definition: A fall in core body temperature to 95°F (35°C) or less. (FMST.07.11a)

b. Hypothermia is potentially lethal. The mortality rate is greater than 50% in severe cases and cases complicated by trauma where the mortality can approach 100%. The most common cause of death in hypothermia is ventricular fibrillation.

c. When speaking of hypothermia, core temperature values must be used. In a controlled environment of an MTF or ER the gold standard is a core temperature taken by rectal or esophageal thermometry. The term "exposure" has often been used to speak of hypothermia but this term is inexact and has numerous other connotations. It should not be used as a synonym for hypothermia.

d. Incidence.

(1) Classically, hypothermia has been thought to occur in extremely cold environments. However, it can occur in temperate climates and at Southern latitudes. In fact, greater than 50% of cases of hypothermia occur at ambient temperatures of 50°F or greater. Contrary to popular belief, the highest incidence of hypothermia is seen in urban areas.

(2) Hypothermia tends tends to occur more frequently in males with a male to female ratio of 2.8:1.0. It's also more common in the elderly, those with drug and alcohol dependency and those with psychiatric diagnoses.

Many standard clinical thermometers don't register a low enough temperature to accurately diagnose hypothermia. Currently, a low-reading rectal thermometer is available through the Federal Stock System (NSN 6515-00-139-4593). All medical personnel attached to cold weather units or deploying to cold weather environments should carry this thermometer. In a high altitude cold weather environment medical providers should be adept at correlating signs and symptoms with specific temperature ranges so as not to unduly expose the patients to the elements. The more accurate, and invasive temperature determination should only be performed when the patient can be adequately protected from the elements.

2. **PATHOPHYSIOLOGY**:

a. Heat Conservation/Loss.

(1) The body is a furnace that generates heat. The ability of the body to thermo regulate depends on the ratio of heat production to heat loss. An imbalance of these two processes can lead to hypothermia. About 75% of food energy consumed goes toward generating body heat while the other 25% goes toward maintaining basic metabolic function.

(2) Thermogenesis can be divided into two main categories:

(A) Shivering thermogenesis creates heat by muscle activity with subsequent breakdown of glycogen. Shivering usually begins at a core temperature of around 97°F (36°C) and finally ceases at core temperatures of 85°- 90°F (30°- 34°C).

(B) Non-shivering thermogenesis creates heat by hormonal interaction with target organs such as the heart. Epinephrine, norepinephrine, growth hormone and thyroid hormone have all been implicated in playing a role in non-shivering thermogenesis.

(3) Changes in hormone secretion patterns are extremely important in acclimatization to cold weather.

(4) Once heat is generated it is retained by three main mechanisms: Peripheral vasoconstriction, abolition of sweating and piloerection. Peripheral vasoconstriction is by far the most important and efficient mechanism of heat conservation. The main internal temperature sensors in the body are located in the hypothalamus in the brain. When these areas sense a drop in core temperature they initiate peripheral vasoconstriction at the surface of the skin to prevent a loss of core body heat.

(5) The body loses heat by five main mechanisms: (FMST.07.11b)

(A) **Radiation** is heat transferred to the surrounding environment via infrared radiation.

(B) **Convection** is the loss of heat via air/water current.

(C) **Conduction** is the direct contact-transfer of heat from a hot to a cold object.

(D) **Evaporation** is the conversion of water from a liquid phase to a gaseous phase (perspiration).

(E) **Respiratory heat loss** takes place when warm humidified air from the lungs is expelled into the atmosphere.

(F) The mechanism accounting for the greatest percentage of heat loss is highly variable and depends on ambient conditions, i.e., a nude man standing in a cold room will lose approximately 65% of his heat by radiation. The same man standing in calm cold water will lose most of his heat by conduction and at a rate 25 times faster.

**NOTE: The liver and brain are the most thermogenic organs on a per weight basis. However, due to their greater mass, skeletal muscle and skin produce the largest percentage of body heat.

(6) Numerous factors predispose to the development of hypothermia. These include but are not limited to:

(A) Malnutrition.
(B) Lack of adaptation.
(C) Inactivity.
(D) Sedatives.
(E) Physical exhaustion.
(F) Burns.
(G) Extremes of age.
(G) Dehydration.
(H) Comorbid illness.

3. **SIGNS AND SYMPTOMS**:

a. Clinical manifestations of hypothermia reflect CNS, and cardiorespiratory involvement.

(1) 98°-95°F (37°-35.6°C): Mild shivering, cold sensation and impaired fine motor coordination.

(2) 95°- 92°F (35°-32.8°C): Violent shivering, difficulty speaking, sluggish thinking, and amnesia. Large muscle in coordination.

(3) 92°-86°F (32.2°-30°C): Shivering is replaced by muscular rigidity. Exposed skin is blue or puffy. Movements are jerky. Dulled sensorium, but victim may be able to maintain posture and the appearance of being in contact with surroundings. Possible atrial arrhythmias.

(4) 85°-81°F (29.4°-27.2°C): Coma, lack of reflexes, possible ventricular arrhythmias. Cessation of shivering below 86oF.

(5) Below 78°F (25.6°C): Failure of cardiac and respiratory centers, pulmonary edema, ventricular fibrillation, and asystole.

b. Some patients with hypothermia will exhibit a phenomenon that is called, "paradoxical undressing."

(1) In these instances the hypothermia victim is found nude with their cloths beside them. If the victims are women they are often misdiagnosed as sexual assault cases. This phenomena results from loss of hypothalamic control over peripheral vasoconstriction which leads to flushing of the periphery with relatively warm blood causing the individual to feel hot. Subsequently the individual undresses to cool down.

c. Again it must be stated that the only way to correctly diagnose hypothermia is with a core temperature reading.

4. **FIELD TREATMENT**:

a. ABC's, C-Spine precautions, remove wet clothing, replace with dry clothes and insulate with a vapor barrier system. (Wrapping the victim in two plastic garbage bags or bubble wrap is ideal. Further insulation may be necessary). Handle the victim carefully to prevent ventricular fibrillation caused by unnecessary jarring and rough handling. (FMST.07.11c) CASEVAC.

b. Victims with altered consciousness should be NPO. Under no circumstances should ETOH be given. This is a peripheral vasodilator and will only enhance heat loss.

c. If CASEVAC is not feasible for a long period of time, or if the victim is only mildly hypothermic, it may be necessary to rewarm in the field by using a proper sleeping bag rewarming technique. Additionally, insulated hot packs placed at the high heat loss areas will help.

d. The hypothermic victim often has no discernible vital signs.

(1) Before beginning CPR thoroughly assess pulses for 30-45 seconds to detect viable cardiovascular activity.
(2) Severely hypothermic victims in ventricular fibrillation and asystole have been successfully resuscitated even after periods as long as 4 hours.

5. **REWARMING**: The re-warming of a hypothermia casualty in a BAS setting is less than optimal since the AMAL for a BAS does not include monitoring devices, laboratory facilities, etc., but rewarming of a casualty can be successfully conducted even in these fairly primitive conditions. The following are three basic techniques of re-warming:

- Passive external re-warming: a sleeping bag or blanket that is not pre-warmed.
- Active external re-warming: a pre-warmed sleeping bag, bear hugger or warm water bath.
- Active internal or core re-warming: IV fluids, warmed/humidified O2, lavage, extracorporal re-warming.

a. Four ideal re-warming strategies that may be available in the field include: (FMST.07.11d)

(1) Sleeping bag rewarming.

(2) Water bath/counter Current rewarming.

(3) Warm/humidified airway rewarming.

(3) Warmed I.V. solutions. NS is preferable as a cold liver does not like to metabolize Lactate.

6. **REWARMING TECHNIQUES (BAS)**:

a. Sleeping Bag Rewarming. (FMST.07.11e) This is the simplest, but least effective method.

(1) Make the diagnosis. Take a rectal temperature to determine if the casualty is clinically hypothermic (95°F or less). Use this temperature as a baseline to determine stabilization.

(2) Warm sleeping bag (pre-warm sleeping bag with two volunteers). Placing a hypothermic casualty into a cold sleeping bag will cause further heat loss by conduction.

(3) Strip casualty. Remove all wet clothing avoiding unnecessary handling. Muscular movement will pump cold blood to the core.

(a) Place the casualty between the two volunteers, if space permits. Their body heat is transferred from the two volunteers to the cold casualty.

(4) Monitor core temperature frequently throughout transport to ensure that the casualty's core temperature is not continuing to drop.

(5) Adequate insulation is required to prevent further heat loss to the environment.

(6) Augment heat by placing insulated heating pads in the high heat loss areas: head, axilla, groin, popliteal region and antecubital fossa. (Be careful not to cause burns)

**NOTE: It should be noted that an evacuation bag can be used; this will provide more room for the casualty and volunteers.

b. Water Bath Re-warming. (FMST.07.11f) This method has a long history of success in rewarming even severely hypothermic casualties. The casualty is placed in a portable/field expedient bathtub or a life raft. The extremities and head are kept out of the water to avoid significant afterdrop. The water bath temperature should be 104°-108°F. An anesthesia temperature probe should be used to constantly monitor the core temperature. Since the casualty will cool the bath water around him, it will be necessary to continue to add warm water to maintain the proper water bath temperature. Keep in mind however that this technique has definite drawbacks. In the event of ventricular fibrillation caused by re-warming shock, unsynchronized cardioversion can be very dangerous in a wet environment. Furthermore, if the casualty in the water is not on a firm surface CPR, if needed, can be very difficult to perform ineffective.

**NOTE: Individual in heat distress can be rapidly cooled using this method with "cool" water instead.

c. Warm Airway Re-warming. A warmed and humidified air/oxygen mixture is used to provide a warming media within the lungs. These devices consist of a method of generating warm humidified air and are usually portable. The warm humidified air modestly increases the amount of heat that can be delivered to core. There should be a thermometer in the airway tubing to monitor the temperature. To prevent injury to the bronchi, the temperature should not exceed 115°F (108° to 115°F is ideal). Warm airway rewarming should be used in conjunction with other rewarming methods. This method, by itself, will probably not deliver sufficient heat to rewarm a severely hypothermic casualty by itself and may in fact blunt the shivering response robbing the body of an important endogenous mechanism for producing heat.

Caveat: When possible, active external re-warming should always be combined with active core re-warming techniques so as to minimize the effects of afterdrop. This occurs as a result of peripheral vasodilatation causing warm core blood to move to the surface while relatively cold surface blood moves to the core and may lower the fibrillation threshold.

**NOTE: Hypothermic submersion incident casualties cannot tolerate humidified air for any length of time.

d. Warmed I.V. Solutions. Hypothermia is a common response to I.V. therapy. Solutions that have been prewarmed have been shown to prevent this complication. There are various methods for warming I.V. solutions:

(1) Crystalloid solutions can be warmed by warm water bath and microwaved with no adverse changes to its integrity.

(2) Fresh Frozen Plasma (FFP; -20°C): Studies have shown that microwaving for 30 seconds followed by gentle manipulation of the bag for 10 seconds, repeated five times will result in only 2.6% cell destruction.

(3) Packed Red Blood Cells (PRBC 4°C): Using conventional rewarming methods, studies have shown that the outer five millimeters of the bag can have isolated areas of hemolysis. Until a device is made that can uniformly shake up PRBC during warming, we do not recommend microwaving of blood. Rewarming of blood by dilution with warm, calcium-free crystalloid solution is useful.

WIND CHILL CHART

WIND	AMBIENT TEMPERATURE (FAHRENHEIT/CENTIGRADE)													
CALM F	35	30	25	20	15	10	5	0	-5	-10	-15	-20	-25	-30
CALM C	27	-1	-4	-7	-9	-12	-15	-18	-21	-23	-26	-29	-32	-34
	EQUIVALENT WIND CHILL FACTOR (FAHRENHEIT/CENTIGRADE)													
5 MPH F	33	27	21	16	12	7	1	-6	-11	-15	-20	-26	-31	-35
8 KPH C	1	-3	-6	-9	-11	-14	-17	-21	-24	-26	-29	-32	-35	-37
10 MPH F	21	16	9	2	-2	-9	-15	-22	-27	-31	-38	-45	-52	-58
16 KPH C	-6	-9	-13	-17	-19	-23	-26	-30	-33	-35	-39	-43	-47	-50
15 MPH F	16	11	1	-6	-11	-18	-25	-33	-40	-45	-51	-60	-65	-70
23 KPH C	-9	-12	-17	-21	-24	-28	-32	-36	-40	-43	-46	-51	-54	-57
20 MPH F	12	3	-4	-9	-17	-24	-32	-40	-46	-52	-60	-68	-76	-81

32 KPH C	-11	-16	-20	-23	-27	-31	-36	-40	-43	-47	-51	-56	-60	-63
25 MPH F	7	0	-7	-15	-22	-29	-37	-45	-52	-58	-67	-75	-83	-89
40 KPH C	-14	-18	-22	-26	-30	-34	-38	-43	-47	-50	-55	-59	-64	-67
30 MPH F	5	-2	-11	-18	-26	-33	-41	-49	-56	-63	-70	-78	-87	-94
48 KPH C	-15	-19	-24	-28	-32	-36	-41	-45	-51	-53	-57	-61	-66	-70
35 MPH F	3	-4	-13	-20	-27	-35	-43	-52	-60	-67	-72	-83	-90	-98
56 KPH C	-16	-20	-25	-31	-33	-37	-42	-47	-51	-55	-58	-64	-68	-72
40 MPH F	1	-4	-15	-22	-29	-36	-45	-54	-62	-69	-76	-87	-94	-101
64 KPH C	-17	-20	-26	-30	-34	-38	-43	-48	-52	-56	-60	-66	-70	-74

7 REWARMING TECHNIQUES (HOSPITAL):

a. Optimal re-warming techniques for higher echelon care facilities still tend to employ rather simple therapeutic modalities. The three safest and most effective are listed below:

- Warmed IVF.
- Warmed/humidified oxygen.
- Bear Hugger.

b. Many texts advocate the use of extracorporal blood re-warming. However, this technique has some disadvantages. It requires a high level of technical and medical support, which are often not available. Since patients undergoing bypass must be anti-coagulated, patients with multiple traumatic injuries may not be good candidates. Lastly, research indicates that anticoagulation with heparin may actually worsen the severity of frostbite that may occur in patients with hypothermia.

c. Lavage: peritoneal, thoracic and mediastinal lavage have all been suggested for use in severe hypothermia. However, the use of these techniques in inexperienced hands tends to carry a much higher complication rate. If performed correctly, peritoneal lavage can be very beneficial with a lower complication rate than thoracic lavage. Gastric, colonic and bladder lavage are indeed safer but are of limited efficacy, as they do not offer access to areas with significant surface area.

d. ACLS: At core temperatures below 86°F attempt three progressive unsynchronized shocks at 200J, 300J, and 360Js. If no response continue CPR and REWARM THE VICTIM. Re-attempt counter-shocks when core temperature rises above 86°F. Above 86°F, perform counter shocks in 2° increments above 86°F, i.e.88°F, 90°F and 92°F. Employ ACLS drug protocols above 93°F along with conventional normothermic ACLS protocol.

UNITED STATES MARINE CORPS
Mountain Warfare Training Center
Bridgeport, California 93517-5001

FMST.07.12
03/21/02

STUDENT HANDOUT

SUBMERSION INCIDENTS

TERMINAL LEARNING OBJECTIVE: Given a casualty in a cold weather environment and the necessary equipment and supplies, manage a submersion incident casualty to prevent death or further injury per the reference. (FMST.07.12)

ENABLING LEARNING OBJECTIVES:

(1) Without the aid of references, given a list of the events, which take place to a submersion victim, select the correct sequence of events, in accordance with the references. (FMST.07.12a)

(2) Without the aid of references, select from a given list, three of the five methods that can be used to increase survival time during cold water immersion, in accordance with the references. (FMST.07.12b)

(3) Without the aid of references, select from a given list, three of the six favorable prognostic signs post submersion incident, in accordance with the references. (FMST.07.12c)

(4) Without the aid of references, select from a given list, three of the four unfavorable prognostic signs post submersion incident, in accordance with the references. (FMST.07.12d)

(5) Without the aid of references, select from a given list the correct field management priorities of a submersion victim, in accordance with the references. (FMST.07.12e)

10

OUTLINE.

1. **TERMINOLOGY**.

a. Definition: A submersion incident is that medical problem that occurs after a casualty has been submersed under water.

(1) Drowning refers to death by suffocation following submersion in water.

(2) Near drowning refers to submersion in water with at least temporary survival (greater than 24 hours).

NOTE: An effort is being made to standardize the terminology used regarding submersion incidents, in order to clarify communication between care-givers. Terms like secondary drowning and other definitions are being discouraged.

2. **PATHOPHYSIOLOGY**.

a. Sequence of events: (FMST.07.12a)

(1) Panic initially, followed by a violent struggle.

(2) Gulping and swallowing air and water to avoid aspiration.

(3) Breath holding until hypoxia leads to unconsciousness.

(4) Once consciousness is lost, the gag reflex relaxes and passive influx of water into the lungs occurs. A small percentage of victims (10-15%) have significant laryngospasm, which prevents any appreciable volume of water to enter the lungs. This is referred to as "dry drowning".

(5) Aspiration is more likely in submersion victims who enter very cold water (10°C). An involuntary gasp reflex occurs, with a marked increase in respiratory rate and depth. This individual is at increased risk of aspiration, and subsequent drowning.

b. Submersion in very cold water will quickly results in hypothermia.

(1) Rapid induction of total body hypothermia (core temperatures of 89.6°F or less) may be protective in submersion incidents. Significant decreases in oxygen requirement result, especially in the case of the brain itself. Marked decreases in cerebral blood flow and oxygen demand have been demonstrated (see Hypothermia lesson). This protective effect of hypothermia has been used to explain the many cases of complete recovery after extended periods of submersion, which have only occurred in cold water.

(2) The "Mammalian Diving Reflex" slows the heart rate, shunts blood to the brain and closes the airway when more primitive marine mammals are exposed to extremely cold water. General consensus holds that this reflex is not a significant response in human beings. However, it has been found to be more prominent in new-born human infants. It has also been speculated that this "Reflex" may be more significant in those individuals who have survived extended submersion times.

c. Fatalities in submersion victims associated with hypothermia are primarily due to three known mechanisms:

(1) Immersion syndrome. Sudden death due to cardiac arrest secondary to massive vagal stimulation. Usually associated with sudden total body exposure to frigid water.

(2) Excessive fatigue and confusion leading to fatal errors of judgment.

(3) Direct hypothermic effects. Thermal conductivity in water is approximately 25 times that of air. Despite the evidence showing hypothermia's protective effects, the complications associated with hypothermia (cardiac arrhythmias, CNS depression, loss of motor control) can lead to submersion and drowning.

d. Some authorities emphasize distinguishing between "wet" and "dry" drowning. In wet drowning, aspiration of water into the lung occurs. (85-90% of casualties). In dry drowning, laryngospasm prevents aspiration of water (10-15% of cases). Ultimately, the end result is the same: obstruction to respiration, oxygen depletion, and asphyxiation. The morbidity and mortality associated with the two types of drowning are not significantly different.

e. Fresh-water and salt-water drowning are somewhat different.

(1) In fresh-water aspiration, the osmotic difference between hypotonic fresh-water and the relatively hypertonic blood leads to net water movement into the vasculature.

This flow across the alveolar membrane also leads to "washout" of surfactant, which leads to collapse of the alveolar spaces. Hypoxemia follows.

(2) In sea-water aspiration, the osmotic shift is reversed, leading to fluid outpouring from the blood stream flooding the alveoli and the interstitium. This leads to decreased lung compliance as well as direct damage to pulmonary capillaries. Again, hypoxemia results.

(3) Hypoxemia is the end result in both types of drowning. No significant difference in morbidity in survivors has been shown between fresh-water and salt-water drowning. The only bodies of water shown to have a worsening effect on prognosis are the hyper-saline bodies, like the Dead Sea and the Great Salt Lake. Otherwise, prognosis is similar.

f. Hypoxemia is the primary insult after submersion. Efforts to correct this problem should be vigorous and sustained. All clinically dead submersion casualties should receive CPR, or ACLS if available, immediately upon removal from the water, regardless of water temperature or duration of submersion. Since the overwhelming majority of naturally occurring water bodies in the world have temperatures below 92°F, all submersion victims should be treated aggressively as hypothermic casualties. Survivals after submersion times up to 66 minutes have been documented. Remember: No one is dead until they are warm and dead.

g. In the case of prolonged submersion or significant aspiration, pulmonary derangements occur rapidly. In lesser insults, the onset of symptoms may be delayed for as long as 24 hours. Physiological changes include:

(1) Cardiovascular. Casualties develop respiratory acidosis and hypoxemia, which may trigger dysrhythmias. Electrolyte changes are not usually significant in survivors of submersion incidents and should correct once spontaneous respirations are re-established.

(2) CNS. C-Spine injuries are common in submersion incidents, most commonly a result of shallow-water diving. Injury patterns include wedge, tear-drop and burst fractures of the upper C-Spine, frequently resulting in tetraplegia. Concussive head injuries also occur. Drug and alcohol use should always be considered.

(3) Metabolic. Respiratory acidosis is the initial problem. Acid/base derangement should correct after resuscitation of the casualty without specific intervention.

3. **SURVIVAL TECHNIQUES IN COLD WATER.**

a. Swimming motion increases heat loss more than it increases heat production. The direct hypothermic effects of cold-water immersion make it impossible to swim for long periods of time. The average swimmer is unlikely to swim more than a kilometer in 50°F water. Swimming is advised only when a rescue vehicle is nearby, or the distance to shore is less than ½ mile.

b. Treading Water and Drown Proofing.

(1) Treading water results in reduced survival time due to increased heat loss secondary to increased activity.

(2) "Drown Proofing" is a technique whereby the swimmer relaxes in the water, floating facedown, arms spread, allowing his face to submerge between breaths. "Drown proofing" is the least desirable survival technique in cold water since the head and neck (high heat loss areas) are maximally exposed.

c. Personal Floatation Device (PFD). Without the aid of the PFD, the swimmer caught in cold water is forced to use less desirable methods to stay afloat.

(1) If wearing a PFD, the individual can utilize the "Heat Escape Lessening Posture" (HELP). This position minimizes the exposure of the swimmer's high heat loss areas because the arms are folded across the chest and pressed to the sides, the knees are drawn up to the chest and the legs are crossed at the ankles. Unfortunately, this position is fatiguing and becomes difficult to hold.

(2) If there is more than one person in the water, the "Huddle" position should be used. Similar to a football huddle, the chest, abdomen and groin area should be pressed together. Children or injured individuals should be placed in the middle of the huddle, since they become hypothermic more quickly.

d. Increasing Survival Time: (FMST.07.12b)

(1) Try to enter the water in a lifeboat or raft.

(2) Locate and don a PFD as quickly as possible.

(3) Wear several layers of clothes, especially to protect high heat loss areas.

(4) If you must enter the water, enter gradually rather than jumping or diving.

(5) If afloat in the water, avoid unnecessary movement. Assume the HELP or Huddle position.

(6) Don't try to swim.

(7) Remember the situation is not hopeless. Have the will to SURVIVE!

4. **RESCUE TECHNIQUES**.

a. Reaching the Casualty. The basic order for water rescue is: Reach and Pull, Throw, Tow and, as a last resort, Go.

(1) Reach and pull. When the casualty is responsive and close to the shore, begin by holding out an object for him to grab then pull him from the water. Floating line is considered the best choice, but if it is not available, a tree branch, fishing pole, oar or other object will do as well.

(2) Throw. Should the casualty be alert, but too far away for you to reach and pull, throw an object that will float to them. A PFD or ring buoy is best, if available. The primary course of action is to throw anything that will float, and to do it as quickly and as accurately as possible.

(3) Tow. Once the conscious casualty has an object to hold onto, try to find a way to tow him to shore. From a secure position, tow the casualty ashore.

(4) Go. Swim to the casualty as a last resort. You should be a strong swimmer, trained in water rescue, in order to avoid the dangers associated with this form of rescue. If the casualty is unconscious or unresponsive, you must go to the casualty.

5. **SIGNS AND SYMPTOMS**.

a. History of Incident. Several questions need to be answered, as they will help determine the approach to the casualty.

(1) When a person is found floating face down on the water surface or is brought to the surface by a rescuer, there usually is little question about the cause of unconsciousness.

(2) When near-drowning is unwitnessed or occurs after a dive or other water activity (surfing), a head injury and a C-spine injury should be presumed, and C-Spine precautions should be instituted.

(3) Determine how long the casualty was submerged. This helps determine prognosis.

(4) How cold was the water? Very cold water (less than 70°F/21°C) may be protective.

(5) What was the casualty's health prior to submersion? Drowning could have been secondary to alcohol ingestion, cardiac arrest or other illness/injury.

b. Casualties may be symptomatic or asymptomatic at the scene:

(1) Respiratory signs and symptoms may include dyspnea, tachypnea (SOB), sore throat, substemal burning, pleuritic chest pain, persistent cough, frothy sputum, cyanosis, rales, rhonchi and wheezes.

(2) Cardiovascular signs may include tachycardia and hypotension.

(3) Neurological findings may include anxiety, restlessness, lethargy, confusion, convulsions, incontinence, hyporeflexia and coma.

(4) Abdominal distention is common as is fever and hypothermia.

c. Favorable prognostic signs: (FMST.07.12c)

(1) Casualty is alert.

(2) Basic or advanced life support is received at the scene.

(3) Good response to initial resuscitation.

(4) Hypothermic.

(5) Older child or adult.

(6) Brief submersion time.

d. Unfavorable prognostic signs: (FMST.07.12d)

(1) Age under three years.

(2) Fixed dilated pupils.

(3) Submersed five minutes or longer.

(4) No resuscitation attempts for the first ten minutes.

(5) PH less than 7.10.

(6) Cardio-pulmonary arrest as initial presentation.

(7) Resuscitation time greater than 30 minutes.

6. **FIELD MANAGEMENT**.

a. The scene of a submersion incident is frequently chaotic and first consideration should be the safety of the rescuers, then retrieval of the casualty from the water.

b. Priorities: (FMST.07.12e)

(1) Establish an airway and check breathing and circulation (ABC's). Rescue breathing, can be performed by a trained rescuer, while the casualty is still in the water.

(2) Observe C-Spine precautions, since cervical spine and head injuries are common, especially in diving accidents.

(3) Begin CPR as soon as possible, if necessary, when the casualty reaches the shore, or suitable rescue platform.

(4) The symptomatic patient should be given high-flow 02 and IV access should be established, if available.

(5) Medevac ASAP.

(6) It is recommended that all victims of submersion be observed in a medical facility for a minimum of 6-8 hours. It is not uncommon for asymptomatic near-drowning victims to refuse further care. All that can be done is attempt to get a contact number in order to follow-up with the victim via telephone 8 hours later.

c. Vomiting is common and should be anticipated by turning the patient's head and having suction ready, if available. Frequently, the cause of vomiting is gastric distention.

(1) The American Heart Association recommends that gastric distention be relieved only when it interferes with CPR.

(2) Gastric decompression can be achieved by "breaking" the patient. This maneuver is performed by placing the patient in the prone position and lifting upwards at the abdomen.

(3) Another method is to turn the patient to a lateral recumbent position and compress the abdomen with the palm of one hand.

(4) The preferred method of gastric drainage is to pass an NG tube and relieve distention with gentle suction.

(5) Gastric drainage should not be attempted without suction equipment available, due to the risk of aspirating vomitus.

7. **PROGNOSTIC PREDICTORS**.

a. 0 - 2 "Unfavorable Prognostic Indicators" yield a greater than 90% survival rate (see above).

b. 3 or more "Unfavorable Prognostic Indicators" yields a less than 5% survival rate.

c. Corm's Criteria: Based on the status of the patient 1 hour after initial resuscitation.

(1) A - Awake	All survive (99+%)
(2) B - Blunted	Most survive (95+%)
(3) C – Comatose	
C1 - Flexor Response	Good survival with ICU care (80+%)
C2 - Extensor Response	Good survival with ICU care (80+%)
03 - Flaccid	Poor survival
C4 - Arrest	Dismal survival

UNITED STATES MARINE CORPS
Mountain Warfare Training Center
Bridgeport, California 93517-5001

FMST.07.14

04/02/02

STUDENT HANDOUT

WILDERNESS ORTHOPAEDIC/ TRAUMA INJURIES

TERMINAL LEARNING OBJECTIVE: Given a casualty in a mountainous enviornment and, necessary equipment and supplies, manage common orthopedic injuries in a mountainous enviornment to prevent death or further injury per the reference. (FMST.07.14)

EABLING LEARNING OBJECTIVES:

(1) Without the aid of references, select from a given list the phases of clearing a cervical spine in the field in accordance with the references. (FMST.07.14a)

(2) Without the aid of references, select from a given list, the correct treatment for an open book pelvic fracture in accordance with the references. (FMST.07.14b)

(3) Without the aid of references, select from a given list, six considerations for orthopedic patient evacuation in accordance with the references. (FMST.07.14c)

OUTLINE

1. INTRODUCTION.

Orthopedic injuries occuring in the wilderness are not significantly different from those suffered in garrison. Bones break, ligaments are torn and strained, and joints are dislocated in both settings. The key to preparing for and managing wilderness orthopedic injuries is understanding that the provider will have significant limitations with regard to diagnostic tools and initial management resources at his/her disposal. The personal risks each expedition member is exposed to can rise dramatically if the team is needed to transport a victim out of the back-country. Determination of a casualty's mobility becomes a major consideration for the wilderness medical provider. Assessment of the terrain, creative use of available natural resources, and effective leadership are all necessary components of caring for orthopedic injuries in the wilderness.

←

2. GENERAL: Making the diagnosis.

The wilderness "doc", whatever his/her training background, needs to have good physical exam skills. NO X-RAY and NO LABS will be available in the back-country. These basic tools are not available in the field. The provider will need to rely on his/her observation, palpation, and auscultation skills to make a diagnosis. This person will also need to use the patient, if possible, to uncover the likely injury by taking a thorough history which includes careful review of the mechanism of injury.

A. HISTORY: Critical information can be gained from talking to your patient.

- Mechanism of Injury - Determine:
    1. High vs. low-velocity accident
    2. Direction of force
    3. Nature of activity
    4. Loss of consciousness

- Ask the casualty:
    1. Did they hear a pop, snap, crack, or breaking sound?
    2. How long ago did the injury occur?
    3. Can they move the injured body part?
    4. Was there swelling within 15 minutes of injury or did it take longer?
    5. Was there immediate disability or was there a gradual loss of function?
    6. Was there immediate discoloration of a joint?
    7. Can they bear weight on the injured extremity?

- Other important information:
    1. Evaluate the environment. (i.e. Hot/Cold, Wet/Dry)
    2. Allergies.
    3. Medications currently being taken.
    4. Previous injuries/surgeries to the site in question.

B. PHYSICAL EXAMINATION

Your powers of observation become crucial.

- Look for:
    1. Swelling.
    2. Discoloration and bruising.
    3. Obvious deformity/angulation.
    4. Open wounds. (with or without protruding bone fragments)
    5. Differences right vs. left.

6. Muscle spasm surrounding injured site.
7. Point tenderness.
8. Crepitus.

- Beware:

Consider the environment while assessing your patient
- Don't undress the casualty in the cold!
- Protect your patient from the elements. (Sun, Rain, Cold)
- Talor your physical exam to meet the constraints of the environment.
    * palpate under clothing.
    * visualize one region at a time, then re-dress.
    * set up temporary shelter from the elements.

3. Head Injuries:

a. A blow to the head may lead to increased intracranial pressure (ICP) or intracranial bleeding neither of which are managable in the wilderness. The job of the clinician is to diferentiate a serious life threatening injury from a minor one.

1. Minor injury: No loss of consiousness (LOC), or LOC of less than 15 seconds with immediate return to full alertness. The casualty can not be on medications which increase risk of bleeding or have a history of bleeding disorders. Patient may be monitored every two hours for mental status changes, lethargy, irritibility, persistent nausea and vomiting, changes in speech or visual changes.

2. Serious Injury: LOC greater than 15 seconds, and/or persisitent confusion or memory loss; signs or symptoms of increased ICP: Debilitating headache, mental status changes, persistent nausea and vomiting, appearance of clear fluid in external auditory canal, Battle sign, raccoon eyes, or seizures.

3. Field treatment of serious head injury: Suspect injury of C-spine, manage airway-be able to clear vomit, elevate head 30 degrees. Evacuate to treatment facility ASAP.

4. Spinal Injuries.

A. Cervical Spine

1. High Risk Activities:

C-spine injuries in the wilderness usually occur after either a fall from a significant height or high-velocity accident from skiing. Common winter activities predisposing participants to C-spine injuries are skiing, snow-boarding, and snow-mobiling.

2. Anatomy:

The cervical spine consists of seven cervical vertebrae interposed between the base of the skull and the thoracic spine. The cervical spine has a great deal of mobility to allow maximal range of motion for the skull. This increased mobility comes at a high price; this portion of the spine is less stable than the rest of the vertebral column, and hence more subject to injury. The C-Spine is stabilized by three longitudinal ligaments: The anterior longitudinal ligament (ALL) runs longitudinally along the anterior surface of the vertebral bodies. It is broad and very strong and helps to prevent hyper-extension of the head and neck. The posterior longitudinal ligament (PLL) also runs longitudinally, this time along the posterior surface of the vertebral bodies, within the vertebral canal itself. This ligament is relatively narrow and somewhat weaker than the ALL, and helps to prevent hyper-flexion at the neck. Third is the supraspinous ligament, connecting the spinous processes. This, too, is a strong ligament, acting to inhibit hyper-flexion.

The peripheral nerves innervating the muscles and the sensory nerves of the upper extremities originate from the cervical portion of the spinal cord and these are some of the nerves likely to be affected in the event of a C-Spine injury. Hence, it is important to assess the neurologic status of the upper extremities if a cervical injury is suspected.

3. C-Spine Injury Statistics:

-Most common injury: Flexion injury at C5/6
(so look for a deficit in the C6 distribution)

-28% of C-Spine injuries have another spinal fracture associated with it.
(so examine entire spine)

-10% of Head injuries/Facial fractures also have a C-Spine injury, especially if there was LOC.

4. Remember:

a. Head injury -> Assume C-Spine injury.
b. Perform complete Neurovascular exam
-Motor, Sensory, Reflexes, Pulses, Babinski reflex.

B. Clear a C-Spine. (FMST.07.14a)

When faced with a possible C-Spine injury in the Wilderness, clearing it without X-ray should be considered only in extreme situations. However, there are situations which demand that every effort be made to make a casualty ambulatory, because the movement of the individual would demand herculean effort from the rest of the team, potentially placing more individuals at risk of injury or prolonging exposure to extreme environmental conditions.

Clearing a C-Spine in the field should follow a three-phased assessment. If the patient fails the assessment at any one of the phases, then the team is obligated to maintain spinal precautions and transport the patient. The phases are outlined below, and expanded upon in the lecture.

PHASE ONE:

-ensure no alteration of mental status.
-maintain in-line traction.
-perform complete neurovascular exam (focusing on the upper extremities)
-palpate spinous processes one-by-one while asking casualty whether palpation causes pain (supraspinous ligament)
-distinguish between pain over spinous process and muscle soreness associated with paraspinal musculature.
-only if palpation was completely pain-free can you move on to ...

PHASE TWO:

-continue to maintain in-line traction
-palpate anteriorly to the left and right of trachea. along the anterior vertebral bodies (anterior longitudinal ligament).
-Assess for bony tenderness under the angle of the jaw to the clavicle.
-Distinguish between tenderness of the sternocleidomastoid muscle and actual vertebral pain.
-only if palpation was pain-free can you progress to...

PHASE THREE:

-while loosely maintaining in-line traction, have patient move head through active range of motion

*first flexion/extension
*followed by rotational movement

-stop test at the first indication of pain with movement

Note: If you are able to move the patient's head through a full range of motion without pain, you have effectively cleared his C-Spine. Have the patient resume activity slowly and cautiously; evaluate any other complaints noted.

B. Thoracic and Lumbar Spine:

-Look for a T-L Spine injury with calcaneal fractures. Approximtely 10% of patients with a calcaneal fracture will also have an associated lumbar fracture.

C. Pelvis:

-Think HEMORRHAGE/SHOCK if faced with a Pelvic Fracture
Up to 6 liters of blood loss possible – (internal iliac areteries)

Place gentle constricting wrap around pelvis if open book fracture is suspected. (FMST.07.14b)

4. FRACTURES.

Some common wilderness fractures:

a. Metacarpal/Phalangeal	-the hands are your means of interfacing with the environment, sometimes it's a forceful meeting.
b. Distal Radius	-Very common wilderness fracture.
c. Scaphoid	-check for snuffbox tenderness, pain with axial load to thumb.
d. Lunate	-also common, pain dorsally at base of 3rd MC.
f. Clavicle	-need good NV exam, check for pneumothorax.
g. Long bone	-Radius/Ulna: if proximal, check elbow -Humerus: radial nerve runs in spiral groove -Tib/Fib: think compartment syndrome- Fib fx may ambulate with cane -Femur: think Hemorrhage, think traction.
f. Ankle	-commonly a fracture/dislocation, check NV
g. Hip	- Leg will typically be externally rotated

REMEMBER:

- Always think HEMORRHAGE with Long Bone fractures
- Always perform NEUROVASCULAR Exam before and after treatment
- Always SPLINT AND PAD for stability, comfort, function and transport.
- Indications for REDUCTION are: NV deficit , severe angulation, severe pain, if angulation prediposes to open fx or makes transport difficult.
- OPEN FRACTURES: Have 6-8 hours to get to surgery. Field Tx- Gently wash off with betadine/ iodine solution to get off obvious dirt- don't scrub. Wrap in sterile gauze. Don't place exposed bone back under skin. Give antibiotics: Augmentin, 2nd or 3rd generation cephalosporin, a quinolone or tetracycline.

THE REASONS WE PLACE FEMUR FRACTURES UNDER TRACTION.

1. Re-establishing length tightens fascia and tamponades bleeding.
2. Dramatic pain relief.
3. Helps prevent open fractures.

4. Helps reduce secondary soft tissue damage.
5. Increases ease and safety of transport.

SOME BASIC RULES OF EXTREMITY SPLINTING:

- Splint all fractures before moving the casualty, unless in imminent danger.
- Splint all fractures as they are found, unless severe angulation complicates transport or is causing neurovascular deficits.
- Splint to include the joint above and below.
- Construct splints using uninjured extremity, then splint in the position of function, comfort or stability.

5. DISLOCATIONS.

a. Rapid Diagnosis and Reduction is imperative, if evacuation time is > 1-2 hours.

- Easier to reduce immediately after injury. (muscle spasm)
- Makes transport much easier. (increased patient comfort)
- Dramatic pain relief.
- Early reduction reduces risk of Neurovascular injury.
- Reduction could make difference between an ambulatory vs. litter patient
  the safety of the entire expedition could be placed at risk

b. Signs and Symptoms: Remember, NO X-RAYS!

- Decreased Range of Motion.
- Obvious Deformity. (compare right vs. left)
- Usually NO Crepitus.
- Typical Posture.

c. Reduction Techniques:

- Phalanges	Linear traction, buddy tape. Can not reduce volar displacement of 1st phalanyx.
- Shoulder	Bedsheet method- counter traction. Prone with weights method. External rotation. Snowbird technique.
- Patella	Extension with medially directed pressure.
- Ankle	Linear traction with knee flexion.

- Hip — Anchor pelvis, hip flexed at 30 degrees, upward traction.

- Sternoclavicular — Roll between scapulae, and apply forceful downward pressure.

-Knee — Should only be reduced as an absolute last resort. Apply anterior traction gently.

REMEMBER: Analgesia and Sedation should be utilized when available before attempting joint or fracture reduction, unless multiple traumatic injuries are present.

6. EVACUATION CRITERIA:

Conditions demanding expeditious evacuation: (FMST.07.14c)

1. Suspected cervical, thoracic, and/or lumbar spine injuries.
2. Pelvic injury with instability and/or significant blood loss.
3. Any open fractures. (6-8 hour window, antibiotics early)
4. Compartment syndromes. (pain with passive motion, pallor, pulseless, paresthesia)
5. Hip or Knee dislocations.
6. Any other injury the medical provider feels he/she is not prepared to manage.

7. OTHER TRAUMA.

Heavy bleeding: Usually can be treated by direct pressure, if a tourniquet is required,release pressure every five minutes to reassess.

Contusion: Apply ice 1st 48 hours which will provide pain relief and limit expansion of hematoma.

Lacerations and avulsions: High pressure irrigation with clean water, may remove debris with flame treated forceps. Use steri-strips for minor clean lacs. Keep dirty or puncture wounds open and give antibiotics.

Animal Bites: Copious irrigation, and antibiotic, do not close. Consider rabies if bite was unprovoked and animal was acting strange. (bats, raccoons, skunks, canines, and felines) Has never been seen in livestock, rabbits, squirrels, rats or mice. Most common zoonotic source in North America is the skunk.

Severed part:	Can be reattached up to 24 hours after injury if the cut is clean. Gently clean part, wrap in gauze and keep cool without freezing.

UNITED STATES MARINE CORPS
Mountain Warfare Training Center
Bridgeport, California 93517-5001

FMST.07.38
04/02/02

STUDENT HANDOUT

REPTILE AND ARTHROPOD ENVENOMATION BITES AND STINGS

TERMINAL LEARNING OBJECTIVE: Given a simulated casualty, in a wilderness environment, treat reptile and arthropod envenomation, in accordance with the references. (FMST.07.38)

ENABLING LEARNING OBJECTIVES:

1) Without the aid of references, select from a given list the first aid measures used for pit viper envenomation, in accordance with the references. (FMST.07.38a)

2) Without the aid of references, select from a given list the first aid measures used for coral snake envenomation, in accordance with the references. (FMST.07.38b)

3) Without the aid of references, select from a given list the first aid measures used for widow spiders (lactordectus species) bites, in accordance with the references. (FMST.07.38c)

4) Without the aid of references, select from a given list the first aid measures used for brown recluse spider (loxosceles species) bites, in accordance with the references. (FMST.07.38d)

5) Without the aid of references, select from a given list choose the first aid measures used for bee, wasp, and hornet stings, in accordance with the references. (FMST.07.38e)

12

OUTLINE.

1. **PIT VIPERS**. Crotalidae have a triangular head, cat-like vertical pupils, hinged fangs and a heat-sensitive "pit" on each side of the head between the tip of the nose and the eye. Rattlesnakes have a variable number of rattles depending upon age and number of molts. They sometimes strike without rattling. About 60% of this country's venomous bites are attributed to rattlesnakes. Cottonmouths (water moccasins) and copperheads are the other two commonly encountered North American pit vipers. Copperhead and cottonmouth venom are quite similar but weaker than most rattlesnake venom.

a. Anatomy of Pit Vipers.

(1) Venom apparatus: Venom gland; Compressor glandular muscle; Primary duct; Accessory gland; Secondary duct; Fang sheath; and fangs.

(2) Facial pit organs: Highly sensitive paired receptors of infrared radiation that can detect temperature changes of less than 0.2°C; used to detect warm-blooded prey/predators and aim strike; also pit organs may have a role in determining volume of venom injected.

(3) Rattles or buttons: Interlocking keratin rings at tip of tail in rattlesnakes; a new rattle is added with each molting cycle (every 50-400 days); some are lost due to trauma.

(4) The strike: Aimed primarily by facial pits; rarely strike farther than 1/2 their body length; speed - approximately 8 feet/second.

(a) Rattlesnakes generally inject 25-75% of their venom when they bite humans.

(b) Can bite without injecting any venom ("dry bite").

1. 20% - no envenomation.
2. 20-30% - mild envenomation.
3. 30-50% - moderate to severe envenomation.

b. Signs and Symptoms of Snake Venom Poisoning:

(1) Pit Vipers:

(a) Local:

1. Puncture wounds/scratches:

A. Pattern can be misleading (venomous vs. nonvenomous).

B. Must differentiate from other animal/insect bites or plant puncture wounds- usually venomous snakebite wounds are larger, with more bleeding.

C. May be one or many fang marks present; range from a few mm to 4cm apart.

2. Pain, edema, erythema:

A. Usually presents within 30 minutes if envenomation.

B. More severe envenomation generally yields more rapid progression and more severe pain.

C. Pain is usually characterized as burning and immediate in onset.

D. Pain usually confined to bite site.

E. Edema limited to subcutaneous tissues (with no increase in intra-compartmental pressures).

F. Usually starts within 5 minutes (if none in 10 minutes, probably no significant envenomation unless dealing with Mojave rattlesnake).

G. Gradually spreads up extremity for 36 hours.

H. Danger of airway obstruction in bites to head and/or neck.

3. Ecchymosis:

A. Ecchymosis starts within several hours.

B. May involve entire extremity.

4. Lymphangitis:

A. Venom absorbed rapidly through lymphatic system results in lymphatic inflammation.

5. Petechiae, vesicles, hemorrhagic bullae:

A. Will occur in untreated rattlesnake envenomation.

B. Rarely seen when treated early with adequate antivenin.

C. Usually occurs 6-36 hr after bite.

6. Necrosis, tissue destruction:

A. Due to direct venom effects.

B, Can be prevented if adequate antivenin is given within 2 hours of bite.

(b) Systemic:

1. Nausea, vomiting:

A. Common (early onset may indicate severe envenomation).

2. Weakness, diaphoresis, chills, dizziness/vertigo.

A. Syncope is common with all pit viper bites.

B. Frequency is proportional to severity of bite.

3. Change in taste:

A. May complain within minutes of rubbery or metallic taste.

4. Increased salivation.

5. Fever.

6. Tingling, numbness in scalp/face/fingers/toes:

A. Can occur within 10 minutes and indicates moderate to severe envenomation.

7. Fasciculations in face/neck/back/ or other involved extremity:

A. Can occur early and indicates severe envenomation.

8. Visual disturbances:

A. Blurred vision, yellowing of vision and blindness.

9. Tachycardia, bradycardia.

10. Hemorrhage, hemolysis, thrombosis and DIC:

A. Bleeding from wounds, gingival bleeding, epistaxis, Hematuria, hematemesis, Melena, lower GI bleeding, hemoptysis, peritoneal hemorrhage and cerebral hemorrhage.

B. Can occur as early as 6 hours; more common 12-72 hours after bite.

C. Systemic bleeding only occurs in moderate to severe envenomation.

D. Most coagulopathic effects secondary to proteases acting at various sites in the coagulation cascade.

11. Hypotension, shock:

A. Can occur rapidly in severe envenomation.

B. Early, due to pooling of blood.

C. Late may be due to loss of volume.

D. Generally little effect from decreased cardiac contractility.

12. Pulmonary edema:

A. Common in severe envenomation.

B. Secondary to toxin and inflammatory injury of pulmonary capillary membranes and pooling of blood in major vessels and capillary beds of the lung.

C. Compounded by direct cardiodepressant factor in some venoms.

13. Oliguria, anuria:

A. Hypotension with resultant decreased GFR. (#1).

B. Hemoglobinuria and myoglobinuria with renal tubular obstruction.

C. Direct toxic effect of venom on kidneys.

D. Acute tubular necrosis from rhabdomyolysis.

14. Paresis, Paralysis:

A. Seen with Eastern Diamondback and Mojave rattlesnakes.

B. Uncommon with other rattlesnakes.

15. Respiratory failure:

A. Uncommon with pit viper bites (except Mojave).

B. Complicated by cardiovascular failure.

16. Seizures:

A. Probably due to hypertension and hypoxia.

17. Coma:

A. Usually sensorium is clear.

B. When coma occurs, it is secondary to cerebral anoxia.

18. Death:

A. Generally occurs in 6 to 48 hours in untreated severe envenomation.

B. Can occur sooner with IV envenomation.

c. First Aid Measures for Pit Viper Envenomation. (FMST.07.38a)

(1) Do not try to kill the snake and bring it in for identification.

(a) Only vital to identify the snake when Mojave rattlesnake or coral snake bites are a possibility, as management is altered.

(b) Risk of a second bite to the victim or rescuer.

(c) Should never delay transportation of the victim.

(2) Remove any jewelry, which could become a tourniquet with progression of edema.

(3) Calm and reassure patient to decrease heart rate and circulation of venom.

(4) Constriction band.

(a) Most widely accepted first aid measure in the literature.

(b) Should be more than 1/2 inch wide.

(c) Apply to proximal end of extremity affected and wrap to approximately 5 cm above swelling (site of envenomation). Then wrap back towards proximal end of extremity from where the wrap started.

(d) Apply only tight enough to occlude lymphatic and superficial venous return.

(e) Probably no benefit when applied 30 min or more after bite.

(f) Maintain until antivenin is started.

(5) Incision and suction.

(a) Very controversial; should only be used by experienced medical personnel.

(b) Do not incise wound.

(c) Suction best applied by mechanical means (Sawyer extractor).

(d) Do not use mouth (unknown oral lesions may expose rescuer to risk).

(e) Should be started within 30 minutes of envenomation.

(f) Continue suction for 30 to 60 min.

(6) Rest and immobilization.

(a) Put victim at rest and splint extremity in position of function, at or just below heart level.

(b) Allow room for swelling in splint.

(7) Ice.

(a) Avoid any method of cooling.

(8) Watch for adverse reactions.

(a) Rare documented cases of anaphylaxis caused by snake venom in patients previously bitten or otherwise exposed to snake venom.

(b) Some evidence that venom can activate the alternate-complement system directly and cause a similar reaction.

(9) Rapid transport to a medical facility.

(a) The most important measure in first aid.

(b) No measures should delay this; antivenin is the only proven definitive therapy.

2. **CORAL SNAKES**. The elapidae species of coral snakes are brightly colored, with black noses and alternating red-yellow-red-black bands around their bodies (remember "red on yellow, kill a fellow"). They have relatively small mouths with fixed fangs. From southern Mexico through tropical South America, the rules for distinguishing coral snakes are highly unreliable. Unless you are a knowledgeable herpetologist, it is best not to pick up colorful snakes in the tropics.

a. Signs and Symptoms of Coral Snake Envenomation:

(1) Fang marks may be hard to see.

(2) Frequent delay in onset of symptoms, followed by extremely rapid progression.

(3) Little or no pain at bite site.

(4) No local necrosis.

(5) Earliest evidence may be drowsiness or euphoria.

(6) N/V, increased salivation, paraesthesias at bite site.

(7) Bulbar paralysis progressing to peripheral paralysis.

(8) Paraesthesias, and fasciculation.

(9) Occasional seizures.

(10) Mild Hypotension.

(11) Death is due to respiratory and cardiovascular failure.

b. First Aid Measures For Coral Snake Envenomation. (FMST.07.38b)

(1) None of proven benefit except rest, reassurance and rapid transportation to medical facility for antivenom administration.

3. **GILA MONSTERS**. No direct injection mechanism for their venom, but have powerful jaws that allow them to chew and tear at their victims. Drooling venom produces a substantial amount of pain and possible complications resembling pit viper envenomation. Follow the recommendations for pit viper envenomation. Fatal Gila monster encounters are extremely rare.

4. **VENOMOUS ARTHROPODS**. Arthropods represent the largest phylum in the animal kingdom. It contains spiders, scorpions, insects, ticks, kissing bugs, water

bugs, caterpillars, moths, butterflies, grasshoppers, centipedes, and millipedes, among others. Some arthropods sting (bees, ants, scorpions, etc.), others bite (spiders, centipedes, kissing bugs, etc.), while still others discharge a secretion that is toxic (millipedes, caterpillars, etc.). Although the number of arthropods that are sufficiently venomous to endanger humans is not known, these animals are implicated in far more bites and stings than all other phyla combined. Almost all of the 20,000 species of spiders are venomous, but luckily for, man, only relatively small numbers have fangs long and strong enough to penetrate the human skin. There are some 500 species of scorpions and all are venomous, although only a small number pose a significant danger to humans.

a. The number of deaths from arthropod stings and bites is not known, nor do most countries keep records of the incidence of such injuries. In Mexico, parts of Central and South America, and in North Africa, deaths from scorpion stings may exceed several thousand per year. Spider bites probably account for no more than 200 deaths per year, worldwide. The number of deaths from arthropod bites or stings in the temperate countries is far greater than the number of deaths from snakebite. Almost all of these deaths, however, are from anaphylactic reactions.

5. **SPIDERS**. -There are at least 200 species of spiders that have been implicated in significant bites to humans (Russell, 1988a). Spiders got the misplaced blame for the bites seen in one large series of case studies. Of some 600 suspected spider bites, 80% were found to be caused by other arthropods. The arthropods most frequently involved in these mis-diagnoses were ticks (including their embedded mouth parts), kissing bugs, mites, bedbugs, fleas, flies, beetles, water bugs and various other stinging arthropods.

a. Lactordectus species (Widow spiders): It is most commonly referred to as the Black Widow, Brown Widow or Frog-legged spider depending on the species. There are many other commons names such as: hourglass, poison lady, deadly spider, red-bottom spider, T-spider, Gray Lady spider, or shoe-button spider. Although both male and female widow spiders are venomous, only the latter have fangs large and strong enough to penetrate the human skin. Mature females range in body length from 10 to 18mm, whereas males are from 3 to 5mm.

(1) Signs and Symptoms of spider envenomation.

(a) Sharp, pinprick-like bite, but in some cases the incident is so minor that it goes unnoticed.

(b) Initial bite-pain is often followed by dull-numbing pain in the affected extremity, and cramps in one of the large muscle masses.

(c) Muscle fasciculations can be seen within 30 minutes of the bite.

(d) Sweating is common.

(e) Complaints of weakness and pain in the regional lymph nodes.

(f) Pain in the lower back, thighs, or abdomen is common and rigidity of the abdominal muscles is seen in severe envenomations.

(g) Severe paroxysmal muscle cramps may occur accompanied by arthralgias.

(h) Tissue reaction around the bite may vary, depending on the species involved. In most cases in the United States, there is no reaction at the bite site and the puncture wounds may be impossible to find. There may be an immediate 2 to 4 mm blanched area around the puncture wounds, surrounded by a slightly erythematous area, which in time becomes pallid. This may persist for 20 to 30 minutes.

(i) Hypertension is a characteristic finding in moderate to severe poisonings. It usually appears 2 to 3 hours post envenomation.

(j) Elevation of body temperature.

(k) Increased salivation.

(1) Anorexia.

(2) Treatment for Widow Spider envenomation.
(FMST.07.38c)

(a) First aid measures are supportive in nature. If the pain is intense, ice can be placed over the wound until the patient arrives at a medical facility.

(b) Antivenin is given in severe cases at medical facilities.

b. Loxosceles species (Violin spiders): Are known in the United States as fiddle-back, or brown recluse spiders. There are over 100 species of Loxosceles. The abdomen of these spiders varies in color from gray to orange or reddish-brown to dark brown. The violin on the cephalothorax is brown to blackish and distinct from the pale yellow to reddish-brown background of the cephalothorax. This spider has three pairs of eyes, forming a curved row. Females average 8 to 15 mm in body length whereas males average 6 to 10 mm. Both are venomous.

(1) Signs and Symptoms of Violin spider envenomation.

(a) Pain; about the same degree of a bee sting but patients may be completely unaware of the bite.

(b) Localized burning sensation around the site of injury.

(c) Pruritus is often present and the area begins to appear red, with a small-blanched area around the immediate bite site.

(d) The reddened area enlarges during the subsequent 1- 8 hours. It often becomes irregular in shape and, as time passes, hemorrhages may develop throughout the area.

(e) A small bleb or vesicle forms at the bite site and increases in size.

(f) Vesicle subsequently ruptures and a pustule forms.

(g) The whole area becomes swollen and painful.

(h) Lymphadenopathy may develop.

(i) Necrosis can develop from superficial to deep layers.

(j) Rare systemic signs and symptoms include:

1. Malaise.
2. Stomach cramps.
3. Nausea and vomiting.
4. Jaundice.
5. Spleen enlargement.
6. Hemolysis.
7. Hematuria.
8. Thrombocytopenia.

(2) Treatment for Loxosceles spider envenomation: (FMST.07.38d)

(a) There are no first-aid measures of value. In fact, all first-aid measures should be avoided, because the natural appearance of the lesion is most important in determining the diagnosis.

(b) An ice pack may be placed over the wound if the pain is severe.

(c) Excision of the bite site is no longer recommended as this has been demonstrated to cause worsening of necrosis and tissue damage. Definitive therapy includes use of dapsone, antibiotics, tetanus prophylaxis and possible steroid use.

c. There are many other species of spiders that include Steatoda species (cobweb spiders), Phidippus species (jumping spiders), Chiracanthium species (running spiders), Lycosa species (wolf spiders), and the newest member to the United States is the Hobo spider. There have been several reported deaths in the Northwestern United States. Originally a European species of brown spider, the Hobo spider was introduced into the U.S. in 1936, and little research has been done on it to date.

6. **SCORPIONS**. Approximately 75 of the 650 species of scorpions can be considered of sufficient importance to warrant medical attention. In the United States, members of the genera Hadrurus and Vejovis are capable of inflicting a painful and erythematous, but non-lethal lesion. However, species of the genus Centruroides are sufficiently dangerous to warrant definitive medical care. Most stings are inflicted by the Vejovis species.

a. Centruroides species. There are approximately 30 species of this genus confined to the New World. Of these, eight are of considerable medical importance, and most of these species are found in Mexico. In the United States, they are commonly referred to as "bark scorpions", because of the preference for hiding under the loose bark of trees, in dead trees, or in logs. Their general color is straw to yellowish-brown or reddish-brown, and they are often easily distinguishable from other scorpions in the same habitat by their long thin tail with its "telson" on the end, the pedipalps, or pincer-like claws. They can reach lengths up to 55 mm.

(1) Signs and symptoms of envenomation.

(a) Initial pain, rarely severe.

(b) Area becomes sensitive to touch (tap test).

(c) Tachycardia may ensue after one hour.

(d) Respiratory rate increases.

(e) Fasciculations may be seen over the face or large muscle masses.

(f) Complaints of generalized muscle weakness.

(g) Respiratory distress may progress to paralysis.

(h) Slurring of speech may be present.

(i) Convulsions have been reported.

(2) Treatment for scorpion envenomation.

(a) There is no first-aid measure of value for scorpion stings.

(b) Ice may be placed over the wound to reduce pain.

(c) In severe cases, antivenin may be recommended.

b. Vejovis species is closely related to the Urectonus. Both genera are ground scorpions. Whereas Urectonus usually inhabits mountain habitats from southern California to Oregon, Bejovis has a wide distribution from the southern portions of Canada, south through Wyoming and Colorado to Texas and west to California. Some species measure up to 85mm, but most are from 30 to 35mm. Vejovis spinigerus, a common western U.S. species, is often called the striped-tail scorpion.

c. Hadrurus species is native to North America. These are the longest and most stout-bodied of our scorpions, generally referred to as the giant hairy scorpions because of their size and the conspicuous bristles on their legs, pedipalps and caudal segments. Adults may can measure up to 135 mm. These are burrowing species and may be found as deep as 2 feet in sandy soil. They are native to Arizona, California and parts of Utah, Nevada and Idaho, as well as Mexico.

7. **BEES, WASPS, HORNETS AND ANTS**. While it may take over 100 bees to inflict a lethal dose of venom in most adults, one sting can cause a fatal anaphylactic reaction in a hypersensitive person. There are 3 to 4 times more deaths in the United States from bees than from snakebites. In the few fatalities that have resulted from multiple bee stings, death has been attributed to acute cardiovascular collapse.

a. Treatment for bee stings: (FMST.07.38e)

(1) The stingers of many bees may remain in the skin and should be removed by teasing or scraping rather than pulling, to avoid even more venom being injected into the patient.

(2) Ice may be applied to the sting site.

(3) Persons with known hypersensitivity to stings should carry a Bee-Sting Kit, containing an antihistamine and epinephrine when in endemic areas.

(4) Desensitization can be carried out using whole-body antigens or, preferably, whole-venom antigens.

8. **TICKS AND MITES**. Carry many diseases. In North America, some species of Dermacentor and Amblyomma cause tick paralysis.

a. Signs and symptoms of Tick Paralysis.

(1) Restlessness and irritability.

(2) Lethargy and anorexia.

(3) Generalized muscle weakness.

(4) Incoordination and ataxia.

(5) Nystagmus.

(6) Ascending flaccid paralysis with loss of deep tendon reflexes.

(7) Bulbar or respiratory paralysis may develop.

b. Treatment for Tick Paralysis:

(1) Remove the tick.

(2) Treat symptomatically (ventilator support may be necessary).

(3) Lesions should be cleaned.

(4) Antivenin and/or corticosteroids are used in severe cases.

9. **LEPIDOPTERA (CATERPILLARS)**. Caterpillars commonly cause envenomation in humans. There are at least ten families of caterpillars that are venomous.

a. MEGALOPYGE OPERCULARIS (Puss Caterpillar or Woolly Slug) The most commonly seen venomous caterpillar in the United States.

(1) Description: Hairy, flat, ovoid shaped: 30-35 millimeters in length.

(2) Location: Texas north to Maryland and Missouri.

(3) Stinging apparatus: Consists of spines intermingled with the hairs of the body:

b. SIGNS AND SYMPTOMS.

(1) Needling pain (Can be intense).

(2) Redness.

(3) Swelling.

(4) Nausea.

(5) Headache.

(6) Fever.

(7) Vomiting.

(8) Lymphadenopathy.

(9) Hypotension.

(10) Shock.

(11) Bleeding of the mucus membranes.

a. Treatment

(1) Apply contact tape to affected area, and remove tape. This will effectively remove spines from affected area.

(2) Group I corticosteroid cream and ointment.

(3) Codeine or Demerol may be required for pain.

(4) Antiemitics may be needed to control vomiting.

Note: Anaphylactic reactions have been reported, but are rare.

UNITED STATES MARINE CORPS
Mountain Warfare Training Center
Bridgeport, California 93517-5001

FMST.07.10

04/02/02

STUDENT HANDOUT

FREEZING AND NEAR FREEZING TISSUE INJURIES

TERMINAL LEARNING OBJECTIVE: Given a casualty in a cold weather environment, and necessary equipment and supplies manage common cold weather injuries in a cold weather environment to prevent death or further injury per the reference. (FMST.07.10)

ENABLING LEARNING OBJECTIVES.

1) Without the aid of references, from a given list select the correct definition of frostbite, in accordance with the references. (FMST.07.10a)

2) Without the aid of references, from a given list select the correct two mechanisms of injury for frostbite, in accordance with the references. (FMST.07.10b)

3) Without the aid of references, from a given list select the correct signs and symptoms of frostbite, in accordance with the references. (FMST.07.10c)

4) Without the aid of references, from a given list select the correct favorable post thaw signs of a frostbite injury, in accordance with the references. (FMST.07.10d)

5) Without the aid of references, from a given list select the correct unfavorable post thaw signs of a frostbite injury, in accordance with the references. (FMST.07.10e)

6) Without the aid of references, from a given list select the correct field treatment for frostbite, in accordance with the references. (FMST.07.10f)

12

7) Without the aid of references, from a given list select the correct primary field treatment consideration when treating frostbite in the field, in accordance with the references. (FMST.07.10g)

8) Without the aid of references, from a given list select the correct four methods to reduce the chance of becoming a frostbite casualty, in accordance with the references. (FMST.07.10h)

9) Without the aid of references, from a given list select the correct definition of immersion foot, in accordance with the references. (FMST.07.10i)

10) Without the aid of references, from a given list select the correct mechanism of injury for immersion foot, in accordance with the references. (FMST.07.10j)

11) Without the aid of references, given a list of categories and characteristics match the category of immersion foot to its characteristics, in accordance with the references. (FMST.07.10k)

12) Without the aid of references, from a given list select the correct field treatment of immersion foot, in accordance with the references. (FMST.07.10l)

OUTLINE.

1. **FROSTBITE**

a. Definition: Frostbite is the actual freezing of tissue. (FMST.07.10a)

(1) Frostbite usually occurs as a result of exposure to subfreezing temperatures for a long period of time. However, brief exposure to high wind-chill or very cold temperatures can also cause frostbite. Frostbite usually occurs after more than twelve hours of exposure to subfreezing temperatures. Skin temperature of the extremities is dependent on heat production and is more a function of underlying muscle mass than of circulation. Thus, the areas of the body most prone to frostbite (hands, feet, ears, nose, chin and molar regions) have small amounts of underlying muscle tissue.

(2) Incidence. Frostbite, unlike hypothermia tends to spare the extremes of age. Thus, it is most prevalent in the middle years of life and like hypothermia, the greatest incidence occurs in urban areas. Other factors predisposing a person to frostbite include: poor nutritional status, dehydration, prior cold injury, individuals from the southeastern part of the United States, diabetics, individuals with atherosclerotic disease, and use of tobacco products.

(3) Although the actual numbers are not known, persons born in the south are 3.7 times more susceptible to frostbite. Persons with a history of prior cold injury are 1.7 times more prone to have a second injury. This includes parts of the body other than the originally frostbitten region.

b. Mechanisms of Tissue Injury. Tissue injury resulting from frostbite is produced in two ways: (FMST.07.10b)

(1) *Actual freezing of the tissues.* Ice crystal formation in the extra-cellular space leads to an osmotic gradient across the cell membrane drawing water out of the cell. This process, known as crenation, causes a marked change in intra-cellular electrolyte and enzyme concentrations, which can lead to cell death or recovery depending on the duration of exposure to freezing conditions and post thaw care of the tissue. Intra-cellular ice crystal formation is much more destructive, however, this requires much faster freezing rates.

(2) *Obstruction of blood supply to the tissues.* Cell injury and death caused by ice crystal formation leads to the release of inflammatory mediators, which damage vascular endothelial cells. The net result is increased vascular permeability. Fluid leaks out of the vascular space and into the intercellular space causing tissue edema and increased blood viscosity. Elevated viscosity combined with platelet aggregation, to the walls of damaged blood vessels, lead to vascular obstruction with resultant tissue damage.

c. Signs/Symptoms. (FMST.07.10c) Initially there is discomfort and pain in the affected tissues, which can progress to loss of sensation. The skin appears white and bloodless. There is decreased or no capillary refill. The skin may feel waxy and will not glide freely over a bony prominence. With minor injury, blisters form, usually in the first 24 hours, and edema may also be present. Deep injury is not associated with substantial formation of blisters. There is also little edema seen in deep injury.

(1) Favorable prognostic signs of frostbitten tissue. These signs indicate probably little or no tissue loss, post thaw. (FMST.07.10d)

(A) Warmth of tissue.

(B) Normal tissue color.

(C) Preservation of sensation.

(D) Blister formation. *A* blister extending the full distal length of the injured digit is a good prognostic sign. However, a blister limited to the line of demarcation between normal and frostbitten tissue is a less favorable sign.

(E) Edema persisting more than 24 hours.

(F) Rapid capillary refill.

(2) Unfavorable prognostic signs. These indicate a poor prognosis with probable tissue loss, post thaw: (FMST.07.10e)

(A) Complete absence of edema.

(B) Cyanotic tissue.

(C) Continued loss of sensation.

(D) Affected areas continue to stay cold.

(3) Since it is extremely difficult to predict ultimate tissue loss from initial examination, any classification must be done retrospectively. Classically, frostbite is divided into four degrees of severity:

(A) First degree - Erythema, edema, transient tingling and/or burning.

(B) Second degree - Blisters, edema, anesthesia and/or paresthesias.

(C) Third degree - Involves the entire thickness of skin and extends into subcutaneous tissue.

(D) Fourth degree - Involves the entire thickness of the skin along with muscle and bone.

(F) Frostnip is a completely reversible injury that is technically not considered frostbite. Frostnip classically presents as a white patch and numbness and is readily warmed in the field by placing the affected extremity in one's axilla or groin. No resultant skin changes occur.

d. Field Management (FMST.07.10f)

(1) The correct treatment for frostbite is as follows:

(A) Once an extremity has suffered frostbite, the primary consideration of treatment is not to rewarm the region if there is a chance of refreezing. (FMST.07.10g) This occurs because the amount of tissue damage is greatly increased if the part is thawed then allowed to re-freeze. Generally, the sooner a frostbitten region is thawed the lower the probability and extent of damage. Rapid rewarming is the preferred method of thawing a frostbitten region. This consists of immersion of an affected part in circulating warm water of 104°-108°F until the tissues are soft and pliable, which usually takes about 40-60 minutes.

(B) If slow rewarming methods (such as allowing a part to sit at room temperature) is used, some refreezing of the melted fluid occurs and the new ice crystals are actually larger than those formed during the original freezing causing further tissue destruction.

(C) Out dated methods such as rubbing with snow or rewarming next to an open fire, etc., only enhance tissue damage.

(D) Once thawing has taken place no benefit of continued rapid rewarming can be demonstrated.

(E) Cold injuries that are old and have thawed spontaneously should not be rapidly re-warmed again.

(F) Nicotine use should be prohibited since it is a vasoconstrictor and may enhance tissue loss.

(G) There is no need for ointments, salves, etc.

(H) Blisters should be allowed to remain intact.

(I) Medevac all cases of frostbite, as soon as the tactical situation permits.

(J) One final note; in military tactical situations such as ours, the victim often must walk down below the snow line to reach a BAS or base camp where rapid rewarming can be carried out. In this situation let the victim walk if he has frostbitten feet. Once his feet have thawed, he will be unable to walk and will have to be carried out.

(K) Hospital and long-term care: After re-warming the frostbitten tissue, attention must turn to prevention of infection and functional restoration. The patient's tetanus status should be determined and meticulous care should be taken to keep the injury clean. Daily whirlpool therapy for 30-45 minutes has been shown to be beneficial. The use of post-thaw medications is controversial but there is evidence that topical aloe vera and NSAIDS to control inflammation are probably beneficial. Some studies have demonstrated benefit from the use of allopurinol as a free radical scavenger. Also, animal studies have shown a definite benefit from the use of 90% DMSO gel, however, this use is not FDA approved on humans and it may be difficult or impossible to obtain this medication. The wide scale use of heparin and dextran 40 has been suggested but a definitive recommendation that supports their use still cannot be made. Studies suggesting the use of thrombolytics are very small and not well controlled therefore there is no current recommendation for use of thrombolytics. Lastly, it has been shown that there is a clear benefit for the use of early and aggressive physical therapy to preserve function and prevent contracture.

e. Prevention (FMST.07.10h)

(1) Some general principles for the prevention of frostbite are listed below

(a) Dress to keep comfortably cool. The head radiates 30-50% or more of all body heat and so goes the old adage, "If your feet are cold put on your hat".

(b) Nutrition. Eat the right type and quantity of foods.

(c) Avoid tight fitting and constrictive clothing. This decreases blood flow and increases the chance of frostbite. Rings, jewelry, etc., conduct heat away from the skin and increase the chance of frostbite.

(d) Wear mittens instead of gloves. They are much warmer. For work involving a high level of dexterity use contact gloves.

(e) Do not touch metals with bare hands at extremely low temperatures. Your flesh will stick to the surface and this will result in subsequent frostbite and tissue loss. If you find yourself in such a predicament pour a warm fluid over the metal to thaw your flesh off the metal.

(f) Avoid nicotine.

2. **IMMERSION FOOT**.

a. Definition. (FMST.07.10i) Immersion foot (AKA trenchfoot) is a non-freezing injury of the extremities in which the tissues are damaged.

b. Usually a disease of the feet (but it can occur in the hands), it occurs when the extremity is exposed to a cold-wet environment (external temperatures above freezing and less than

50°F) for long periods of time (usually greater than 12 hours). Prolonged standing enhances the pathologic potential for injury due to wet and cold conditions.

c. Mechanism of Injury. (FMST.07.10j) Even though this injury is called immersion foot, actual immersion of the feet in water is not necessary. This injury can also occur in the hands.

(1) The circulation of blood to the affected extremity is reduced because the victim's extremity is cold, wet and the peripheral blood vessels have constricted to reduce overall heat loss.

(2) Because prolonged cooling damages sensitive tissues, nerve injuries are also involved. This nerve damage is responsible for the severe pain and paresthesias. Damaged skin and subcutaneous tissues result in erythema, edema and possible blister formation.

d. Signs/Symptoms Early signs include numbness and paresthesias followed by complete numbness and sometimes leg/arm cramps. The skin at first appears erythematous but becomes progressively pale and mottled then grayish-blue as the condition progresses. Often there is a large amount of edema and blister formation. In severe cases the arterial pulses are absent. Immersion foot is classified by severity: (FMST.07.10k)

CATEGORY	CHARACTERISTICS
Minimal	Reddening of skin; slight sensory changes
Mild	Edema; sensory changes (reversible)
Moderate	Edema, redness, blisters, intra-cutaneous hemorrhage; irreversible nerve damage
Severe	Severe edema, massive intra-cutaneous hemorrhage; necrosis, gangrene

e. Field Management (FMST.07.10l) Treatment consists of:

(1) Pat drying of extremity.

(2) Gentle rewarming.

(3) Elevation of affected extremity.

(4) Bed rest.

(5) Treatment after this is supportive. For more severe injuries, debridement or resection of a portion of the foot may be necessary. However, this is best accomplished at a higher echelon of care. Do not be fooled by this injury. These patients can get

extremely ill from wet-gangrene which may necessitate amputation of tissue. However, the initial treatment should be as conservative as possible. DO NOT TREAT THESE PATIENT'S WITH WATER IMMERSION THERAPY. If infection is present obtain surgical consult early, and plan on small sequential amputations.

3. **CHILBLAINS (PERNIO)**.

a. Definition. Chilblains are a non-freezing injury that leads to red inflamed, pruritic skin.

b. Mechanism of Injury. Pernio is caused by intermittent exposure to above freezing temperatures and high humidity. (More extreme cold is usually associated with lower humidity). Classically, chilblains occur in young individuals and most commonly on the anterior surface of the legs and the dorsum of the hands. The cause of pernio is not clearly known, but it is thought to reflect an abnormal vascular response to the environment.

c. Signs/Symptoms. Skin is red, tender, and pruritic, edematous and blanches with pressure. Itching may be severe and hyperhydrosis can occur.

d. Management. The field treatment is massage and gentle heat. This treatment is usually self-limiting and will not recur without re-exposure. Topical steroids are effective for itching and burning.

4. **CORNEAL FROSTBITE**

a. Definition. Actual freezing of the cornea.

b. Mechanism of injury. This occurs when large amounts of cold air pass over and cool the cornea, resulting in corneal clouding and loss of vision. This can occur as the result of even short periods of exposure to activities associated with high wind chill i.e. snowmobiling.

c. Signs/Symptoms. On examination the cornea will be clouded over.

d. Treatment and Prevention. Antibiotic ointment and DO NOT PATCH THE EYES. Prevention is the key.

UNITED STATES MARINE CORPS
Mountain Warfare Training
Center Bridgeport, California
93517-5001

FMST.07.22
04/03/02

STUDENT HANDOUT

LAND NAVIGATION REVIEW

TERMINAL LEARNING OBJECTIVE In a mountainous environment, navigate in mountainous terrain, in accordance with the references. (FMST.07.22)

ENABLING LEARNING OBJECTIVES

1) Given a map and compass, orientate the map, in accordance with the references. (FMST.07.22a)

2) Given a map and a protractor locate a six-digit grid coordinate, in accordance with the references. (FMST.07.22b)

3) Given a map, compass, and protractor perform a resection to locate an unknown point from three known points, in accordance with the references. (FMST.07.22c)

4) Without the aid of references, select from a given list the correct definition for true north, grid north, and magnetic north, in accordance with the references. (FMST.07.22d)

OUTLINE.

1. **THE MAP**. Before you can properly use a map, there is some basic information you will need to understand.

a. Definition. A map is a reduced or scale drawing of the ground and important things on the ground as seen from the air. Essentially it is a picture of the surface of the earth, as it would appear looking at it from an aircraft.

13

b. Marginal Information. As you take a quick look at the borders, you will notice that there is a large amount of information. When using a map, the instructions are placed around the outer edges of the map and are known as marginal information. All maps are not the same, some having more marginal information than others, and the information being located in different places around the border. It is important that you know how to use this information if you plan to get the most out of your map. The different information on the margins is as follows:

(1) Sheet name. The sheet name is found in two places: the center of the upper margin and lower right margin of the map. The map is usually named after it's outstanding cultural or geographical features. Whenever possible, the name of the largest city is used.

(2) The sheet number. The sheet number is found in the upper right margin of your map and is used as a reference number for the map sheet.

(3) Series name and scale.

(a) The series name is found in the upper left margin of your map. A map series usually comprises a group of similar maps on the same scale and on the same sheet lines or format designed to cover a particular geographical area. It may also be a group of maps designed to serve a common purpose. The name given a series is of the most prominent area.

(b) The scale notation represents the ratio of the maps distance to the corresponding (actual) distance on the earth's surface. For example, the scale notation of 1:50,000 indicates that one unit of measure on the map equals 50,000 of the same units of measure on the actual surface of the ground (1 map inch = 50,000 earth inches).

(4) Series number. The series number appears in the upper right margin and lower left margin. It is a comprehensive reference expressed either as a four digit number or as a letter followed by three or four numbers.

(5) Index to adjoining sheets. The index to adjoining sheets is found in the lower right margin of your map.

(6) Bar scales. The bar scale is located in the center lower margin of your map.

(7) Contour interval. The contour interval is located in the center lower margin of your map, just below the bar scales. It states the vertical distance between adjacent contour lines on the map. For example, every contour line represents an increase in elevation as indicated by the distance of the contour interval.

(8) Northern References. There are 3 different northern references located on the declination diagram. It is located in the right portion of the lower margin of your map.

(a) True North. A line from any position on the earth's surface to the North Pole, which is the exact top of the earth.

(b) Grid North. The north direction established by the vertical grid lines on the map.

(c) Magnetic North. The direction to the North Magnetic Pole.

(9) Legend. Is located in the lower left margin. It illustrates and identifies some of the symbols on the map. There is not enough room on a map sheet to show the true outlines of objects and land features so the mapmaker uses a set of standard symbols to represent them. Sometimes the symbols are not the same, or standard types of symbols typically used. This is because of the type of map, the scale, or the origin. Because of this possible difference, to prevent errors in symbol identification, the legend should be referred to every time a map is used. (The example on the legend may state, "unimproved road, hard surface road, jeep trail," etc. The right side of the symbol will have three symbols; the first being "unimproved road", the second "hard surface road", and the third being "jeep trail".)

(10) Map colors. The colors you see on your map are not there to make the map pretty or enjoyable to the user. They have important meaning. The colors vary with different types of maps, but on a standard, large scale map, there are five basic colors:

(a) Black. Represents most of the man-made features.

(b) Red. Represents major roads.

(c) Blue. Represents water.

(d) Green. Represents vegetation.

(e) Brown. Represents all terrain features, such as hills, draws, fingers, depressions, and saddles by the use of contour lines. These contour lines are shown in brown.

(f) Other Colors. Occasionally other colors are used to show special information. These, as a rule, will be indicated in the marginal information. Currently, there are changes being made to the color system, which may result in changes of the marginal information in regards to map colors.

2. **HORIZONTAL SCALES**.

a. Bar Scale. This special ruler, located at the bottom of the map sheet, enables you to measure a distance between 2 points on your map and change that map distance into the actual ground distance.

(1) The primary scale. The primary scale is the part of the scale that starts at the zero (0) and extends to the right.

(2) The extension scale. The second part of the bar scale is the extension scale. It starts at the zero (0) and extends left.

b. Measuring Straight-Line Distance. If you are given a mission that requires you to move from your present position to a different position some distance away and you plan to travel in a straight line between these two points, the following method is used to determine the distance you will have to travel:

(1) First locate your present position and your destination on your map. Then you will lay a straight edged piece of paper on the map so that the edge of the paper touches the location of your present position and the location of your destination. Make a tick mark on the edge of the paper at each point. Now move the paper down to the bar scale. Place the left tick mark on the zero (0) of the scale so that the other tick mark will fall somewhere on the primary scale.

c. Measuring a Curved Line Distance. If you are given a mission that requires you to move from your present position to a different position and this time you choose to move along a road to reach your destination, the following method is used to work out the ground distance you will have to travel:

(1) Find your present position, your destination, and the road you will follow, on your map. Position a straight edged piece of paper at your location and make a tick mark,. Now, rotate the paper so that it runs along the road until the road turns, and make a tick mark on both the paper and the map. Keeping these 2 tick marks together, rotate the paper again until the road turns, and again mark the paper and the map. Continue this procedure until you have reached your destination. Now using your bar scale as previously stated, find the actual ground distance your squad will be traveling.

3. **ELEVATION AND RELIEF**. In order to begin learning the subject of elevation and relief, let's first look at some definitions.

a. Contour lines. Of the several ways to indicate elevation and relief on maps, the most common is through the use of contour lines. A contour line represents an imaginary line on the ground along which all points on this line are the same elevation above sea level.

b. Characteristics of Contour Lines. There are a few characteristics of a contour line that you should know that will help you understand what they represent on your map. These characteristics are as follows:

(1) Contour lines never cross.

(2) Contour lines connect on themselves.

c. Types of Contour Lines. In addition to knowing the definition, purpose, and characteristics of a contour line, you need to realize that there are different types of contour lines.

(1) Index contour lines. Starting at zero elevation, every 5th contour line is drawn with a heavier line.

(2) Intermediate contour lines. These are the contour lines falling between the Index contour lines.

(3) Supplementary contour lines. These lines are shown as dashed brown lines.

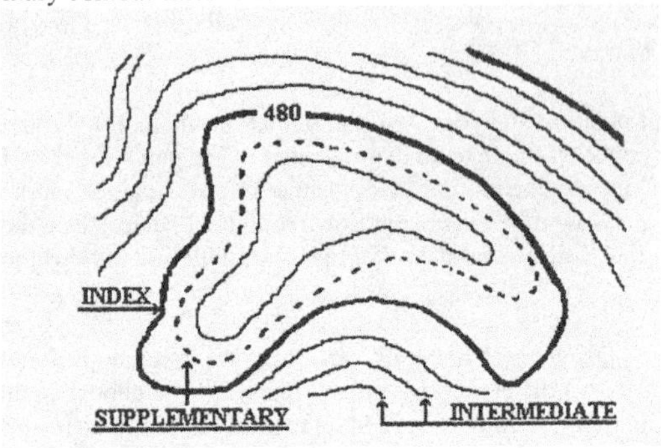

d. Determining Elevation of a Point. Now, you are ready to learn how to determine the elevation of any point of the ground represented on your map.

(1) Given elevation nearest the point, examine the area around your point until you find the nearest point that has a marked elevation. There are a few different ways that contour lines are marked, and they are:

(a) Labeled index contour line. Index contour lines are broken with the elevation of the line printed in the break.

(b) Spot elevations. At various places on your map, usually on prominent landforms such as road junctions or hilltops, a spot elevation will be given. It will be printed in either brown or black.

(c) Bench marks. Similar to Spot elevations, but more accurate. They are marked on your map by a black "x", with the letters "BM", and the elevation.

(d) A point on a contour line. If a point is located on a contour line, its elevation will be the same as that contour line.

(e) A point between contour lines. If a desired point is less than 1/4 the distance between 2 lines, its elevation is considered to be the same as the closest contour line. If the point is from 1/4 to 3/4 the distance, its elevation is considered to be in the middle of the 2 lines.

(f) A point on top of a hill. To estimate the elevation on the top of an unmarked hill, add 1/2 of the contour interval to the elevation of the highest contour line around the hill.)

4. **TERRAIN FEATURES**. There are many types of terrain features that you must be able to identify. By interpreting the arrangement of contour lines, you should be able to get a mental picture of how these terrain features would appear if you were actually out there on the ground looking at them. Some of the features you need to be able to recognize are as follows:

a. Hills. Defined as a point or small area of high ground. A hill is represented on your map by a number of contour lines each of which circles around and connects onto itself. These rough circles get smaller as they progress toward the top of the hill. You need to carefully compare the arrangement of these contour lines to the actual appearance of the hill.

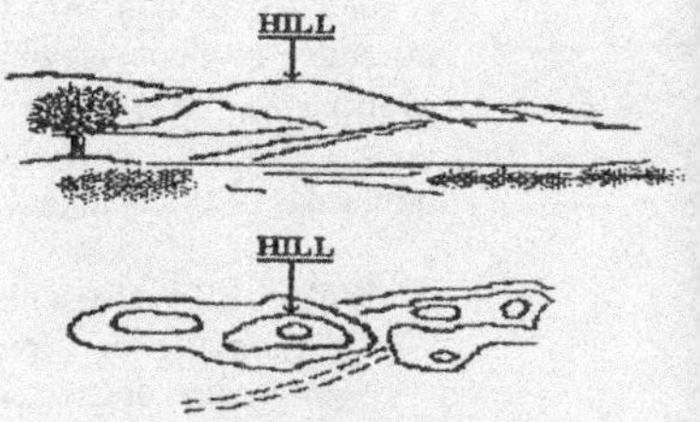

b. Spur. Defined as a line of high ground. The points along the top or crest of the spur are higher than the ground on both sides. A spur is represented on the map by a number of contour lines, each of which form a U-Shaped or V-Shaped design. The end or bottom of the U or V-Shaped lines point down to a lower elevation.

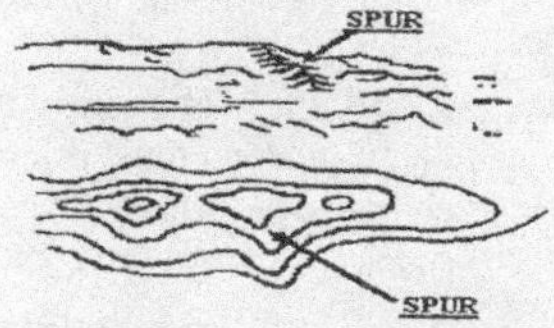

c. Draw. Defined as a line of low ground. The points located in the bottom of a draw are at a lower elevation than the ground to either side. A draw usually runs between 2 spurs. It is sometimes difficult to tell the difference between the contour lines that represent a draw from those that represent a spur because they are both U or V-Shaped lines. Remember that the curved portion representing a spur points down to ground of lower elevation, and the curved portion representing a draw points to ground of higher elevation. Now by looking closely at the area on your map, you will be able to tell which of the 2 you are dealing with. In many cases, draws will contain a stream, while spurs usually never have a stream running down their centers.

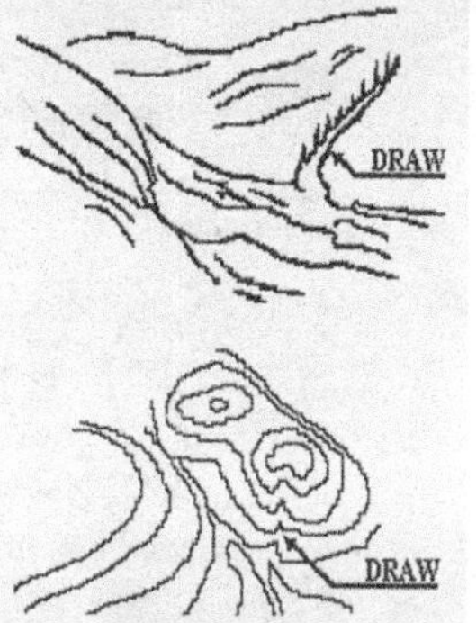

d. Saddle. Defined as the low ground between 2 hilltops, or a dip along the crest of a spur. On your map a saddle is represented by contour lines that form 2 hilltops that are located closely to each other. The area between these 2 hilltops being the location of the saddle.

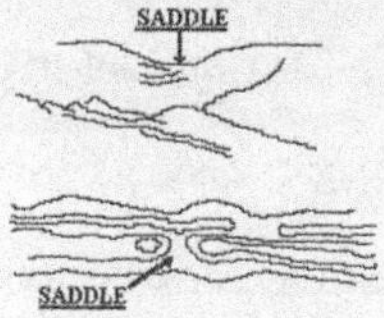

e. Depression. Defined as a point or area of low ground surrounded on all sides by higher ground. A depression is represented by contour lines forming circles within a small area on the map. These rough circles look identical to hilltops except for the presence of hachures. These are short stubby lines that are connected to the contour line at one end and point away from the contour lines at right angles in a downhill direction.

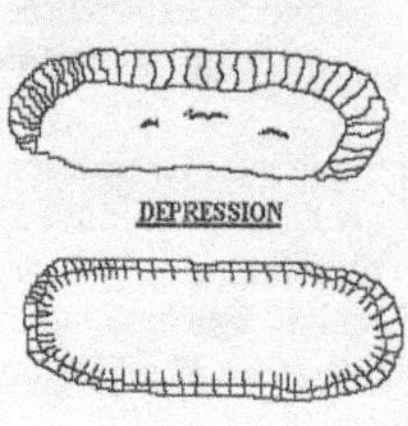

f. Cut. A cut is a man-made feature by which the bed of a road or railroad is leveled by cutting through high ground.

A cut is represented by a sudden ending of contour lines, the ends being connected by a straight line. Hachures are connected to this straight line and point downhill toward the roadbed. This sudden ending of contour lines occurs on both sides of the road bed.

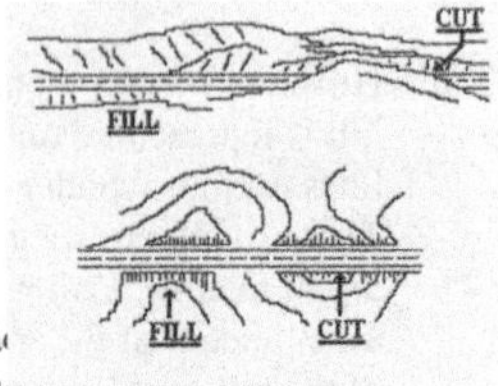

g. Fill. A fill is also a man-made feature by which the bed of a roa[d] ... filling in the low area. A sudden ending of contour lines can also represent a fill. A straight line connects these ends with Hachures pointing downhill away from the road bed. Again, this occurs on both sides of the road bed.

5. **SLOPES.** The rate of rise or fall of a ground form is known as its slope. In combat, this factor of slope must be considered when making plans for an operation. You must have the ability to recognize the ground forms in your area of operation by studying your map and the spacing between the contour lines on your map.

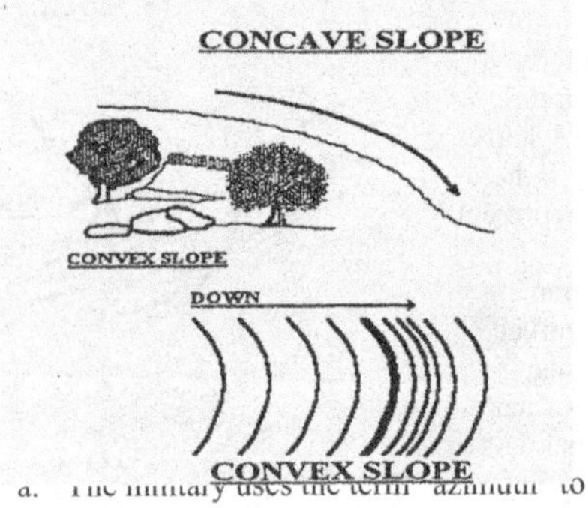

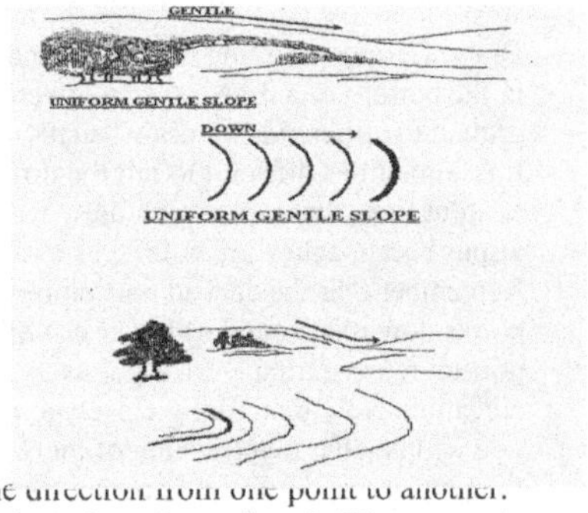

a. The military uses the term "azimuth" to describe the direction from one point to another. There are two base direction lines used to measure the value of an azimuth. They are as follows:

(1) Grid north. It is used when measuring direction from your map.

(2) Magnetic north. Used when measuring direction from your compass.

b. Units of Measurement. Now that you have learned about the two base direction lines that are used to measure direction, we need to learn the units of measurement that are used to measure the angle from one of these direction lines to the direction line.

(1) Degrees. The degree is the unit of measurement that is used most often. Anytime you need an azimuth to move from one point to another, you will use degrees as your unit of measurement. The base direction line, grid north or magnetic north, will be zero (0) degrees and measuring clockwise from this base direction line, the angle increases until you reach 360 degrees which brings you back around in a full circle to the base direction line.

(2) Mils. The mil is also used as a unit of measurement for direction. It is used when working with indirect fire weapons such as mortars and artillery. The azimuth you send will have to

be in mils, not in degrees. Remember, you are still measuring the angle between the base direction line and the line of sight from you to the distant object only you are using a different unit to measure the angle. Moving clockwise from the base direction line, which would be zero (0) mils, the angle back to the direction line. Now you know there are 360 degrees and also 6400 mils in the circle. This means that each degree equals approximately 17.5 mils.

c. Converting Map and Compass Directions. (WMC.5.2d) Earlier you learned that an azimuth was always measured from a base direction line. You also learned that we can use two different base direction lines. Grid north is the base direction line used when you are working with the map and getting a map direction or a grid azimuth. Magnetic north is the base direction line used when you are using the compass to get a compass direction or a magnetic azimuth. Because these two different base direction lines are used and because inmost cases there is a difference of one or more degrees between them, it is necessary for you to learn how to convert from a grid azimuth to a magnetic azimuth or from a magnetic azimuth to a grid azimuth. In order to learn this procedure we must take another look at the declination diagram.

(1) Declination diagram. A diagram that shows the relation between three different norths true north, grid north and magnetic north. It is the difference between grid north and magnetic north that we need to look at. The difference between these two north's is known as the grid-magnetic angle or G-M angle.

(a) Grid-Magnetic angle. The grid-magnetic angle is the angular difference between grid north and magnetic north. As you can see on this diagram a line that connects grid north and magnetic north represents the G-M angle. The size of the angle is stated off to the left side of the diagram in both degrees and mils. The G-M angle on your maps is 18 degrees or 320 mils. It is important to know that the G-M angle will change depending on where you are located on the earth's surface.

(2) Converting. When you receive a mission as a squad leader to move your squad from your present position to some distant point, you take out our map, a protractor and figure out the map direction or the grid azimuth to that distant point. However, before you can use your compass to direct you to your destination, you will have to convert the grid azimuth you have from your map, to a magnetic azimuth that you can set on your compass.

(a) The first step in making this conversion is to examine the declination diagram on your map. The G-M angle on your map is 18 degrees. You know that you will have to add the G-M angle to, or subtract it from the azimuth you measured on your map, or shot on your compass.

(b) The instructions as to whether you will add or subtract this angle are written on the map depending on whether you are going from grid to magnetic or vise-versa.

d. Orientation of a Map. (WMC.6.14a) In order to gain the best use of your map and to avoid confusion while navigating across terrain, you will need to know how to orient your map to the ground. *A* map is oriented when it is lying flat and level with it's north and south pointing in the same directions as north and south on the ground. There are two methods you can use to orient a map.

(1) Inspection. The first method used to orient the map is by inspection. By carefully examining your map and the ground around the area in which you are located to find linear features such as roads, railroads and powerless, you can align the feature on the map with the same feature on the ground. If there is only one linear terrain feature in the area in which you are located, you will have to be careful not to reverse your directions. Even though you have a road running north-south on your map aligned with the same road on the ground, the north end of your map may be pointed in a southerly direction down the road. To avoid this from occurring, find another terrain feature in the area that you can also identify on the map. Once you have aligned your map along the linear feature, check to insure that the second feature is positioned on the same side of the linear feature on your map as it is on the ground.

(2) Compass. This is the second method of orienting a map. You begin by placing your map on the ground so it lies flat and level. Open your compass so that the cover and eyepiece are completely open. Align the graduated straight edge of the compass along one of the north-south grid lines on your map with the cover end of the compass pointing toward the top of the map. This will place the stationary index line parallel to the grid north. Since the north seeking arrow on the dial of your compass will always point to magnetic north, you set your compass up just like a declination diagram. Grid north being the stationary index line and magnetic north being your north seeking arrow. Now rotate your map, insuring that the graduated straight edge of your compass remains aligned with the north-south grid line, until the face of the compass appears exactly as the declination diagram in the lower margin of the map. Your map is now oriented. For the Sonora Pass map sheet you would rotate the map until the north seeking arrow on the compass lies 16.5 degrees to the right of the stationary index line.

7. **GRID SYSTEM.** (WMC.6.14b) In most combat situations it is very important to be able to pinpoint your location, the location of another unit, or the location of an objective. In order to do this, it is necessary to have a location system that will allow you to accurately determine a specific location and state that location in such a way that will allow others to find that location quickly. This system is known as a grid system.

a. Grid Box. This box gives basic instructions on reading grids in determination of specific points on the map.

b. Grid Lines. The straight black lines that run across, and up and down the surface of your maps form the 1000 meter grid squares. These lines are called grid lines. If you look around the border of your map you will see that both the horizontal and vertical grid lines are identified by a two digit number.

c. Grid Square Identification. Each 1000 meter grid square on your map has it's own unique number. This number is called a grid coordinate. A four digit number is the result of combining the two digit number identification of the vertical grid line and the two digit number identification of the horizontal grid line that cross to form the lower left corner of the grid square. To find the grid coordinate of any grid square on your map you will always read your map right and up. You find the two digit number of that vertical grid line and you will have the first 2 digits of your coordinate. Next, find the two digit number of the horizontal grid line and place it after your vertical number and you now have the 4 digit grid coordinate that identifies your grid square.

d. Locating Terrain Features Within a Grid Square. If you are given a 4 digit grid coordinate to find a certain point on the earth's surface, that grid coordinate would put you within 1000 meters of that point. If you are told to find a certain feature within that grid square and you are good at reading your map, a 4-digit coordinate may get you close enough to your point to enable you to find it. However, there are many situations in which a 4 digit number will not get you close enough to a given position to allow you to accomplish your task. It is necessary to be able to locate points within a grid square and using a six-digit grid coordinate does this.

(1) Six Digit Grid Coordinate. The 1000-meter grid square will have to be broken down further into 100-meter grid squares. Because the surface of your map would become too cluttered by all these additional lines, a special instrument has been made to help you divide up a 1000-meter grid square. This instrument is called a coordinate scale or protractor.

8. **RESECTION**. (WMC.6.14c)

a. You will face a situation where you need to pinpoint the location of your position. This may not be possible by associating the terrain in your area with the map because you may have a poor map or the terrain in your immediate area is flat and has no distinguishing features. When you find yourself in this situation you will need to know how to perform a resection.

(1) Examine the terrain around your position and find two terrain features or objects that you can also identify on your map.

(2) Use your compass to find the magnetic azimuth to one of the distant terrain features you have identified.

(3) Convert the magnetic azimuth you found into a grid azimuth.

(4) Change the grid azimuth you found into a back azimuth.

(5) Drawing a line on your map from a distant point you identified in step one back toward your present position. Place a small mark on your map at this point. Using a straight edge, draw a line from where you plotted the distant object to the mark you made by the edge of your protractor. This line should be extended to insure it continues through the area where you are located.

(6) Repeat, finding the magnetic azimuth to the second distant object. That is at least 60 degrees from the first.

(7) When you have drawn your second line on the map, the point where the second line crosses the first line is the location of your position. The process that has just been covered is called a two-point resection.

(8) If your position is located on a linear line feature such as a road, power line, railroad or river, then it is only necessary to perform a one-point resection in order to pinpoint your location. Instead of identifying two distant points, you will only need to select one. Where the single line that you drew on your map crosses the linear feature on which you are located will be the location of your position.

9. **DETOURING AROUND AN OBSTACLE**.

a. There will be occasions when it will become necessary to move off your plotted route in order to avoid an obstacle such as a small lake or swamp. You must be able to move around these obstacles and return to the route you were following before you took your detour. This can be accomplished by using one of three detour methods: Far side landmark method, nearside landmark method, and the 90-degree offset method.

(1) Far side landmark. Used during hours of daylight when you are able to identify some feature on the far side of the obstacle that lies on your line of march and is obvious enough so that you can find it once you have moved around to the far side.

(2) Nearside landmark. Used during hours of daylight when you cannot find a landmark on the far side. However, you can identify a feature on the nearside that is on your line of march and you will still be able to see once you have moved around to the far side. Again, let us use the example of being on an azimuth of 198 degrees, and it brings you to a small lake. From the nearside you cannot locate any features on the far side. There is, however, a small pier on the nearside that is located on your line of march. You can now move around the lake using the easiest and safest route. Move to an area on the far side that is as close to being opposite of the nearside landmark as possible. You then figure out the back azimuth of your route azimuth, which in this case would be 18 degrees. Shoot an azimuth back across the lake to the pier. Move to the right or left until you are in a position where the pier lies at an azimuth of 18 degrees. This puts you back on your line of march and you can continue on following your original azimuth of 198 degrees.

(3) 90 degree offset. Used during hours of reduced visibility or during daylight hours when there are no visible landmarks along the line of march on either side of the obstacle. This time you are moving along a 360 degree azimuth and you encounter a large, impassable area of thick brush. The brush is too high to see any landmarks from either side of the thicket. The following procedure should be taken to detour around the thicket.

(a) Stop the pace count you have been making for movement along your assigned route.

(b) Decide which way you will detour around the thicket. Let's say in this case the right flank appears to be the easiest and safest. Add 90 degrees to your original azimuth of 360 degrees in order to move your patrol to the right.

(c) Move on this 90 degrees azimuth until you have reached a point where you can clear the right flank of the thicket. You will keep a new pace count starting from the point where you made your 90 degrees turn to the right to the point where you can clear the right flank of the thicket.

(d) Make another change of direction. This time you want to return to a 360 degrees azimuth so you will parallel your original course and move by the right flank of the thicket. In order to return to the 360 degrees azimuth you will now subtract 90 degrees.

(e) Move on the 360-degree azimuth until you have by passed the thicket. At the point of your second 90 degree turn, you will pick up the pace count you were making before you began making your detour around the obstacle. Even though you are not on your original line of march, you are moving parallel to your original line of march and therefore must include the distance you cover during this leg of the offset in your pace count for your route.

(f) Once you have bypassed the thicket, you will make another change of direction. This time, you want to move back toward your original line of march. Since that means you will be turning toward the left, you will subtract 90 degrees from the 360 degrees azimuth you have just completed. Your new direction will be 270 degrees.

(g) Pace back toward the original line of march following the 270-degree azimuth, the same number of paces as you made off the original line of march. In doing this you will have returned to your original line of march.

(h) Once you have moved back to your original line of march you will make your last change of direction. By adding 90 degrees to the original azimuth of the last leg of the offset, which in this case was 270 degrees, you will be back on the original azimuth of 360 degrees.

(i) Return to your original line of march pace count.

UNITED STATES MARINE CORPS
Mountain Warfare Training Center
Bridgeport, California 93517-5001

FMST.07.39
04/03/02

STUDENT HANDOUT

SEARCH AND RESCUE MANAGEMENT

TERMINAL LEARNING OBJECTIVE: In a mountainous environment, execute a search and rescue, in accordance with the references. (FMST.07.39)

ENABLING LEARNING OBJECTIVES:

(1) Without the aid of references and from a given list, choose the definition of a search, in accordance with the references. (FMST.07.39a)

(2) Without the aid of references and from a given list, choose the definition of a rescue, in accordance with the references. (FMST.07.39b)

(3) Without the aid of references and from a given list, choose one the four methods used to determine a search area, in accordance with the references. (FMST.07.39c)

(4) Without the aid of references and from a given list, choose one of the five methods for confining a search area, in accordance with the references. (FMST.07.39d)

(5) Given a lost individual scenario in a mountainous environment, plan a search and rescue in accordance with the references. (FMST.07.39e)

OUTLINE.

1. **GENERAL**. When discussing a search and rescue situation we must first be able to differentiate between the two facets of the operation. A "search" means that you are looking for an overdue missing person (FMST.07.39a). A "rescue" 'is the extraction of a known subject from a known location (FMST.07.39b). In either of these situations time is critical for several reasons:

a. Subject may need emergency medical care.

b. Size of the search area is limited.

c. Time and weather destroys clues.

d. Subject may need protection from the environment or from himself.

2. **SEARCH METHODS**. When conducting a search you will be searching for clues, instead of searching for the victim. There are two reasons for this:

a. There are more clues than victims. Anyone moving through an area will leave signs.

b. Clue detection greatly reduces search difficulty by reducing the search area.

c. This determination will assist the search manager in the use of assets, the justification for requesting outside agency assistance, and other management decisions. There are eight points that must be considered in determining the search urgency:

1. Subject profile.
2. Weather profile.
3. Equipment available to subject.
4. Subject's training, experience, and mobility.
5. Terrain hazards.
6. History of incidents in the area.
7. Time.
8. Political sensitivity.

3. **SUBJECT PROFILE**. The subject profile is based on the subject's capabilities and mobility in the terrain and conditions of the search area. The three points are age, medical condition, and number of subjects. Discussing this will assist us is rating the relative urgency:

a. Age: Very young/old, other.

b. Medical condition: Injured, ill, mental problems, healthy, or dead.

c. Number of subjects: Single or group.

d. Subject experience profile: No experience/no knowledge of area, no experience/knows area, experienced/no knowledge of area, or experienced/knows area.

e. Weather profile: Past or current bad weather, bad weather in next 8 hours, bad weather in next 24 hours, or no bad weather predicted. Temperature and visibility.

f. Equipment profile: Inadequate, questionable, or adequate for environment or weather.

g. Terrain/Hazard profile: Known hazardous terrain or, hazards or few or no hazards.

h. History of incidents in the area: Outcome of other searches for individuals with similar profiles in similar situations.

i. Time: Time since the subject went missing and the effects of time on clues.

j. Political sensitivity: External sources; VIPs, politicians, relatives, media, or higher authority in your chain of command.

4. **DETERMINING THE SEARCH AREA**. There are four methods used to determine the search area by rescue managers. The best application is to use a combination of these methods by personnel with search experience. (FMST.07.39c)

a. Theoretical: This is the distance that the victim could have traveled in the elapsed time since he/she was reported missing. (PI x R Squared)

b. Statistical: Historical data on searches for similar individuals in similar conditions.

c. Subjective: Evaluation of limiting factors for this specific incident and location.

d. Deductive reasoning: Sherlock Holmes version, based on analysis of circumstances involving the specific incident.

5. **BEHAVIOR ASPECTS OF A LOST PERSON**. General aspects of lost person behavior will give you an awareness of the actions of the individuals you are searching for by separating them into different categories. Some factors that could effect the search strategy are the general health of the subject, his/her past experiences, physiological effects of the environment, and the victims' mental state or biological cycle. Also, looking at the subject's ability to assist in his/her own survival with simple skills such as fire, shelter, signaling, and land navigation will determine strategy. The psychological impact on the victim should also be taken into account. Will the subject be overcome by Stress? Fear? Panic? Embarrassment? Or will the subject be calm and develop a positive "I'll be OK attitude? A further point to consider is the circumstances under which the subject became lost. Was the subject at a known location such as a campground? Enroute to a specific location or on a trail? Traveling in a wilderness area with no specific location or destination in mind? Or does the subject fall under one of the following categories:

a. Children age 1 to 3: Unaware of concept of being lost, no navigation skills or sense of direction, wander aimlessly, might seek out a convenient place to take a nap. i.e. hollow log, under bush/rock/picnic table.

b. Children age 3 to 6: More mobile, more inquisitive and easily distracted, will attempt to "go home" once they decide that they are lost, taught to avoid strangers and may hide from searchers.

c. Children 6 to 12: Good navigation skills and sense of direction in familiar terrain but become disoriented in unknown surroundings, may have intentionally run away and do not want to be found, may not answer when called, same fears as adults but no experiences to fall back on, once night sets in more likely to respond to searchers and except assistance.

d. Mentally retarded (all ages): Act similar to children 6 to 12, generally will not respond to name, will hide from searchers or seek shelter, once shelter found might remain in place, will do little or nothing to assist in their own survival or rescue.

e. Elderly (above 65): Possibly suffering from senility or Alzheimer's, easily confused or attracted to something that strikes their fancy, may not be oriented to the present, likely to over-extend themselves physically, and often have hearing impairments.

f. Despondent: Often are seeking solitude and will not respond to searchers, usually found near prominent terrain/scenic views, within sight and sound of civilization, rarely found in the underbrush.

g. Hikers: Rely on trails and have a set destination in mind, have poor land navigation once they loose the trail, often get confused at trail intersections or in springtime when trails are covered with snow, hiking parties of different ability levels get separated, will miss trails and head down wrong drainages. Hunters: Concentrate on game more than navigation, often pursue game into drainages/deadfalls/underbrush/deep snow, tend to over-extend themselves, typically not prepared for foul weather.

h. Cone pickers/Berry pickers/Photographers/ rockhounds/etc: Intention to stay or go to a specific location, carry little provisions or survival gear, go in good weather so they don't carry foul weather gear, often are not aware of surrounding as they are intent on pictures or the ground if pickers, these people are an extremely high risk group.

i. Fishermen: Well oriented once they arrive at lake or river, most often late due to accident, these are often recovery operations.

j. Climbers: Generally well equipped and self sufficient, tend to remain on marked routes and leave good plans with friends, primary factors are weather and hazardous conditions such as falling debris or avalanches, once located these searches often become technical rescues. 1. Marines: Subject profile will depend on individual's MOS, mission, training, and abilities learned as a civilian.

6. **BEGINNING THE SEARCH**. Once we have determined that a subject is missing we must determine whether or not he/she is in our search area. The sheriffs department will initiate a "Bastard Search", this is to ensure that the subject is not simply at the local tavern, shacked up with an old friend, or was not in an automobile accident enroute to the area. Next we will begin our search at one of the four following area:

a. Last Known Point (LKP): This is the last place that you have physical evidence to support that the subject was there. i.e. Subject's vehicle at the trailhead, name in a logbook, campsite with gear, etc.

b. Place Last Seen (PLS): The last place he was physically seen. i.e. On the trail by other hikers, at the hunting cabin, at a restaurant or gas station, or back home in San Diego.

c. Departure Point: The place where the subject was supposed to enter the search area. Usually a trailhead or specific point relating to the individual's activity.

d. Clues: As clues are discovered, and you confirm that they belong to the subject, you will update or adjust the search area.

7. **MATTSON**. Once we have determined that we do have an overdue subject and can assume that he/she is in our search area we need to break the search area into manageable sized segments. Depending upon the area to be searched, special training required to search high hazard areas, and assets available, we will divide the area by hazard or by terrain features. The lines of division should follow terrain features so that the searchers on the ground can easily navigate and stay within their zones. We will then determine the POA using the Mattson approach. This is done with each of the five or six members of the management team with the most search experience assigning a percentage to each of the segments using his individual judgment on the likelihood of the subject being in that area. Always include an extra column to represent the "rest of the world". Once all personnel have made their determinations, divide the totals and this will give you your areas to search in order of precedence:

NAME	A	B	C	D	E*	TOTAL
DUNSTON	40%	20%	10%	20%	10%	100%
KEOGH	50%	30%	5%	10%	5%	100%
KOFFEL	40%	25%	15%	10%	10%	100%
AYRES	60%	20%	10%	5%	5%	100%
TOTAL	190%	95%	40%	45%	30%	400%
AVERAGE	47.50%	23.75%	10.00%	11.25%	7.5%	
PRIORITY	1 ST	2ND	4TH	3RD	5TH	

8. **COMMAND SYSTEM**. The Office of Emergency Service (OES), has developed a command system that they use for all emergency situations. This system is called the

Incident Command System (ICS). The breakdown is very similar to the military chain of command and the duties correspond to those of the various "S" shops in the battalion. The personnel in each duty area will complete their assigned duties as you would in a military unit with the Incident Commander (IC) being overall responsible for the entire action. Normally, in California, the IC is the County Sheriff.

a. Incident Commander. CO

b. Operations-Tactics, coordination, insert resources. S-3

c. Plans-Strategy, intelligence, briefing and debriefing teams. S-2

d. Logistics-Beans, Band-Aids, beds, heads, motor-T, Special equipment. S-4

e. Communication-COC and field, landlines, frequency coordinator. S-6

f. Finance-Documentation of all man-hours, damages, legal issues.
S-1

g. Liaisons-inter agency, media, safety, family. PAO/Chaplain

9. **CONFINING A SEARCH AREA**. There are several reasons to confine the search area. The most important is that the longer the subject has free movement, the farther he/she can travel. (Remember the Theoretical Method) Additionally, this will ensure that the subject does not walk out without you being aware of it. It's real dumb to be conducting a search for a subject that is on his/her way home. The use of confinement will not interfere with the other searchers, decreases the chances of a more massive search and if promptly initiated has been shown to be highly effective. There are several methods that we can use. (FMST.07.39d)

a. Sign-cutters - Searchers that hike or ski the perimeter of the entire search area, then the perimeter of the segments. This will let you know if he/she moves from one area to another and allows you to concentrate limited assets in a specific area.

b. Trail-watchers/Road Blocks/Camp-ins - Searchers that camp at trailheads. They can alert you if the subject walks out and interview other individuals concerning possible sightings of the subject.

c. Look-outs - These personnel are already in place in fire towers. Simply notify them of the situation.

d. Track Traps - Clearing areas of the trail that are checked periodically for fresh tracks from the subject.

e. String-Lines - Spools of line with paper arrows attached that point to a safe area of a trail-watcher camp.

10. **USING CLUES**. The use of clues cannot be underestimated. They can tell you if a subject was or was not in the area, by freshness you can tell how long ago the subject was there, by the type of clue you can determine physical and mental condition, and by following clues you can determine a direction and speed of travel. Once you have located a CONCLUSIVE clue, you will re-mattson the area, using the clue as the LKP. The clues we will be looking for fall under four categories:

a. Physical - Footprints, equipment, candy wrappers, cigarette butts, etc.

b. Recorded - Summit logs or trail registers.

c. People - Witnesses who have seen the subject.

d. Event - Mirror flashes, camp fires, whistle signals, etc.

11. **ADDITIONAL ASSETS**. There are a variety of assets available to you in the search and rescue business. Some are volunteers, some charge money and some (Like the military) require higher authority to grant permission for their use. Remember that in addition to active searchers you can also use passive measures such as lights, noise, balloons, fires, smoke generators, sirens, or P.A. systems. Some examples of active searchers are:

a. Scent dogs.

b. Tracking dogs.

c. Human trackers.

d. Hasty teams.

e. Grid teams.

f. Helicopters.

g. Civil Air Patrol.

h. Snowmobile clubs.

i. Horse teams.

UNITED STATES MARINE CORPS
Mountain Warfare Training Center
Bridgeport, California 93517-5001

FMST.07.16
04/03/02

STUDENT HANDOUT

TRIAGE

TERMINAL LEARNING OBJECTIVE: Given simulated casualties, in a wilderness environment, triage casualties, in accordance with the references. (FMST.07.16)

ENABLING LEARNING OBJECTIVES:

1) Without the aid of references and from a given list, choose when actual triage most often begins, in accordance with the references. (FMST.07.16a)

2) Without the aid of references and from a given list, identify one of the four-triage categories used by NATO, in accordance with the references. (FMST.07.16b)

3) Without the aid of references and from a given list, choose the primary cause of death in a mass casualty situation, in accordance with the references. (FMST.07.16c)

OUTLINE.

1. **DEFINITION**: Triage or sorting (often used interchangeably) is the evaluation and classification of casualties for the purpose of establishing the priority for treatment and evacuation.

NOTE: Casualty sorting is one of the most important tasks for the entire medical service. No other job requires more informed judgment, hard work and courage. The Officer and the Corpsman responsible for triaging wounded Sailors and Marines must exercise sound judgment so that casualties receive proper, expedient care. Triage is a continuing process and any individual assigned should be the most capable and experienced available.

a. Decision Factors. The factors concerning casualty triage are based upon the following:

(1) The need for immediate resuscitation.

(2) The need for emergency surgery.

(3) The futility of surgery due to obviously lethal wounds.

(4) Time needed to perform surgery compared with:

(a) Probability of success.

(b) Number of other casualties needing treatment.

b. The actual triage begins most often with the casualty himself. (FMST.07.16a). This usually occurs when the injured individual determines that "Buddy-Aid" or "Self-Aid" will not be adequate.

(1) The corpsman, once presented with the casualty, must make several determinations:

(a) Is the casualty a walking-wounded or a litter case?

(b) If severely wounded, are resuscitative efforts required immediately?

(c) Does the casualty need any other immediate treatment (bleeding, asphyxia or pain control)?

(2) Once the casualty has been transported to a Battalion Aid Station (BAS), triage continues. The dental officer will commonly perform the duties of the Triage Officer at the echelon I or echelon II level, if present.

2. **NATO TRIAGE CATEGORY CODES**. (FMST.07.16b)

a. At each medical treatment facility (MTF) in the area of operations (AO), incoming casualties are classified by level of treatment required. Four triage category groups have been universally adopted for use by both United States and NATO Forces. These categories are listed and defined as they appear in standardization agreement (STANAG) No 2879:

(1) Immediate Treatment (Group T1) (RED Tag). To include those requiring emergency life-saving surgery. These procedures should not be time consuming and should concern only those patients with high chances for survival. Some examples are:

(a) Respiratory obstruction.

(b) Accessible hemorrhage.

(c) Emergency amputation.

(2) Delayed Treatment (Group T2) (YELLOW Tag). To include those badly in need of time-consuming major surgery, but whose general condition permits delay in surgical treatment without unduly endangering life. To mitigate the often, critical effects of delay in surgery, sustaining treatment, such as stabilizing I.V. fluids, splinting, administration of antibiotics, catheterization, gastric decompression, and relief of pain will be required. Examples are:

(a) Large muscle wounds.

(b) Fractures of major bones.

(c) Intra-abdominal and/or thoracic, head or spinal injuries.

(d) Uncomplicated major burns.

(3) Minimal Treatment (Group T3) (GREEN Tag). To include those with relatively minor injuries who can effectively care for themselves or who can be helped by untrained personnel. Some examples are:

(a) Minor lacerations.

(b) Abrasions.

(c) Fractures of small bones.

(d) Minor burns.

(4) Expectant Treatment (Group T4) (BLUE Tag). This group comprises patients who have received serious and often multiple injuries, and whose treatment would be time-consuming and complicated with a low chance of survival. If fully treated, they may make heavy demands on medical manpower and supplies. Until the mass casualty situation is under control, they will receive appropriate supportive treatment. The extent of treatment depends on available supplies and manpower, and may involve the use of large doses of narcotic analgesics. These patients should not be abandoned, and every effort should be devoted to their comfort. The possibility of their survival, despite alarming injuries, must always be kept in mind. Examples are:

(a) Severe multiple injuries.

(b) Severe head or spinal injuries.

(c) Large doses of radiation.

(d) Widespread severe burns.

3. **TREATMENT CONCEPTS**.

a. The triage officer must be familiar with all aspects of combat casualty care (airway management, hemorrhage control, use of anesthesia, surgical procedures, fluid resuscitation and post-operative care, to name a few), but ever-present in his mind should be the following concepts:

(1) Asphyxia and hemorrhage are initially the primary causes of death in any mass casualty situation. (FMST.07.16c)

(2) Once at a definitive care facility, uncontrolled hemorrhage is the leading cause of death.

(3) Life has priority over limb, and preserving function overrides the correction of superficial anatomic defects.

(4) A casualty is in a constant state of change until the wound or injury has been repaired.

(5) Systemic disturbances caused by the wound continue until healing is complete.

(6) Combat conditions will dictate the modifications necessary in triage and care of mass casualties.

b. Priorities of Treatment. With the six concepts of treatment in mind, we can now address three basic priorities for treatment.

(1) First Priority:

(a) Asphyxia:

1. Mechanical obstruction of the airway.

2. Sucking chest wound.

3. Tension pneumothorax.

4. Maxillofacial trauma.

5. Massive hemo-pneumothorax.

(b) Shock:

1. Exsanguinating hemorrhage.

2. Major internal hemorrhage (thoracic, abdominal, and pelvic).

3. Visceral injuries/evisceration.

4. Cardio-pericardial injuries.

5. Massive muscle injury leading to Fat Embolization to the lung.

6. Full-thickness or deep-partial burns of 10 % or greater.

(2) Second Priority:

(a) Visceral injuries:

1. Perforation of G.I. Tract.

2. Pancreatic and Biliary system.

3. Genitourinary tract.

4. Thoracic wounds w/o asphyxia.

(b) Vascular injuries requiring repair:

1. Tourniquet cases.

(c) Closed head injuries:

1. Increasing loss of consciousness.

(d) Major burns:

1. Partial-thickness burns of greater than 25% TBSA in the low risk group, 20% TBSA in the high risk group.

2. Burns involving the poor risk group.

3. Burns involving the face, hands, feet, genitalia/perineum.

4. Circumferential limb burns.

(3) Third Priority:

(a) Brain and spinal injuries - requiring burr hole decompression.

(b) Soft tissue injuries.

(c) Debridement necessary with less than major muscle damage.

(d) Lesser fractures and dislocations.

(e) Eye injuries.

(f) Maxillofacial injuries w/o asphyxia.

(g) Burns under 20% in other locations other than priority 2.

4. **TRIAGE TOOLS**.

a. Trauma Score/Glasgow Coma Scale. In recent years, many methods have been proposed to speed up the classification and sorting of casualties. One method that has gained wide acceptance is the Trauma Score/Glasgow Coma Scale. Several studies have been performed using various parameters, i.e., respiratory rate, chest wall expansion, BP, capillary refill, etc. By assigning a number scale to each specific response, the likelihood of survival as a percentage can be calculated.

1. Glasgow Coma Scale.

a. Eye opening:

1) Spontaneous = 4.
2) To voice = 3.
3) To pain = 2.
4) None = 1.

b. Verbal Response.

1) Oriented = 5.
2) Confused = 4.
3) Inappropriate words = 3.
4) Incomprehensive words = 2.
5) None = 1.

c. Motor Response.

1) Obeys Command = 6.

2) Localized Pain = 5.
3) Withdraw Pain = 4.
4) Flexion (pain) = 3.Extension (pain) = 2.
5) None = 1.

d. Total the points up and they equal:

1) 13-15= 4.
2) 9-12 = 3.
3) 6-8 = 2.
4) 4-5 = 1.
5) <3 = 0

2. Trauma Scale.

a. Respiratory Rate:

1) 12-29 min = 4.
2) 10-12 min = 3.
3) 6-9 min = 2.
4) 1-5 min = 1.
5) None = 0.

b. Systolic BP:

1) >89 mmHg = 4.
2) 76 mmHg = 3.
3) 50-75 mmHg = 2.
4) 1-49 mmHg = 1.
5) None = 0.

c. Total the points.

3. Total trauma Scale + GCS = % survival (1-12)

12 = 99
10 = 88
8 = 67
6 = 63
4 = 33
2 = 29

1 = 25
0 = 4

UNITED STATES MARINE CORPS
Mountain Warfare Training Center
Bridgeport, California 93517-5001

FMST.07.08
04/03/02

STUDENT HANDOUT

PREVENTIVE MEDICINE AND WATER PURIFICATION

TERMINAL LEARNING OBJECTIVE. Given a unit in a mountainous environment and the necessary equipment and supplies, perform preventive medicine, in accordance with the references. (FMST.07.08)

ENABLING LEARNING OBJECTIVES.

1) Without the aid of references, from a given list select the five important areas of personal hygiene, in accordance with the references. (FMST.07.08a)

2) Without the aid of references, from a given list select the correct steps taken to ensure proper immunization prior to deployment, in accordance with the references. (FMST.07.08b)

3) Without the aid of references, from a given list select the three methods of water purification, in accordance with the references. (FMST.07.08c)

4) Without the aid of references, from a given list select the four forms of halogens used for water purification, in accordance with the references. (FMST.07.08d)

5) Without the aid of references, from a given list select the definition of giardia, in accordance with the references. (FMST.07.08e)

6) Without the aid of references, from a given list select the most common drug used to treat giardia infection in the United States, in accordance with the references. (FMST.07.08f)

7) Without the aid of references, from a given list select the three important reasons for proper waste disposal, in accordance with the references. (FMST.07.08g)

8) Without the aid of references, from a given list select the correct methods of waste disposal, in accordance with the references. (FMST.07.08h)

16

1. **PERSONAL HYGIENE** Your body's defense mechanisms break down due to the physical demands of the arduous terrain, environment, stress and metabolic changes. Personal

hygiene becomes the key to prevent the spread of communicable diseases. Five important areas are: (WMC.6.17a)

a. Body.

(1) The body should be washed frequently in order to minimize the chances of small cuts and scratches developing into full blown infections and as a defense against parasitic infections.

(2) A daily bath or shower consisting of soap and hot water is ideal. However, when this is not possible, you should:

(a) Give yourself a "sponge bath" using soap and water, making sure particular attention is given to body creases, i.e., armpits, groin area, face ears, hands and feet.

(b) If water is in extremely short supply, you should take an "air bath" by:

(1) Remove all clothing and hang it up to air.

(2) Expose the body for two hours to the sunlight, which is ideal, but the effects will basically be the same if done indoors or during an overcast day. BE CAREFUL NOT TO SUNBURN!

(3) Shaving. Shaving at night before bedtime is preferable because shaving will remove facial oils, which aid in protecting the face from a harsh cold environment during the day.

b. Hair.

(1) Clean frequently.

(2) Inspect at least once a week for parasites. Have your mountain buddy inspect the areas not within your view. This is particularly important during the summer season. Any time the area you are operating in an area that contains wild life there is the possibility of parasites like fleas and ticks.

c. Hands. Hand washing with soap and water will cut down in the transmission of viral and parasitic infections via the person-person and fecal-oral route.

(1) Keep fingernails trimmed and clean to prevent accidentally scratching yourself and to prevent harborage for bacteria.

d. Feet. Are your prime sources of transportation, so take care of them!

(1) Inspect frequently for:

(a) Blisters.

(b) Infections, bacterial and fungal.

(2) Preventive measures:

(a) Keep your feet dry, change socks frequently.

(b) Use antifungal foot powder.

(c) Antiperspirants. Pretreatment for 2 weeks prior to field with an antiperspirant containing Aluminum Chlorhydrate. This will aid in the prevention of athlete's foot.

e. Oral Hygiene. The mouth and teeth should be cleaned at least daily to prevent tooth decay and gum disease.

(1) Ideally with:

(a) Tooth brush.

(b) Tooth paste.

(c) Dental floss.

(2) If these items are not available:

(a) A "chew stick" made from a clean twig about 8" long and finger width. Chew one end until it becomes frayed and brush-like, then use as a toothbrush.

(b) Field expedient dental floss can be pulled from the center lines of paracord.

(c) The gums should be stimulated at least once daily by rubbing them vigorously with a clean index finger.

2. **IMMUNIZATIONS (NAVMEDCOMINST 6230.3)**. In addition to being required, immunizations are your best front line defense against communicable disease.

a. Eight weeks prior to deployment: (WMC.6.17b)

(1) Ensure Basic Series Completed (BSC) on required immunizations.

(2) Determine if additional immunizations/chemophylaxis will be required in that geographical area of operations.

(3) Identify and flag health risk personnel: i.e. TB converters (NAVMEDCOMINST 6224.1), hepatitis (NAVMEDCOMINST 6230.1A), mononucleosis, parasitic infections and anemic personnel.

(4) Review medical intelligence reports:

(a) EPMU, Navy Environmental Preventive Medicine Unit.

(b) G-2, USMC.

(c) NAVDISVECTOECOLCONCEN, Navy Disease Vector Ecology and Control Centers.

(d) CDC - Center For Disease Control.

(e) WHO - World Health Organization.

3. **WATER.** A satisfactory source is one where quantity and quality is enough to supply the needs of all the troops. Water or ice taken from the environment should be considered contaminated and must be purified with available means. Be aware that cold weather slows the chemical reaction time of purification measures.

a. Sources.

(1) Rivers

(2) Streams

(3) Ponds

(4) Lakes

(5) Ice

(6) Snow

b. Water Source Selection. The choice of a water source is influenced by:

(1) Not contaminated by: (Check all to one mile upstream).

(a) Sewage

(b) Enemy pollution

(c) Chemicals

(d) Biologicals

(e) Radiologicals

NOTE: Check condition of vegetation or for dead animals.

(2) Quantity.

(3) Ease of procurement.

(4) Ease of purification.

(5) Freedom from turbidity.

(6) Excessive organic or non-organic contamination.

(7) Should be easily protectable.

4. **PURIFICATION.** Simply consists of removing or destroying enough impurities to make water safe and pleasant to drink.

a. The first step in purifying is to select a treatment method. (All figures are based on one quart of water.) The three different purifying methods are: (WMC.6.17c)

(1) Boiling. The standard recommendation is to bring the water to a rolling boil for 1 minute at sea level for complete sterilization of the water. Recent research showed that additional boiling time for sterilization was not necessary and that boiling for two minutes at any altitude rendered water safe for consumption.

(2) Halogens. The number of tablets and contact time must be considered. All charts based on one-quart canteen. (WMC.6.17d)

(a) Iodine tablets.

TYPE WATER	TABLETS	CONTACT TIME
CLEAR	1	25 MIN
CLOUDY	2	25 MIN
COLD	2	25 MIN

(b) Chlorine bleach. The strength of the solution, number of drops and contact time must be considered.

PERCENT	CLEAR WATER	CLOUDY WATER	COLD WATER
1%	10GTTS/30 MIN	20GTS/30 MIN	10GTTS/60MIN
4-6%	2GTTS/30 MIN	4GTTS/30 MIN	2GTTS/60 MIN
7-10%	1 GTTS/30 MIN	2GTTS/30 MIN	1 GTTS/60 MIN

(c) Iodine solution. The strength, number of drops and contact time must be considered.

PERCENT	CLEAR WATER	CLOUDY WATER	COLD WATER
2%	5GTTS/30 MIN	10GTTS/30 MIN	8GTTS/60 MIN

(d) Betadine solution.

CLEAR WATER	CLOUDY WATER	COLD WATER
8GTTS/30 MIN	16GTTS/30 MIN	8GTTS/60 MIN

(1) Filtration Most filters are not small enough to completely disinfect water. They remove particulate debris, which allows lower halogen dose and improved appearance and taste of "dirty" water. Filters clog quickly if the water is dirty or has a lot of suspended particles. NEVER USE FILTERS AS YOUR SOLE METHOD OF WATER PURIFICATION.

(a) Maximum Filter Pore Size:

1) Parasitic eggs and larvae: 20 micrometers (um).

2) Giardia, E. histolytica: 5 um.

3) Cryptosporidium: 1 um

4) Enteric bacteria: 0.2 um.

5) Viruses: 0.01 um (too small for field water filtration).

5. **GIARDIA**

a. Definition. (WMC.6.17e) An intestinal flagellate protozoa that can cause clinical illness in humans.

(1) Found worldwide in tropical and temperate areas, the infectious cysts are passed by humans and animals and can be transmitted by direct fecal-oral route or waterborne contamination (cysts maintain viability for months in cold water). Due to these factors, Giardia is both a common cause of diarrhea in travelers to foreign countries, and campers in the United States where it has become endemic and is now considered the most common human parasite.

(2) It is found with increasing frequency in formerly "pristine" mountain streams, rivers and lakes and outbreaks have occurred in remote recreational areas and resorts throughout the western mountain states.

(3) Giardiasis is currently a risk for anyone drinking untreated surface water in the western mountains and for communities with water treatment techniques inadequate for eliminating the cysts.

b. Pathophysiology. The organism has two forms: Trophozoite and cyst.

(1) These may be passed in the stool of an infected person. The trophozoites die within one hour outside the body, however the cysts are very hardy, retaining infectivity more than two months in cold water. If the cysts are ingested, excystation is encouraged by stomach acid, and the motile trophozoite attaches by suction disc to the wall of the duodenum and proximal jejunum, where they may cause symptoms of illness.

(2) Water, ice, or food prepared with contaminated water, are the major sources of infection, but direct person-to-person transmission can occur from sexual contact or hand-to-mouth contamination.

(3) Incubation time: 5 - 25 days (7 - 10 medium). Most infections are thought to be asymptomatic carrier states, however, a wide array of clinical syndromes may occur.

c. Symptoms.

(1) A number of people have abrupt onset of explosive diarrhea accompanied by abdominal cramps, foul gas, vomiting, fever and malaise. This commonly lasts three to four days before developing the more common subacute syndrome. In most individuals, the onset is more insidious, and symptoms are persistent or recurrent.

(2) Stools become mushy, greasy and malodorous. Over a period of weeks, watery diarrhea may alternate with soft stools and even constipation. Upper GI symptoms accompany stool changes, mid and upper abdominal cramping, substernal burning, acid indigestion, sulfurous belching, nausea and foul flatus, distention with an early "full" feeling are all typically increased after a meal.

(3) Symptoms of anorexia, fatigue and weight loss are common, but fever and vomiting are infrequent except during an acute onset of illness.

d. Laboratory Parameters.

(1) The "typical" victim should suggest a diagnosis of giardiasis. This individual may have a history of camping, vacationing at a mountain resort, or foreign travel. Symptoms of upper abdominal discomfort, foul gas and soft stools arose one or two weeks after a short trip or during a lengthy one and have fluctuated for weeks. Sometimes improving, the individual held off visiting the doctor, only to have the symptoms recur. Work or school is not interrupted, however, physical discomfort and embarrassment from the gas are concerns.

(2) Laboratory confirmation of giardiasis can be difficult. Stool samples are the first means of diagnosis. Trophozoites may be found in fresh, watery stools, but disintegrate rapidly. Cysts remain in fresh stools for at least 24 hours. Cysts passage, however, is extremely variable and not related to clinical symptoms. Three samples, taken every other day, should be examined for a definite diagnosis.

(3) Average duration of symptoms has varied from three to ten weeks. Acquired resistance has been shown. Immunity is suspected, but not proven and human reinfection definitely occurs.

e. Treatment. (WMC.6.17f) Complete cure of most Giardia infections can be achieved 90% of the time with *Metronidazole* (flagyl) 250 mg three time s a day for 5 days.

f. Prevention There is no safe or effective chemoprophylactic drug and prevention should be practiced by travelers and campers. The purity of American wild waters can no longer be assumed and in developing countries, even tap water should be suspected. In addition to choosing a technique for individual field water disinfection, individuals must become more responsible for prevention of water contamination. This requires attention to personal latrines and camper trailer waste disposal, a subject that most people do not want to touch.

6. **WASTE**. The importance of proper waste disposal cannot be over emphasized.

a. It serves to: (WMC.6.17g)

(1) Eliminate harborage for rodents and vermin.

(2) Prevents attraction of rodents and vermin.

(3) Prevents a pathogenic contamination source.

b. Two basic types of waste are organic and non-organic. The basic ways to dispose of them are burning, burying, or hauling it away. (WMC.6.17h)

(1) Organic wastes:

(a) Human waste - Burn or haul away to a designated waste pit area.

(b) Urine - Use only assigned marked areas.

(c) Edible garbage - Burn. It should not be left exposed for animals, vermin or the enemy.

(d) Paper -Burn.

(e) Contaminated bandages - Sterilize/burn.

(2) Non-organic waste:

(a) Metals - Haul away or bury.

(b) Liquids - Burn or bury.

UNITED STATES MARINE CORPS
Mountain Warfare Training Center
Bridgeport, California 93517-5001

FMST.07.19
04/03/02

STUDENT OUTLINE

REQUIREMENTS FOR SURVIVAL

TERMINAL LEARNING OBJECTIVE Given a survival situation in any type of environmental condition and minimal equipment and resources, apply the requirements for survival to facilitate individual or group survival and recovery per the references. (FMST.07.19)

ENABLING LEARNING OBJECTIVES

1. Without the aid of references and from a given list, choose one of the survival stressors, in accordance with the references. (WMC.A1.a)

2. Without the aid of references and from a given list, choose one of the priorities of work in a survival situation, in accordance with the references. (WMC.A1.b)

1. REQUIREMENTS FOR SURVIVAL

a. This mental "mind-set" is important in many ways. We usually call it the "will to survive" although you might call it "attitude" just as well. This basically means that, if you do not have the right attitude, you may still not survive.

A1

b. A guideline that can assist you is the acronym " SURVIVAL".

(1) **S**ize up.

(a) Size up the situation.

1. Conceal yourself from the enemy.

2. Use your senses to hear, smell, and see to determine and consider what is developing on the battlefield before you make your survival plan.

(b) Size up your surroundings.

1. Determine the rhythm or pattern of the area.

2. Note animal and bird noises and their movement.

3. Note enemy traffic and civilian movement.

(c) Size up your physical condition.

1. Check your wounds and give yourself first aid.

2. Take care to prevent further bodily harm.

3. Evaluate condition of self and unit prior to developing survival plan.

(d) Size up your equipment.

1. Consider how available equipment may affect survival senses; tailor accordingly.

(2) **U**ndue haste makes waste.

(a) Plan your moves so that you can move out quickly without endangering yourself if the enemy is near.

(3) **R**emember where you are.

(a) If you have a map, spot your location and relate it to the surrounding terrain.

(b) Pay close attention to where you are and where you are going. Constantly orient yourself.

(c) Try to determine, at a minimum, how your location relates to the following:

1. The location of enemy units and controlled areas.

2. The location of friendly units and controlled areas.

3. The location of local water sources.

4. Areas that will provide good cover and concealment.

(4) Vanquish fear and panic.

(a) Realistic and challenging training builds self-confidence and confidence for a unit's leadership.

(b) The feeling of fear and panic will be present. The survivor must control these feelings.

(5) Improvise and improve.

(a) Use tools designed for one purpose for other applications.

(b) Use objects around you for different needs. (i.e. use a rock for a hammer)

(6) Value living.

(a) Place a high value on living.

(b) Refuse to give into the problem and obstacles that face you.

(c) Draw strength from individuals that rise to the occasion.

(7) Act like the natives.

(a) Observe the people in the area to determine their daily eating, sleeping, and drinking routines.

(b) Observe animal life in the area to help you find sources of food and water.

NOTE: Remember that animal reactions can reveal your presence to the enemy. Animals cannot serve as an absolute guide to what you can eat and drink.

(8) Live by your wits, **but for now**, learn basic skills.

(a) Practice basic survival skills during all training programs and exercises.

2. **STRESS**. Stress has many positive benefits. Stress provides us with challenges: it gives us chances to learn about our values and strengths. Too much stress leads to distress. While many of these signs may not be self-identified, it remains critical that all survivors remain attentive to each other's signs of distress. Listed are a few common signs of distress found when faced with too much stress:

   a. Difficulty in making decisions (do not confuse this sign for a symptom of hypothermia).

   b. Angry outbursts.

   c. Forgetfulness.

   d. Low energy level.

   e. Constant worrying.

   f. Propensity for mistakes.

   g. Thoughts about death or suicide.

   h. Trouble getting along with others.

   i. Withdrawing from others.

   j. Hiding from responsibilities.

   k. Carelessness.

3. **SURVIVAL STRESSORS**. (FMST.07.19a) **Any** event can lead to stress. Often, stressful events occur simultaneously. These events are not stress, but they produce it and are called "stressors". In response to a stressor, the body prepares either to "fight or flee". Stressors add up. Anticipating stressors and developing strategies to cope with them are the two ingredients in the effective management of stress. It is essential that the survivor be aware of the types of stressors they will encounter.

   a. Injury, Illness, or Death: Injury, illness, and death are real possibilities a survivor has to face. Perhaps nothing is more stressful than being alone in an unfamiliar environment where you could die from hostile action, an accident, or from eating something lethal. Illness and injury can also add to stress by limiting your ability to maneuver, get food and drink, find shelter, and defend yourself.

   b. Uncertainty and Lack of Control: Some people have trouble operating in settings where everything is not clear-cut. The only guarantee in a survival situation is that nothing is guaranteed. This uncertainty and lack of control also add to the stress of being ill, injured, or killed.

c. Environment: A survivor will have to contend with the stressors of weather, terrain, and the variety of creatures inhabiting an area. Heat, cold, rain, winds, snow, mountains, insects, and animals are just a few of the challenges awaiting the Marine working to survive. Depending on how a survivor handles the stress of environment, his surroundings can be either a source of food and protection or a cause of extreme discomfort leading to injury, illness, or death.

d. Hunger and Thirst: Without food and water a person will weaken and eventually die. Getting and preserving food and water takes on increasing importance as the length of time in a survival setting increases. With the increased likelihood of diarrhea, replenishing electrolytes becomes critical. For a Marine used to having his provisions issued, foraging can be a significant source of stress.

e. Fatigue: It is essential that survivors employ all available means to preserve mental and physical strength. While food, water, and other energy builders may be in short supply, maximizing sleep to avoid deprivation is a very controllable factor. Training data collected at MCCDC demonstrates that individuals who lack sleep for 24 hours or more suffer a 25% decrease in effective performance. Further, sleep deprivation directly correlates with increased fear. Effective survival is highly unlikely when fatigue builds to a level where staying awake becomes a stressful evolution.

f. Isolation: Although Marines complain about higher headquarters, we become used to the information and guidance it provides, especially during times of confusion. Being in contact with others provides a greater sense of security and a feeling someone is available to help if problems occur. A significant stress in survival situations is that often a person or team has to rely solely on its own resources.

4. **NATURAL REACTIONS**. Man has been able to survive many shifts in his environment throughout the centuries. His ability to adapt physically and mentally to a changing world kept him alive. The average person will have some psychological reactions in a survival situation. These are some of the major internal reactions you might experience with the survival stresses.

   a. Fear: Fear is our emotional response to dangerous circumstances that we believe have the potential to cause death, injury, or illness. Fear can have a positive function if it encourages us to be cautious in situations where recklessness could result in injury. Unfortunately, fear can also immobilize a person. It can cause us to become so frightened that we fail to perform activities essential for survival.

   b. Anxiety: Anxiety can be an uneasy, apprehensive feeling we get when faced with dangerous situations. When used in a healthy way, anxiety urges us to act to end, or at least master, the dangers that threaten our existence. A survivor reduces his anxiety by performing those tasks that will ensure his coming through the ordeal alive. Anxiety can overwhelm a Marine to the point where he becomes easily confused and has difficulty thinking.

c. Anger and Frustration: Frustration arises when a person is continually thwarted in his attempts to reach a goal. The goal of survival is to stay alive until you can reach help or until help can reach you. To achieve this goal, Marines must complete some tasks with minimal resources. One outgrowth of frustration is anger. Getting lost, damaged or forgotten equipment, the weather, inhospitable terrain, enemy patrols, and physical limitations are just a few sources of frustration and anger. Frustration and anger encourage impulsive reactions, irrational behavior, poorly thought-out decisions, and, in some instances, an "I quit" attitude. If the Marine does not properly focus his angry feelings, he can waste much energy in activities that do little to further either his chances of survival or the chances of those around him.

d. Depression: Depression is closely linked with frustration and anger when faced with the privations of survival. A destructive cycle between anger and frustration continues until the person becomes worn down-physically, emotionally, and mentally. When a person reaches this point, he starts to give up, and his focus shifts from "What can I do" to "There is nothing I can do." If you allow yourself to sink into a depressed state, then it can sap all your energy and, more important, your will to survive.

e. Loneliness and Boredom: Man is a social animal and enjoys the company of others. Loneliness and boredom can be another source of depression. Marines must find ways to keep their minds productively occupied.

f. Guilt: The circumstances leading to your survival setting are sometimes dramatic and tragic. It may be the result of an accident or military mission where there was a loss of life. Perhaps you were the only, or one of a few, survivors. While naturally relieved to be alive, you simultaneously may be mourning the deaths of others that were less fortunate. Do not let guilt feelings prevent you from living.

5. **PRIORITIES OF WORK IN A SURVIVAL SITUATION.** (FMST.07.19b) Each survival situation will have unique aspects that alter the order in which tasks need to be accomplished. A general guideline is to think in blocks of time.

a. First 24 hours: The first 24 hours are critical in a survival situation. You must make an initial estimate of the situation. Enemy, weather, terrain, time of day and available resources will determine which tasks need to be accomplished first. They should be the following:

(1) Shelter.

(2) Fire.

(3) Water.

(4) Signaling.

b. Second 24 hours: After the first 24 hours have passed, you will now know you can survive. This time period needs to be spent on expanding your knowledge of the area. By completing the following tasks, you will be able to gain valuable knowledge.

(1) Tools and weapons: By traveling a short distance from your shelter to locate the necessary resources, you will notice edible food sources and game trails.

(2) Traps and snares: Moving further away from your shelter to employ traps and snares, you will be able to locate your shelter area from various vantagepoints. This will enable you to identify likely avenues of approach into your shelter area.

(3) Pathguards: Knowing the likely avenues of approaches, you can effectively place noise and casualty producing pathguards to ensure the security of your shelter area.

c. Remainder of your survival situation. This time is spent on continuously improving your survival situation until your rescue.

6. **GROUP SURVIVAL.** In-group survival, the group's survival depends largely on its ability to organize activity. An emergency situation does not bring people together for a common goal; rather, the more difficult and disordered the situation, the greater are the disorganized group's problems.

a. Groups Morale: High morale must come from internal cohesiveness and not merely through external pressure. The moods and attitudes can become wildly contagious. Conscious, well-planned organization and leadership on the basis of delegated or shared responsibility often can prevent panic. High group morale has many advantages.

(1) Individual feels strengthened and protected since he realizes that his survival depends on others whom he trusts.

(2) The group can meet failure with greater persistency.

(3) The group can formulate goals to help each other face the future.

b. Factors that Influence Group Survival: There are numerous factors that will influence whether a group can successfully survive.

(1) Organization of Manpower - Organized action is important to keep all members of the group informed; this way the members of the group will know what to do and when to do it, both under ordinary circumstances and in emergencies.

(2) Selective Use of Personnel - In well-organized groups, the person often does the job that most closely fits his personal qualifications.

(3) Acceptance of Suggestion and Criticisms - The senior man must accept responsibility for the final decision, but must be able to take suggestion and criticisms from others.

(4) Consideration of Time - On-the-spot decisions that must be acted upon immediately usually determines survival success.

(5) Check Equipment - Failure to check equipment can result in failure to survive.

(6) Acquiring survival knowledge and skills increases survival Knowledge and Skills - Confidence in one's ability to survive.

UNITED STATES MARINE CORPS
Mountain Warfare Training Center
Bridgeport, California 93517-5001

FMST.
07.20
04/03/02

STUDENT OUTLINE

SURVIVAL KIT

TERMINAL LEARNING OBJECTIVE. In a summer mountainous environment, construct a survival kit, in accordance with the references. (FMST.07.20)

ENABLING LEARNING OBJECTIVES

(1) Without the aid of references, choose from a given list in the criteria for a survival kit, in accordance with the references. (FMST.07.20a)

(2) Without the aid of references, choose from a given list one example of each criteria for a survival kit, in accordance with the references. (FMST.07.20b)

1. COMPONENTS FOR A SURVIVAL KIT

a. The environment is the key to the types of items you will need in your survival kit. How much equipment you put in your kit depends on how you will carry the kit. A kit on your body will have to be much smaller than one carried in a vehicle.

b. Always layer your survival kit, keeping the most important items on your body.

c. In preparing your survival kit, select items that can be used for more than one purpose.

d. Your survival kit does not need to be elaborate. You only need functional items that will meet your needs and a case to hold them. The case might be a first aid case, an ammunition pouch, or another suitable case. This case should be:

(1) Water repellent or waterproof.

(2) Easy to carry or attach to your body.

(3) Suitable to accept various sized items.

(4) Durable.

e. When constructing a survival kit, you should have the following components: (FMST.07.20a)

(1) Fire starting items.

(2) Water procurement items.

(3) Food procurement items.

(4) Signaling items.

(5) First aid items.

(6) Shelter items.

2. **<u>ITEMS CONTAINED WITHIN EACH COMPONENT</u>** (FMST.07.20b)

a. <u>Fire Starting Items</u>.

(1) Matches.

(2) Magnifying glass.

(3) Flint and Steel.

(4) Lighter.

(5) Potassium Permanganate, with a container of sugar or anti-freeze.

(6) Prepackaged Tinder.

(a) Commercially Manufactured

(b) Cotton Balls and Petroleum Jelly

b. Water Procurement Items

(1) Water Disinfecting Chemicals.

(a) Iodine Tablets

(b) Betadine Solution

(c) Iodine Solution

(2) Metal Container. (Serves for boiling water)

(a) Canteen Cup

(b) Survival Kit Container

(c) Any suitable can that contained no petroleum products.

(3) Water Carrying Items.

(a) Canteen

(b) Plastic Bag

(c) Plastic/Metal/Glass Container which contained no petroleum products

c. Food Procurement Items

(1) Fish.

(a) Various sized hooks

(b) Various sized sinkers/weights

(c) Metal leaders and swivels

(d) Small weighted jigs

(e) Fishing line

i. Think about the size of fish for that environment when selecting weights and sizes.

(2) Game.

(a) Snares

i. Commercially Manufactured

ii. Aircraft Cable

iii. Tie Wire

(b) Bait

i. MRE Cheese Spread or Peanut Butter Package

(c) 550 Cord for Gill Net and Trap Construction

i. Engineer/Marking Tape

ii. Sling shot rubber and pouch

d. Signaling Items

(1) Day.

(a) Mirror

(b) Whistle

(c) Pyrotechnics (Smoke, Pen Flares)

(d) Air Panels

(2) Night.

(a) Pyrotechnics (Pen Flares, Star Clusters)

(b) Lights (Flashlight, Strobe, Chemlight)

(c) Whistle

e. Shelter Items

(1) Cordage.

(a) 550 Cord

(b) Wire

(c) Communication wire

(d) Tie wire

(2) Finger Saw.

(3) Sewing Kit with Needles for construction/repair of clothing.

(4) Tentage.

(a) Poncho

(b) Tarp

(c) Space blanket

(d) Plastic trash bags

f. First Aid Items

(1) Band-Aids.

(a) Steristrips

(b) Adhesive Tape

(c) Non-stick pads, 4x4's, Gauze, Battle Dressings

(d) Muslin Bandage

(2) Ointments.

(a) Burn

(b) Anti-septic

(3) Miscellaneous.

(a) Salt

(b) Sugar

(c) Eye Wash

(d) Alcohol prep pads

(e) Suture Kit

(f) Scalpel

(g) Vile of Yarrow

g. Miscellaneous items

(1) Fingernail clippers

(2) Compass

(3) Notebook with pen or pencil

(4) Wood eye screws and nails

(5) Surgical tubing

NOTE: It is assumed that the Marine is always carrying a high quality fixed bladed knife, a multi-tool knife, and a sharpening stone.

UNITED STATES MARINE CORPS
Mountain Warfare Training Center
Bridgeport, California 93517-5001

FMST.07.23
04/05/02

STUDENT OUTLINE

SIGNALLING AND RECOVERY

TERMINAL LEARNING OBJECTIVE Given a survival situation in any type of environmental condition and minimal equipment and resources, signal for aid to facilitate individual or group recovery per the references. (FMST 07.23)

ENABLING LEARNING OBJECTIVES

(1) Without the aid of references and from a given list, choose the audio international distress Signal, in accordance with the references. (FMST.07.23a)

(2) Without the aid of references and from a given list, choose the visual international distress signal, in accordance with the references. (FMST.07.23b)

(3) Without the aid of references and given a list of messages, match in writing the international code symbols to each corresponding message, in accordance with the references. (FMST.07.23c)

OUTLINE

1. **SIGNALING DEVICES**. The equipment listed below are items that may be on your body or items inside an aircraft. Generally, these items are used as signaling devices while on the move. They must be accessible for use at any moment's notice. Additionally, in a summer mountainous environment, Marines may experience areas that are snow covered and must be familiar with the effects that snow will have on specific signaling devices.

    a. Pyrotechnics. Pyrotechnics include star clusters and smoke grenades. When using smoke grenades in snow pack, a platform must be built. Without a platform, the smoke grenade will sink into the snow pack and the snow will absorb all smoke. A rocket parachute flare or a hand flare has been sighted as far away as 35 miles, with an average of 10 miles. Pyrotechnic flares are effective at night, but during daylight their detectability ranges are reduced by 90 percent.

A3

b. M-186 Pen Flare: The M-186 Pen Flare is a signaling device carried in the vest of all crew chiefs and pilots. Remember to cock the gun prior to screwing in the flare.

c. Strobe Light: A strobe light is generally carried in the flight vests of all crew chiefs and pilots. It can be used at night for signaling. Care must be taken because a pilot using goggles may not be able to distinguish a flashing strobe from hostile fire.

d. Flashlight: By using flashlights, a Morse code message can be sent. A SOS distress call consists of sending three dots, three dashes, and three dots. Keep repeating this signal.

e. Whistle: The whistle is used in conjunction with the audio international distress signal. It is used to communicate with forces on the ground.

f. AN/PRC-90 & AN/PRC-112: The AN/ PRC 90 survival radio is a part of the aviator's survival vest. The AN/PRC-112 will eventually replace the AN/PRC-90. Both radios can transmit either tone (beacon) or voice. Frequency for both is **282.8 for voice**, and **243.0 for beacon**. Both of these frequencies are on the UHF Band.

g. Day/Night Flare: The day/night flare is a good peacetime survival signal. The flare is for night signaling while the smoke is for day. A red cap identifies the older version flare with three nubbins while the new generation has three rings around the body for identification during darkness. The flare burns for approximately 20 second while the smoke burns for approximately 60 seconds.

NOTE: Once one end is used up, douse in water to cool and save the other end for future use.

h. Signal Mirror: A mirror or any shiny object can be used as a signaling device. It can be used as many times as needed. Mirror signals have been detected as far away as 45 miles and from as high as 16,000', although the average detection distance is 5 miles. It can be concentrated in one area, making it secure from enemy observation. A mirror is the best signaling device for a survivor; however, it is only as effective as its user. Learn how to use one now, before you find yourself in a survival situation.

(1) Military signal mirrors have instructions on the back showing how to use it. It should be kept covered to prevent accidental flashing that may be seen by the enemy.

(2) Any shiny metallic object can be substituted for a signal mirror.

(3) Haze, ground fog, or a mirage may make it hard for a pilot to spot signals from a flashing object. So, if possible, get to the highest point in your area when flashing. If you can't determine the aircraft's location, flash your signal in the direction of the aircraft noise.

AIMING THE SIGNAL MIRROR

2. **METHODS OF COMMUNICATION**

a. Audio: Signaling by means of sound may be good, but it does have some disadvantages.

(1) It has limited range unless you use a device that will significantly project the sound.

(2) It may be hard to pinpoint one's location due to echoes or wind.

(3) International Distress Signal. (FMST.07.23a) The survivor will make six blasts in one minute, returned by three blasts in one minute by the rescuer.

b. Visual: Visual signals are generally better than audio signals. They will pinpoint your location and can been seen at a greater distances under good weather conditions.

(1) The visual international distress symbol is recognized by a series of three evenly spaced improvised signaling devices. (FMST.07.23b)

3. **IMPROVISED SIGNALING DEVICES**. Improvised signaling devices are generally static in nature. They must be placed in a position to be seen by rescuers. They are made from any resources available, whether natural or man-made.

a. Smoke Generator: The smoke generator is an excellent improvised visual signaling device. It gives the survivor the flexibility to signal in either day or night conditions. This type of signal has been sighted as far away as 12 miles, with an average distance of 8 miles. Smoke signals are most effective in calm wind conditions or open terrain, but effectiveness is reduced with wind speeds above 10 knots. Build them as soon as time and the situation permits, and protect them until needed.

(1) Construct your fire in a natural clearing or along the edge of streams (or make a clearing). Signal fires under dense foliage will not be seen from the air.

(2) Find two logs, 6 - 10 inches in diameter, and approximately five feet long. Place the two logs parallel to each other with 3 - 4 feet spacing.

(3) Gather enough sticks, approximately two inches in diameter and four feet long, to lie across the first two logs. This serves as a platform for the fire.

(4) Gather enough completely dry branches to build a pyramid fire. The pyramid fire should be 4 feet by 4 feet by 2 feet high.

(5) Place your tinder under the platform.

(6) Gather enough pine boughs to lay on top of the pyramid fire. This serves to protect the fire and the tinder.

(7) To light, remove the pine bough and ignite the tinder. If available, construct a torch to speed up the lighting process, especially for multiple fires.

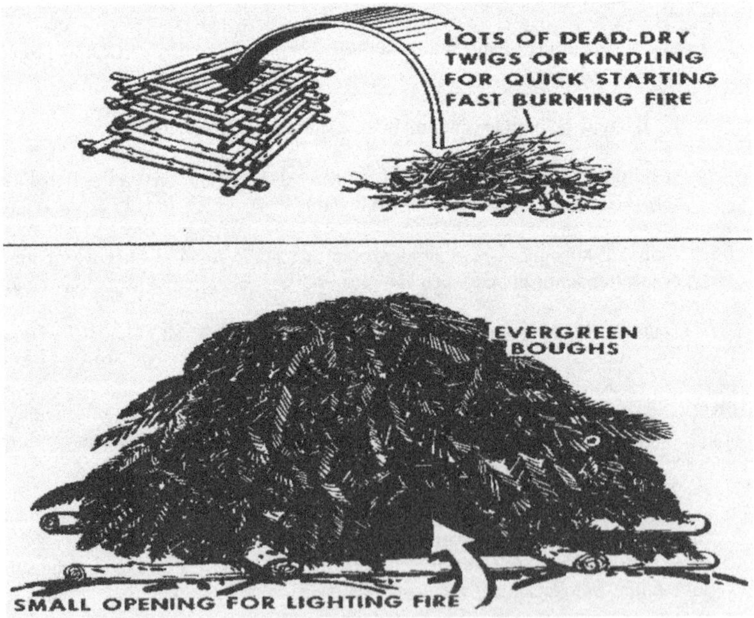

SMOKE GENERATOR

(8) To create a smoke effect during the day light hours, place the pine bough on the ignited fire.

(9) Placing a smoke grenade or colored flare under the platform will change the color of the smoke generated. Remember that you want the fire to draw in the colored smoke, which will create a smoke color that contrasts with the background will increase the chances of success.

b. Arrangement or alteration of natural materials: Such things as twigs or branches, can be tramped into letters or symbols in the snow and filled in with contrasting materials. To attract more attention, ground signals should be arranged in big geometric patterns.

(1) International symbols. The following symbols are internationally known. (FMST.07.23c)

Number	Message	Code symbol
1	REQUIRE ASSISTANCE	V
2	REQUIRE MEDICAL ASSISTANCE	X
3	NO OR NEGATIVE	N
4	YES OR AFFIRMATIVE	Y
5	PROCEED IN THIS DIRECTION	↑

INTERNATIONAL SYMBOLS

(2) Shadows: If no other means are available, you may have to construct mounds that will use the sun to cast shadows. These mounds should be constructed in one of the International Distress Patterns. To be effective, these shadow signals must be oriented to the sun to produce the best shadows. In areas close to the equator, a North—South line gives a shadow anytime except noon. Areas further north or south of the equator require the use of East—West line or some point of the compass in between to give the best result.

(3) Size: The letters should be large as possible for a pilot or crew to spot. Use the diagram below to incorporate the size to ratio for all letter symbols.

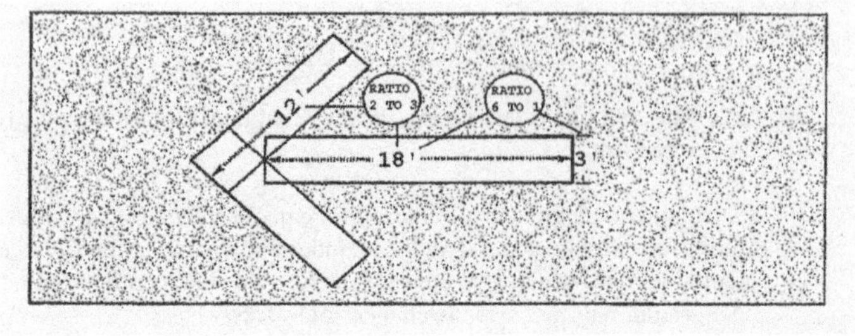

SIZE AND RATIO

(4) Contrast: When constructing letter symbols, contrast the letter from the surrounding vegetation and terrain. Ideally, bring material from another location to build the letter. This could be clothing, air panels, space blanket, etc.

(a) On snow, pile bough or use sea dye from a LPP (Life preserver, personal). Fluorescent sea dye markers have been sighted as far away as 10 miles, although the average detection distance is 3 miles.

4. **AIR TO GROUND COMMUNICATIONS**. Air to ground communications can be accomplished by standard aircraft acknowledgments.

a. Aircraft will indicate that ground signals have been seen and understood by:

(1) Rocking wings from side to side: This can be done during the day or in bright moonlight.

b. Aircraft will indicate that ground signals have been seen but not understood by:

(1) Making a complete, clockwise circle during the day or in bright moonlight.

5. **RECOVERY**. Marines trapped behind enemy lines in future conflicts may not experience quick recovery. Marines may have to move to a place that minimizes risk to the recovery force. No matter what signaling device a Marine uses, he must take responsibility for minimizing the recovery force's safety.

a. Placement Considerations: Improvised signaling devices, in a hostile situation, should not be placed near the following areas due to the possibility of compromise:

(1) Obstacles and barriers.

(2) Roads and trails.

(3) Inhabited areas.

(4) Waterways and bridges.

(5) Natural lines of drift.

(6) Man-made structures.

(7) All civilian and military personnel.

b. Tactical Consideration: The following tactical considerations should be adhered to prior to employing an improvised signaling device.

(1) Use the signals in a manner that will not jeopardize the safety of the recovery force or you.

(2) Locate a position, which affords observation of the signaling device and facilitates concealed avenues of escape (if detected by enemy forces). Position should be located relatively close to extract site in order to minimize "time spent on ground" by the recovery force.

(3) Maintain continuous security through visual scanning and listening while signaling devices is employed. If weapon systems are fire and/or observation should cover available, signaling devices.

(4) If enemy movement is detected in the area, attempt to recover the signaling device, if possible.

(5) Employ improvised signaling devices only during the prescribed times, if briefed in the mission order.

c. Recovery Devices: In mountainous terrain, a helicopter landing may be impossible due to ground slope, snow pack, or vegetation. The survivor must be familiar with recovery devices that may be aboard the aircraft.

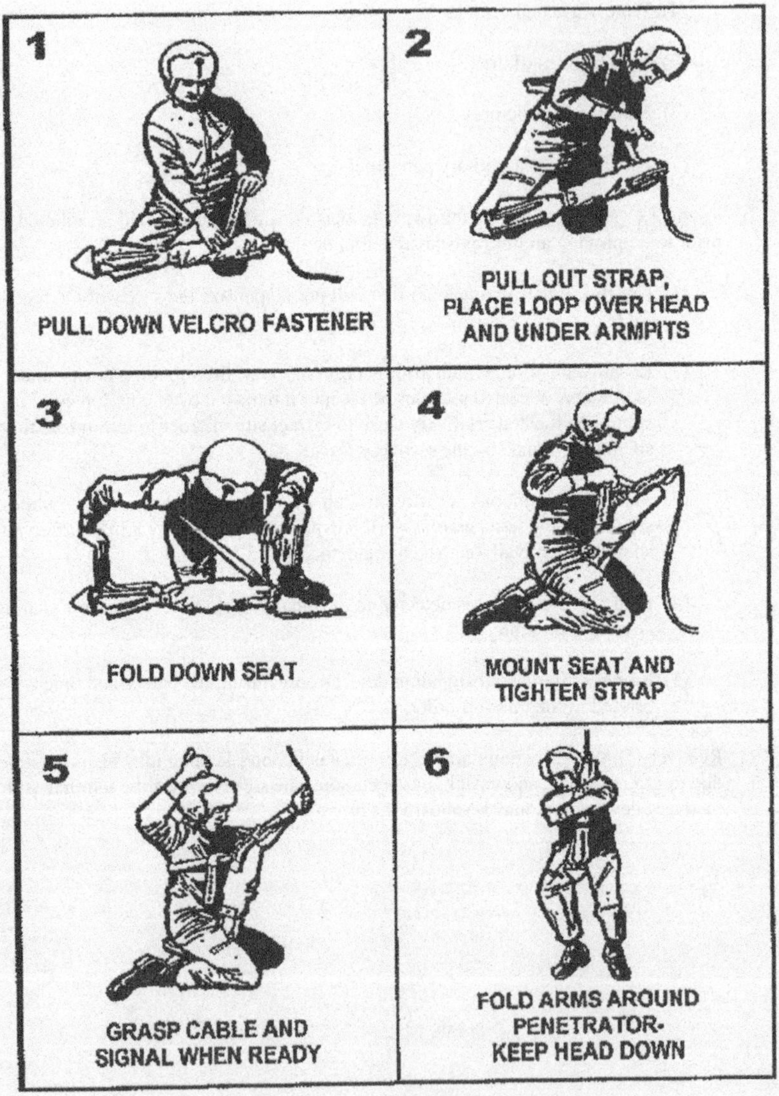

JUNGLE PENETRATOR

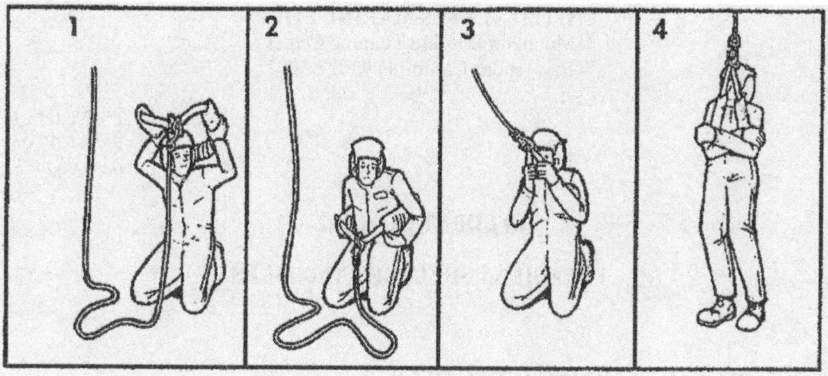

SLING HOIST

d. Recovery by other than aircraft. Recovery by means other than aircraft may occur. Unit Sop's should include signaling and link-up with forces at the following locations:

(1) Border Crossings. The evader who crosses into a neutral country is subject to detention by that country for the duration of the war.

(2) FEBA/FLOT.

(a) Static: Recovery along a static FEBA is always difficult. Under these conditions, enemy and friendly forces can be expected to be densely arrayed and well camouflaged, with good fields of fire. Attempts to penetrate the FEBA should be avoided.

(b) Advancing: Individuals isolated in front of advancing friendly units should immediately take cover and wait for the friendly units to overrun their position.

(c) Retreating: Individuals between opposing forces should immediately take cover and wait for enemy units to pass over their position. After most enemy units have moved on, evaders should try to link up with other isolated friendly elements and return to friendly forces.

(3) Link-up with friendly patrols. Unit authentication numbers and/or locally developed codes may assist the evader to safely make contact in or around the FEBA and when approached by friendly forces.

UNITED STATES MARINE CORPS
Mountain Warfare Training Center
Bridgeport, California 93517-5001

FMST.07.21
FMST.07.04
04/07/02

STUDENT OUTLINE

EXPEDIENT SHELTERS AND FIRES

TERMINAL LEARNING OBJECTIVES

(1) In a summer mountainous environment, construct expedient shelters, in accordance with the references. (FMST.07.21)

(2) In a summer mountainous environment, construct fires, in accordance with the references. (FMST.07.04)

b. **ENABLING LEARNING OBJECTIVES**

(1) Without the aid of references and from a given list, choose the four characteristics of a safe expedient shelter, in accordance with the references. (FMST.07.21a)

(2) Without the aid of references and from a given list, choose the four hazards avoid when using natural shelters, in accordance with the references. (FMST.07.21b)

(3) Without the aid of references and from a given list, choose one of man-made expedient shelters, in accordance with the references. (FMST.07.21c)

(4) Without the aid of references, start a fire using a primitive method, in accordance with the references. (FMST.07.04a)

A4

OUTLINE

1. **BASIC CHARACTERISTICS FOR SHELTER**. (FMST.07.21a) Any type of shelter, whether it is a permanent building, tentage, or an expedient shelter must meet six basic characteristics to be safe and effective. The characteristics are:

   a. Protection from the Elements: The shelter must provide protection from rain, snow, wind, sun, etc.

   b. Heat Retention: It must have some type of insulation to retain heat; thus preventing the waste of fuel.

   c. Ventilation: Ventilation must be constructed, especially if burning fuel for heat. This prevents the accumulation of carbon monoxide. Ventilation is also needed for carbon dioxide given off when breathing.

   d. Drying Facility: A drying facility must be constructed to dry wet clothes.

   e. Free from Natural Hazards: Shelters should not be built in areas of avalanche hazards, under rock fall or "standing dead" trees have the potential to fall on your shelter.

   f. Stable: Shelters must be constructed to withstand the pressures exerted by severe weather.

2. **NATURAL SHELTERS**. Natural shelters are usually the preferred types because they take less time and materials construct. The following may be made into natural shelters with some modification.

   a. Caves or Rock Overhangs: Can be modified by laying walls of rocks, logs or branches across the open sides.

   b. Hollow Logs: Can be cleaned or dug out, then enhanced with ponchos, tarps or parachutes hung across the openings.

   c. Hazards of Natural Shelters: (FMST.04.21b)

      (1) Animals: Natural shelters may already be inhabited (i.e. bears, coyotes, lions, rats, snakes, etc.). Other concerns from animals may be disease from scat or decaying carcasses.

      (2) Lack of Ventilation: Natural shelters may not have adequate ventilation. Fires may be built inside for heating or cooking but may be uncomfortable or even dangerous because of the smoke build up.

      (3) Gas Pockets: Many caves in a mountainous region may have natural gas pockets in them.

(4) Instability: Natural shelters may appear stable, but in reality may be a trap waiting to collapse.

3. **MAN-MADE SHELTERS**. (FMST.04.21c) Many configurations of man-made shelters may be used. Over-looked man-made structures found in urban or rural environments may also provide shelter (i.e. houses, sheds, or barns). Limited by imagination and materials available, the following man-made shelters can be used in any situation.

a. Poncho Shelter.

b. Sapling Shelter.

c. Lean-to.

d. Double lean-to.

e. A-frame Shelter.

f. Fallen Tree Bivouac.

4. **CONSTRUCTION OF MAN-MADE SHELTERS**. To maximize the shelter's effectiveness, Marines should take into consideration the following prior to construction.

a. Considerations.

(1) Group size.

(2) Low silhouette and reduced living area dimensions for improved heat conservation.

(3) Avoid exposed hill tops, valley floors, moist ground, and avalanche paths.

(4) Create a thermal shelter by applying snow, if available, to roof and sides of shelter.

(5) Location of site to fire wood, water, and signaling, if necessary.

(6) How much time and effort needed to build the shelter.

(7) Can the shelter adequately protect you from the elements (sun, wind, rain, and snow). Plan on worst case scenario.

(8) Are the tools available to build it. If not, can you make improvised tools?

(9) Type and amount of materials available to build it.

(10) When in a tactical environment, you must consider the following:

(a) Provide concealment from enemy observation.

(b) Maintain camouflaged escape routes.

(c) Use the acronym BLISS as a guide.

B – Blend in with the surroundings.
L –Low silhouette.
I – Irregular shape.
S – Small.
S – Secluded located.

b. Poncho Shelter: This is one of the easiest shelters to construct. Materials needed for construction are cord and any water-repellent material (i.e. poncho, parachute, and tarp). It should be one of the first types of shelter considered if planning a short stay in any one place.

(1) Find the center of the water-repellent material by folding it in half along its long axis.

(2) Suspend the center points of the two ends using cordage.

(3) Stake the Four Corners down, with sticks or rocks.

c. Sapling Shelter: This type of shelter is constructed in an area where an abundance of saplings are growing. It is an excellent evasion shelter.

(1) Find or clear an area so that you have two parallel rows of saplings at least 4' long and approximately 1 ½' to 2' apart.

(2) Bend the saplings together and tie them to form several hoops, which will form the framework of the shelter.

(3) Cover the hoop with a water-repellent covering.

(4) The shelter then may be insulated with leaves, brush, snow, or boughs.

(5) Close one end permanently. Hang material over the other end to form a door.

d. Lean-to: A lean-to is built in heavily forested areas. It requires a limited amount of cordage to construct. The lean-to is an effective shelter but does not offer a great degree of protection from the elements.

(1) Select a site with two trees (4-12" in diameter) spaced far enough apart that a man can lie down between them. Two sturdy poles can be substituted by inserting them into the ground the proper distance apart.

(2) Cut a pole to support the roof. It should be at least 3-4" in diameter and long enough to extend 4-6" past both trees. Tie the pole horizontally between the two trees, approximately 1 meter off the deck.

(3) Cut several long poles to be used as stringers. They are placed along the horizontal support bar approximately every 1 ½' and lay on the ground. All stringers may be tied to or laid on the horizontal support bar. A short wall or rocks or logs may be constructed on the ground to lift the stringers off the ground, creating additional height and living room dimensions.

(4) Cut several saplings and weave them horizontally between the stringers. Cover the roof with water-repellent and insulating material.

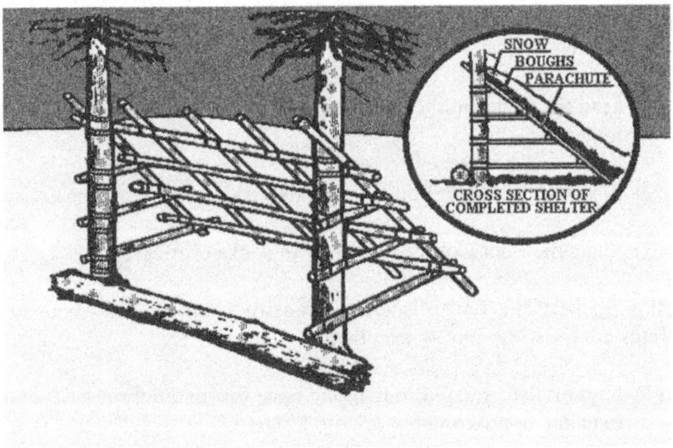

LEAN-TO

e. Double lean-to: The double lean-to shelter is constructed for 2-5 individuals. It is constructed by making two lean-tos and placing them together.

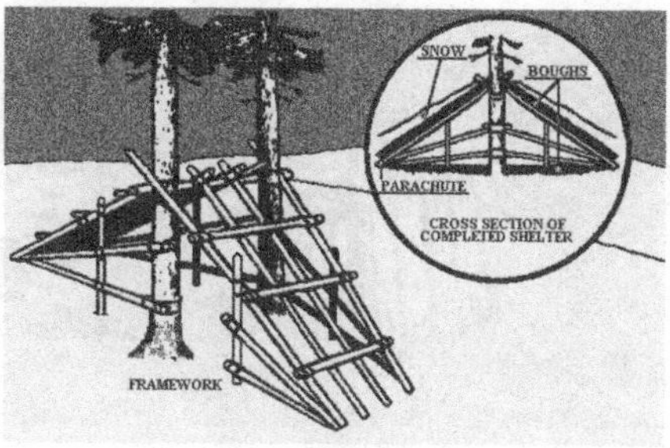

DOUBLE LEAN-TO

f. A-Frame Shelter: An A-Frame shelter is constructed for 1-3 individuals. After the framework is constructed, bough/tentage is interwoven onto the frame and snow, if available, is packed onto the outside for insulation.

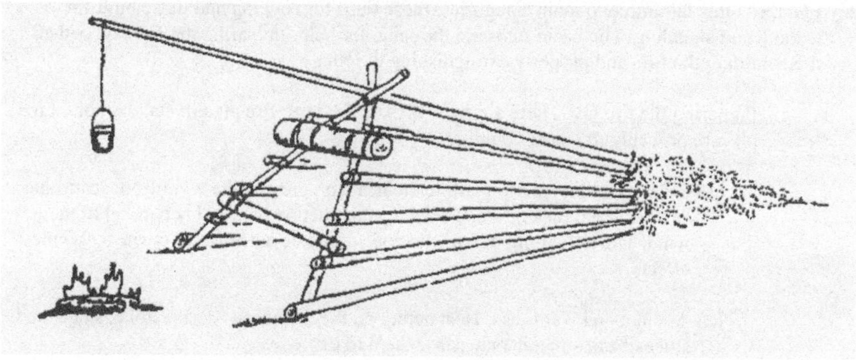

A-FRAME

g. Fallen Tree Bivouac: The fallen tree bivouac is an excellent shelter because most of the work has already been done.

(1) Ensure the tree is stable prior to constructing.

(2) Branches on the underside are cut away making a hollow underneath.

(3) Place additional insulating material to the top and sides of the tree.

(4) A small fire is built outside of the shelter.

<u>**FALLEN TREE BIVOUAC**</u>

5. **REFLECTOR WALLS.** Heating a shelter requires a slow fire that produces lots of steady heat over a long period of time. A reflector wall should be constructed for all open-ended shelters. A reflector wall is constructed with a flat rock or a stack of green logs propped behind the fire. A surprising amount of heat will bounce back from the fire into the shelter.

6. **FIRES**. Fires fall into two main categories: those built for cooking and those built for warmth and signaling. The basic steps are the same for both: preparing the fire lay, gathering fuel, building the fire, and properly extinguishing the fire.

   a. <u>Preparing the fire lay</u>: There are two types of fire lays: fire pit and Dakota hole. Fire pits are probably the most common.

      (1) Create a windbreak to confine the heat and prevent the wind from scattering sparks. Place rocks or logs used in constructing the fire lay parallel to the wind. The prevailing downwind end should be narrower to create a chimney effect.

      (2) Avoid using wet rocks. Heat acting on the dampness in sandstone, shale, and stones from streams may cause them to explode.

      (3) <u>Dakota Hole</u>: The Dakota Hole is a tactical fire lay. Although no fire is 100% tactical, this fire lay will accomplish certain things:

         (a) Reduces the signature of the fire by placing it below ground.

         (b) Provides more of a concentrated heat source to boil and cook, thus preserving fuel and lessening the amount of burning time.

         (c) By creating a large air draft, the fire will burn with less smoke than the fire pit.

         (d) It is easier to light in high winds.

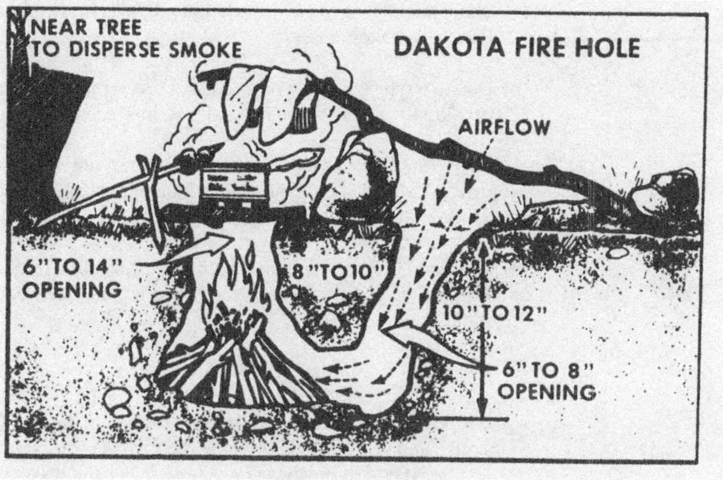

<u>**DAKOTA HOLE**</u>

b. <u>Gather Fuel</u>: Many Marines take shortcuts when gathering firewood. Taking a few extra minutes can mean the difference between ease and frustration when building a fire.

(1) <u>Tinder</u>: Tinder is the initial fuel. It should be fine and dry. Gather a double handful of tinder for the fire to be built and an extra double handful to be stored in a dry place for the following morning. Dew can moisten tinder enough to make lighting the fire difficult. Some examples are:

(a) Shredded cedar/juniper bark, pine needles.

(b) Dry grass.

(c) Slivers shaved from a dry stick.

(d) Hornet's nest.

(e) Natural fibers from equipment supplemented with pine pitch (i.e., cotton battle dressing).

(f) Cotton balls and petroleum jelly or Char-cloth.

NOTE: Sticks used for tinder should be dry and not larger than the diameter of a toothpick.

(2) <u>Kindling</u>: This is the material that is ignited by the tinder that will burn long enough to ignite the fuel.

(a) Small sticks/twigs pencil-thick up to the thickness of the thumb. Ensure that they are dry.

(b) Due to a typically large resin content, evergreen limbs often make the best kindling. They burn hot and fast, but typically do not last long.

(3) Fuel Wood: Fuel Wood is used to keep the blaze going long enough to fulfill its purpose. Ideally, it should burn slow enough to conserve the woodpile, make plenty of heat, and leave an ample supply of long-lasting coals.

(a) Firewood broken from the dead limbs of standing trees or windfalls held off the ground will have absorbed less moisture and therefore should burn easily.

(b) Refrain from cutting down live, green trees.

(c) Softwoods (evergreens and conifers) will burn hot and fast with lots of smoke and spark, leaving little in the way of coals. Hardwoods (broad leaf trees) will burn slower with less smoke and leave a good bed of coals.

(d) Learn the woods indigenous to the area. Birch, dogwood, and maple are excellent fuels. Osage orange, ironwood, and manzanita, though difficult to break up, make terrific coals. Aspen and cottonwood burn clean but leave little coals.

(e) Stack your wood supply close enough to be handy, but far enough from the flames to be safe. Protect your supply from additional precipitation.

(f) If you happen to go down in an aircraft that has not burned a mixture of gas and oil may be used. Use caution when igniting this mixture.

c. Building the Fire. The type of fire built will be dependent upon its intended use, either cooking or heating and signaling.

(1) Cooking Fires. The following listed fires are best used for cooking:

(a) Teepee Fire: The teepee fire is used to produce a concentrated heat source, primarily for cooking. Once a good supply of coals can be seen, collapse the teepee and push embers into a compact bed.

TEEPEE FIRE

(2) Heating and Signaling Fires.

(a) Pyramid Fire: Pyramid fires are used to produce large amounts of light and heat. They will dry out wet wood or clothing.

PYRAMID FIRE

d. Starting Fires. Starting a fire is done by a source of ignition and falls into two categories, modern igniters and primitive methods.

(1) Modern Methods: Modern igniters use modern devices we normally think of to start a fire. Reliance upon these methods may result in failure during a survival situation. These items may fail when required to serve their purpose.

(a) Matches and Lighters. Ensure you waterproof these items.

(b) Convex Lens. Binocular, camera, telescopic sights, or magnifying lens are used on bright, sunny days to ignite tender.

(c) Flint and Steel: Sometimes known as metal matches or "Mag Block". Scrape your knife or carbon steel against the flint to produce a spark onto the tinder. Some types of flint & steel designs will have a block of magnesium attached to the device, which can be shaved onto the tinder prior to igniting. Other designs may have magnesium mixed into the flint to produce a higher quality of spark.

(2) Primitive Methods. Primitive fire methods are those developed by early man. There are numerous techniques that fall into this category. The only method that will be taught at MCMWTC is the Bow & Drill.

(3) Bow & Drill. The technique of starting a fire with a bow & drill is a true field expedient fire starting method, which requires a piece of cord and knife from your survival kit to construct. The components of the bow & drill are bow, drill, socket, and fireboard, ember patch, and bird's nest.

(a) Bow: The bow is a resilient, green stick about 3/4 of an inch in diameter and 30-36 inches in length. The bowstring can be any type of cord, however, 550 cord works best. Tie the string from one end of the bow to the other, without any slack.

(b) Drill: The drill should be a straight, seasoned hardwood stick about 1/2 to 3/4 of an inch in diameter and 8 to 12 inches in length. The top end is tapered to a blunt point to reduce friction generated in the socket. The bottom end is slightly rounded to fit snugly into the depression on the fireboard.

(c) Socket: The socket is an easily grasped stone or piece of hardwood or bone with a slight depression on one side. Use it to hold the drill in place and to apply downward pressure.

(d) Fire board: The fire board is a seasoned softwood board which should ideally be 3/4 of an inch thick, 2-4 inches wide, and 8-10 inches long. Cut a depression 3/4 of an inch from the edge on one side of the fireboard. Cut a U-shape notch from the edge of the fireboard into the depression. This notch is designed to collect and form an ember, which will be used to ignite the tinder.

(e) Ember Patch: The ember patch is made from any type of suitable material (i.e., leather, aluminum foil, and bark). It is used to catch and transfer the ember from the fireboard to the bird's nest. Ideally, it should be 4 inches by 4 inches in size.

(f) Bird's Nest: The bird's nest is a double handful of tinder, which will be made into the shape of a nest. Tinder must be dry and finely shredded material (i.e., outer bark from juniper/cedar/sage brush or inner bark from cottonwood/aspen or dry grass/moss). Lay your tinder out in two equal rows about 4 inches wide and 8-12 inches long. Loosely roll the first row into a ball and knead the tinder to further break down the fibers. Place this ball perpendicular onto the second row of tinder and wrap. Knead the tinder until all fibers of the ball are interwoven. Insert the drill half way into the ball to form a partial cylinder. This is where the ember will be placed.

(4) Producing a fire using the bow & drill.

(a) Place the ember patch under the U-shaped notch.

(b) Assume the kneeling position, with the left foot on the fireboard near the depression.

(c) Load the bow with the drill. Ensure the drill is between the wood of the bow and bow string. Place the drill into the depression on the fireboard. Place the socket on the tapered end of the drill.

(d) Use the left hand to hold the socket while applying downward pressure.

(e) Use the right hand to grasp the bow. With a smooth sawing motion, move the bow back and forth to twirl the drill.

(f) Once you have established a smooth motion, smoke will appear. Once smoke appears, apply more downward pressure and saw the bow faster.

(g) When a thick layer of smoke has accumulated around the depression, stop all movement. Remove the bow, drill, and socket from the fire board, without moving the fireboard. Carefully remove your left foot off the fireboard.

(h) Gently tap the fireboard to ensure all of the ember has fallen out of the U-shaped notch and is lying on the ember patch. Remove the fireboard.

(i) Slowly fan the black powder to solidify it into a glowing ember. Grasping the ember patch, carefully drop the ember into the cylinder of the bird's nest.

(j) Grasp the bird's nest with the cylinder facing towards you and parallel to the ground. Gently blow air into the cylinder. As smoke from the nest becomes thicker, continue to blow air into the cylinder until fire appears.

(5) Trouble Shooting the Bow & Drill

(a) Drill will not stay in depression- Apply more downward pressure and/or increase width/depth of depression.

(b) Drill will not twirl- lessens the amount of downward pressure and/or tighten bowstring.

(c) Socket smoking- Lessen the amount of downward pressure. Wood too soft when compared to hardness of drill. Add some lubrication: animal fat, oil, or grease.

(d) No smoke- Drill and fireboard are the same wood. Wood may not be seasoned. Check drill to ensure that it is straight. Keep left hand locked against left shin while sawing.

(e) Smoke but no ember- U-shaped notch not cut into center of the depression.

(f) Bowstring runs up and down drill- use a locked right arm when sawing. Check drill to ensure that it is straight. Ensure bowstring runs over the top of the left boot.

(g) Birds nest will not ignite- Tinder not dry. Nest woven too tight. Tinder not kneaded enough. Blowing too hard (ember will fracture).

e. Extinguishing the Fire. The fire must be properly extinguished. This is accomplished by using the drown, stir, and feel method.

(1) Drown the fire by pouring at water in the fire lay.

(2) Stir the ember bed to ensure that the fire is completely out.

(3) Check the bed of your fire by feeling for any hot spots.

(4) If any hot spots are found, start the process all over again.

UNITED STATES MARINE CORPS
Mountain Warfare Training Center
Bridgeport, California 93517-5001

FMST.07.22
04/07/02

STUDENT OUTLINE

SURVIVAL NAVIGATION

LEARNING OBJECTIVES

TERMINAL LEARNING OBJECTIVE. In a summer mountainous environment, navigate in a survival situation, in accordance with the references. (FMST.07.22)

ENABLING LEARNING OBJECTIVES

(1)Without the aid of references and from a given list, choose one of the considerations for travel, in accordance with the references. (FMST.07.22a)

(2)Without the aid of references and from a given list, choose two methods for locating the North Star, in accordance with the references. (FMST.07.22b)

OUTLINE

1. **CONSIDERATIONS FOR STAYING OR TRAVELLING**. (FMST.07.22a)

   a. Stay with the aircraft or vehicle if possible. More than likely somebody knows where it was going. It is also a ready-made shelter.

   b. Leave only when:

      (1) Certain of present location; have known destination and the ability to get there.

      (2) Water, food, shelter, and/or help can be reached.

      (3) Convinced that rescue is not coming.

A5

c. If the decision is to travel, the following must also be considered:

(1) Which direction to travel and why.

(2) What plan is to be followed.

(3) What equipment should be taken

(4) How to mark the trail.

(5) Predicted weather.

d. If the tactical situation permits leave the following information at the departure point:

(1) Departure time.

(2) Destination.

(3) Route of travel/direction.

(4) Personal condition.

(5) Available supplies.

2. DAYTIME SURVIVAL NAVIGATION

a. Sun Movement. It is generally taken for granted that the sun rises in the east and sets in the west. This rule of thumb, however, is quite misleading. In fact, depending on an observer's latitude and the season, the sun could rise and set up to 50 degrees off of true east and west.

b. The following diagram and terms are essential to understanding how the sun and stars can help to determine direction:

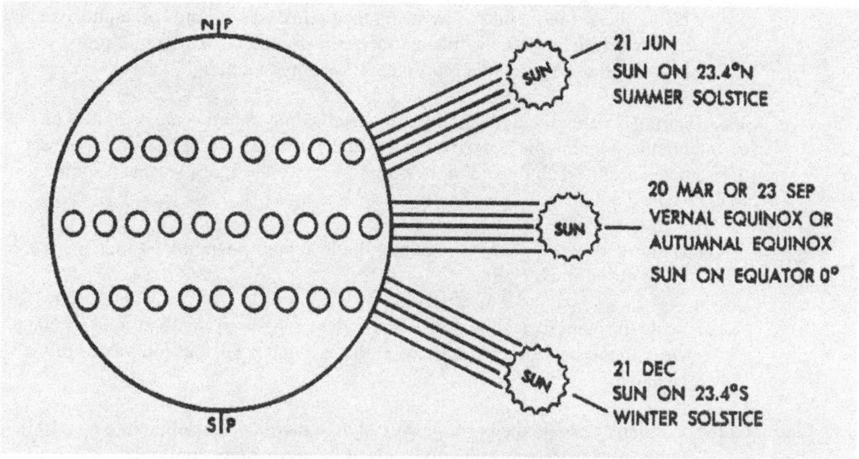

Position of the Sun at Equinox and Solstice

(1) Summer/Winter Solstice: (21 June/21 December) Two times during the year when the sun has no apparent northward or southward motion.

(2) Vernal/Autumnal Equinox: (20 March/23 September) Two times during the year when the sun crosses the celestial equator and the length of day and night are approximately equal.

c. Sun's Movement. As reflected in the diagram above, the sun continuously moves in a cycle from solstice to equinox, throughout each day, however, the sun travels a uniform arc in the sky from sunrise to sunset. Exactly halfway along its daily journey, the sun will be directly south of an observer (or north if the observer is in the Southern Hemisphere). This rule may not apply to observers in the tropics (between 23.5 degrees north and south latitude) or in the Polar Regions (60 degrees latitude). It is at this point that shadows will appear their shortest. The time at which this occurs is referred to as "local apparent noon."

d. Local Apparent Noon. Whenever using any type of shadow casting device to determine direction, "local apparent noon" (or the sun's highest point during the day) must be known. Local apparent noon can be determined by the following methods.

(1) Knowing sunrise and sunset from mission orders, i.e., sunrise 0630 and sunset 1930. Take the total amount of daylight (13 hours), divide by 2 (6 hours 30 minutes), and add to sunrise (0630 plus 6 hours 30 minutes). Based on this example, local apparent noon would be 1300.

(2) Using the string method. The string method is used to find two equidistant marks before and after estimated local apparent noon. The center point between these two marks represents local apparent noon.

e. Sun's Bearing. With an understanding of the sun's daily movement, as well as its seasonal paths, a technique is derived that will determine the true bearing of the sun at sunrise and sunset. With the aid of a circular navigational chart, we can accurately navigate based on the sun's true bearing:

(1) Determine the sun's maximum amplitude at your operating latitude using the top portion of the chart.

(2) Scale the center baseline of the chart where 0 appears as the middle number; write in the maximum amplitude at the extreme north and south ends of the baseline.

(3) Continue to scale the baseline; you should divide the baseline into 6 to 10 tick marks that represent equal divisions of the maximum amplitude.

(4) From today's date along the circumference, draw a straight line down until it intersects the baseline.

(5) The number this line intersects is today's solar amplitude. If the number is left of 0, it is"north" amplitude; if the number is right of 0, it is a "south" amplitude. Use the formula at the bottom of the chart to determine the sun's bearing at sunrise or sunset.

LATITUDE (N OR S)	5	10	15	20	25	30	35	40	45	50	55	60
MAXIMUM AMPLITUDE	24°	24°	24°	25°	26°	27°	29°	31°	34°	38°	44°	53°

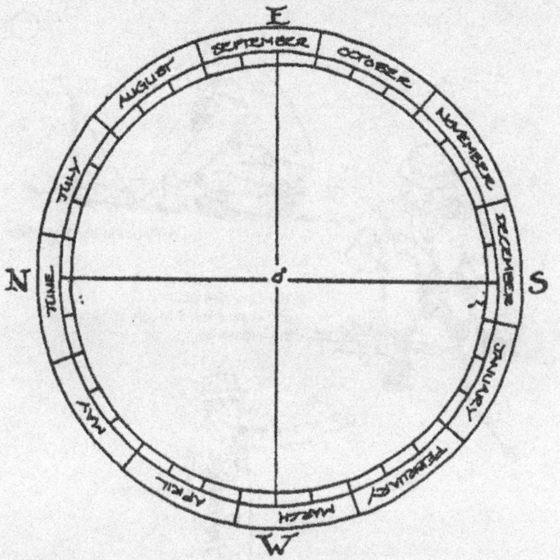

NORTH AMPLITUDES	SOUTH AMPLITUDES
SUNRISE BEARING = 90° - AMPLITUDE	SUNRISE BEARING = 90° + AMPLITUDE
SUNSET BEARING = 270° + AMPLITUDE	SUNSET BEARING = 270° - AMPLITUDE

Circular Navigational Chart

f. Shadow Stick Construction. This technique will achieve a cardinal direction within 10 degrees of accuracy if done within two hours of local apparent noon. Once again, this technique may be impractical near the Polar Regions, as shadows tend to be very long; similarly, in the tropics shadows are generally very small.

(1) Get a straight, 3-6 foot stick free of branches and pointed at the ends and 3-5 small markers: i.e., sticks, rocks, or nails.

(2) Place the stick upright in the ground and mark the shadow tip with a marker.

(3) Wait 10-15 minutes and mark shadow tip again with a marker.

(4) Repeat this until all of the markers are used.

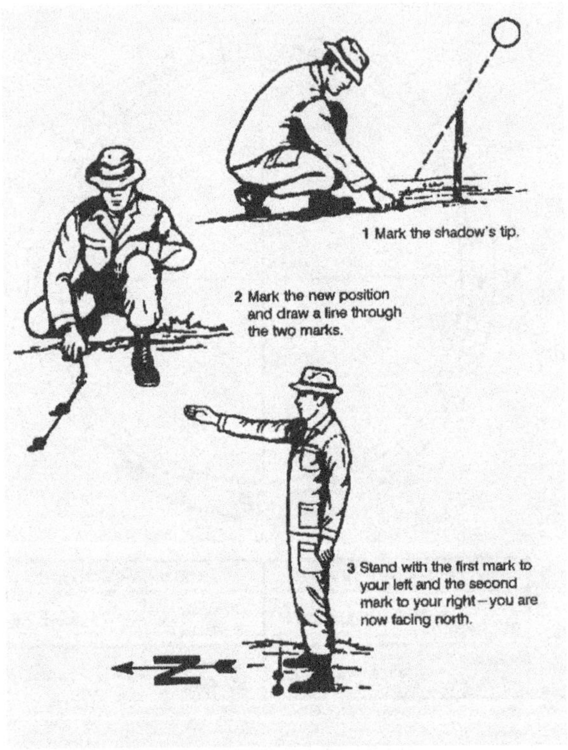

SHADOW STICK METHOD

(5) The markers will form a West—East line.

(6) Put your left foot on the first marker and your right foot on the last marker, you will then be facing north.

3. **POCKET NAVIGATOR** The only material required is a small piece of paper or other flat-surface material upon which to draw the trace of shadow tips and a 1 to 2 inch pin, nail, twig, wooden matchstick, or other such device to serve as a shadow-casting rod.

a. Set this tiny rod upright on your flat piece of material so that the sun will cause it to cast a shadow. Mark the position where the base of the rod sits so it can be returned to the same spot for later readings. Secure the material so that it will not move and mark the position of the material with string, pebbles, or twigs, so that if you have to move the paper you can put it back exactly as it was. Now, mark the tip of the rod shadow.

b. As the sun moves, the shadow-tip moves. Make repeated shadow-tip markings every 15 minutes. As you make the marks of the shadow tip, ensure that you write down the times of the points.

c. At the end of the day, connect the shadow-tip markings. The result will normally be a curved line. The closer to the vernal or autumnal equinoxes (March 21 and September 23) the less pronounced the curvature would be. If it is not convenient or the tactical situation does not permit to take a full day's shadow-tip markings, your observation can be continued on subsequent days by orienting the pocket navigator on the ground so that the shadow-tip is aligned with a previously plotted point.

d. The markings made at the sun's highest point during the day, or solar noon, is the north—south line. The direction of north should be indicated with an arrow on the navigator as soon as it is determined. This north-south line is drawn from the base of the rod to the mark made at solar noon. This line is the shortest line that can be drawn from the base of the pin to the shadow-tip curve.

e. To use your pocket navigator, hold it so that with the shadow-tip is aligned with a plotted point at the specified point. I.e.; if it is now 0900 the shadow-tip must be aligned with that point. This will ensure that your pocket navigator is level. The drawn arrow is now oriented to true north, from which you can orient yourself to any desired direction of travel.

f. The pocket navigator will work all day and will not be out of date for approximately one week.

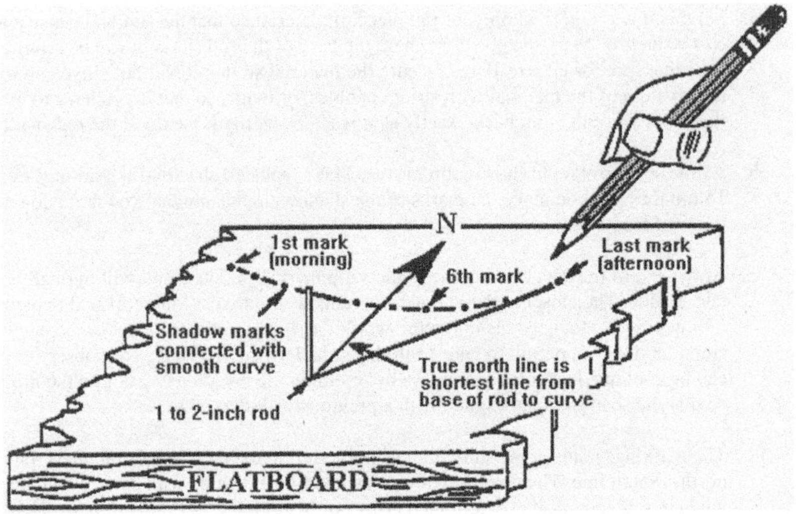

POCKET NAVIGATOR

4. NIGHTTIME SURVIVAL NAVIGATION

a. Mark North. To aid you in navigating at night, it is beneficial to watch where the sun goes down. If you're going to start moving after dark mark the northerly direction.

b. Locating the North Star. There are two methods used in locating the North Star. (FMST.07.22b)

(1) Using the Big Dipper (*Ursa Major*). The best indictors are the two "dippers ". The North Star is the last star in the handle of the Little Dipper, which is not the easiest constellation to find. However, the Big Dipper is one of the most prominent constellations in the Northern Hemisphere. The two lowest stars of the Big Dipper's cup act as pointers to the North Star. If you line up these two stars, they make a straight line that runs directly to the North Star. The distance to the North Star along this line is 5 times that between the two pointer stars.

(2) Using Cassiopeia (*Big M or W*). Draw a line straight out from the center star, approximately half the distance to the Big Dipper. The North Star will be located there.

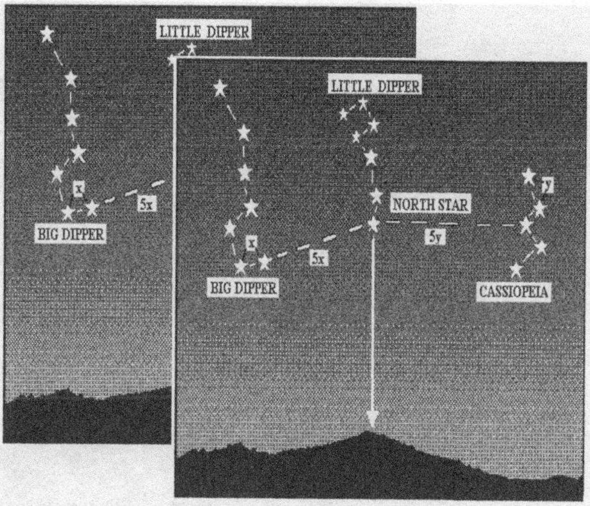

LOCATING THE NORTH STAR

NOTE: Because the Big Dipper and Cassiopeia rotate around the North Star, they will not always appear in the same position in the sky. In the higher latitudes, the North Star is less effective for the purpose of orienting because it appears higher in the sky. At the center of the Arctic Circle, it would be directly overhead and all directions lead south.

c. Southern Cross. In the Southern Hemisphere, Polaris is not visible. There, the Southern Cross is the most distinctive constellation. An imaginary line through the long axis of the Southern Cross, or True Cross, points towards a dark spot devoid of stars approximately three degrees offset from the South Pole. The True Cross should not be confused with the larger cross nearby know as the False Cross, which is less bright, more widely spaced, and has five stars. Two closely spaced, very bright stars that trail behind the crosspiece can confirm the True Cross. These two stars are often easier to pick out than the cross itself. Look for them. Two of the stars in the True Cross are among the brightest stars in the heavens; they are the stars on the southern and eastern arms. The stars on the northern and western arms are not as conspicuous, but are bright.

NOTE: The imaginary point depicted in the picture is the dark spot devoid of stars.

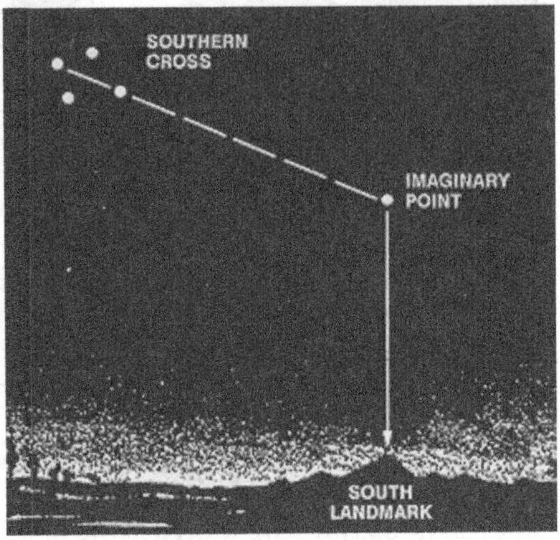

SOUTHERN CROSS

d. Moon Navigator. Like the sun, the moon rises in the east and sets in the west. Use the same method of the shadow stick as you did during the day.

5. **IMPROVISED COMPASSES**. There are three improvised techniques to construct a compass.

a. Synthetic technique. The required items are a piece of synthetic material, (i.e., parachute cloth), and a small piece of iron or steel that is long, thin, and light. Aluminum or yellow metals won't work (only things that rust will do). A pin or needle is perfect, but a straightened paper clip, piece of steel baling wire or barbed wire could also work.

(1) Stroke the needle repeatedly in one direction against the synthetic material. Ensure that you lift the material a few inches up into the air at the end of each stroke, returning to the beginning of the needle before descending for another stroke in the same direction. Do this approximately 30 strokes. This will magnetize the needle.

(2) Float the metal on still water using balled up paper, wood chip, or leaf. Gather some water in a non-magnetic container or a scooped out recess in the

ground, such as a puddle. Do not use a "tin can" which is made of steel. (Aluminum can would be fine.) Place the float on the water, then the metal on it. It will slowly turn to orient itself.

b. Magnet technique. You will achieve the same results by using a magnet. Follow the same steps as you did with the synthetic material. The magnets you are most likely to have available to you are those in a speaker or headphones of a radio.

c. Magnetization through a battery. A power source of 2 volts or more from a battery can be used with a short length of insulated wire to magnetize metal. Coil the wire around a needle. If the wire is non-insulated, wrap the needle with paper or cardboard. Attach the ends to the battery terminals for 5 minutes.

d. Associated problems with improvised compasses. The following are common problems with all improvised compasses.

(1) Soft steel tends to lose its magnetism fairly quickly, so you will have to demagnetize your needle occasionally, though you should not have to do this more than two or three times a day.

(2) Test your compass by disturbing it after it settles. Do this several times. If it returns to the same alignment, you're OK. It will be lined up north and south, though you will have to determine by other means which end is north. Use the sun, stars, or any other natural signs in the area.

(3) Remember that this will give magnetic north. In extreme northern latitudes, the declination angle can be extreme.

6. SURVIVAL NAVIGATION TECHNIQUES

a. Navigator.

(1) Employ a navigation method.

(2) Find the cardinal direction.

(3) Pick a steering mark in the desired direction of travel.

b. Maintain a Log: The possibility may arise when you will not have a map of the area. A log will decrease the chance of walking in circles.

(1) Construction.

(a) Use any material available to you i.e., paper, clothing, MRE box, etc.

(b) Draw a field sketch annotating North, prominent terrain features, and Friendly/enemy position.

(2) Maintenance.

(a) Annotate distance traveled, elevation gained and lost, and cardinal directions.

(b) Maintain and update field sketch as movement progresses.

(c) Ensure readability of your field sketch. (I.e., don't clutter the sketch so much that it can't be read.)

c. <u>During Movement Constantly Refer To</u>.

(1) Log.

(2) Steering marks.

d. <u>Actions If You Become Lost</u>.

(1) Immediate action

(a) Orient your sketch. This will probably make your mistake obvious.

(2) Corrective action

(a) Backtrack using steering marks until you have determined the location of your error.

(b) Re-orient your sketch.

(c) Select direction of travel and continue to march.

UNITED STATES MARINE CORPS
Mountain Warfare Training Center
Bridgeport, California 93517-5001

FMST.07.08
04/07/02

STUDENT OUTLINE

WATER PROCUREMENT

LEARNING OBJECTIVES

TERMINAL LEARNING OBJECTIVE Given a unit in a cold weather environment and the necessary equipment and supplies, perform cold weather preventive medicine, in accordance with the references. (FMST.07.08)

ENABLING LEARNING OBJECTIVES

(1) Without the aid of references from a given list, choose one of the techniques for finding incidental water, in accordance with the references. (FMST.07.08i)

(2) Without the aid of references from a given list, choose one of the four hazardous fluids to avoid substituting for potable water, in accordance with the references. (FMST.07.08j)

(3) Without the aid of references, construct a water-producing device, in accordance with the references. (FMST.07.08k)

OUTLINE

1. WATER INTAKE

a. Thirst is not a strong enough sensation to determine how much water you need.

b. The best plan is to drink, utilizing the OVER DRINK method. Drink plenty of water anytime it is available and particularly when eating.

A6

c. Dehydration is a major threat. A loss of only 5 % of your body fluids causes thirst, irritability, nausea, and weakness; a 10% loss causes dizziness, headache, inability to walk, and a tingling sensation in limbs; a 15% loss causes dim vision, painful urination, swollen tongue, deafness, and a feeling of numbness in the skin; also a loss of more than 15% body fluids could result in death.

d. Your water requirements will be increased if:

(1) You have a fever.

(2) You are experiencing fear or anxiety.

(3) You evaporate more body fluid than necessary. (i.e., not using the proper shelter to your advantage)

(4) You have improper clothing.

(5) You ration water.

(6) You overwork.

2. **INCIDENTAL WATER**. (FMST.07.08.i)

a. During movement, you may have to ration water until you reach a reliable water source. Incidental water may sometimes provide opportunities to acquire water. Although not a reliable or replenished source, it may serve to stretch your water supply or keep you going in an emergency. The following are sources for incidental water:

(1) Dew: In areas with moderate to heavy dew, tying rags or tuffs of fine grass around your ankles can collect dew. While walking through dewy grass before sunrise, the rags or grass will saturate and can be rung out into a container. The rags or grass can be replaced and the process is repeated.

(2) Rainfall: Rainwater collected directly in clean container or in plants that contain no harmful toxins is generally safe to drink without disinfecting. The survivor should always be prepared to collect rainfall at a moment's notice. An inverted poncho works well to collect rainfall.

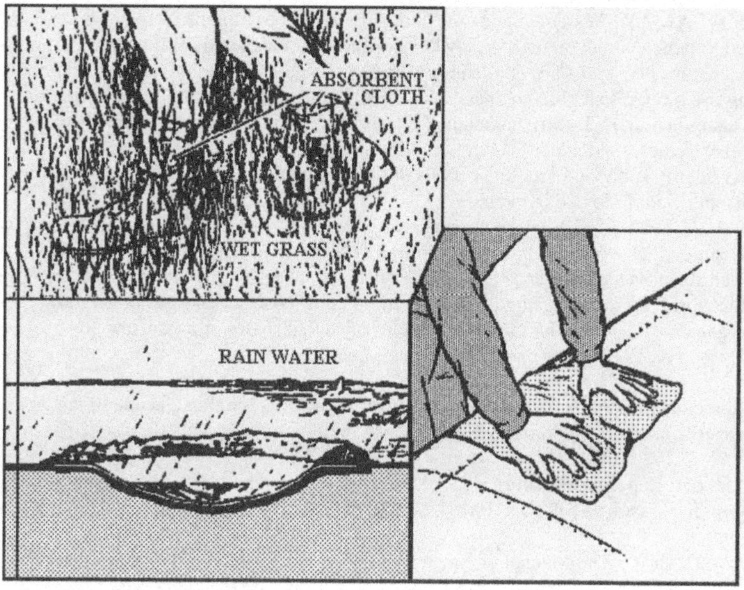

3. **<u>HAZARDOUS FLUIDS</u>** (FMST.07.08j)

a. Survivors have occasionally attempted to augment their water supply with other fluids, such as alcoholic beverages, urine, blood, or seawater. While it is true that each of these fluids has a high water content, the impurities they contain may require the body to expend more fluid to purify them. Some hazardous fluids are:

(1) <u>Seawater</u>: Seawater in more than minimal quantities are actually toxic. The concentration of sodium and magnesium salts is so high that fluid must be drawn from the body to eliminate the salts and eventually the kidneys cease to function.

(2) <u>Alcohol</u>: Alcohol dehydrates the body and clouds judgment. Super-cooled liquid, if ingested, can cause immediate frostbite of the throat, and potential death.

(3) <u>Blood</u>: Blood, besides being salty, is a food. Drinking it will require the body to expend additional fluid to digest it.

(4) <u>Urine</u>: Drinking urine is not only foolish, but also dangerous. Urine is nothing more than the body's waste. Drinking it only places this waste back into the body, which requires more fluid to process it again.

4. **WATER QUALITY**. Water contains minerals, toxins, and pathogens. Some of these, consumed in large enough quantities may be harmful to human health. Pathogens are our primary concern. Pathogens are divided into Virus, Cysts, Bacteria, and Parasites. Certain pathogens are more resistant to chemicals and small enough to move through microscopic holes in equipment (i.e., T-shirt, parachute). Certain pathogens also have the ability to survive in extremely cold water temperatures. Pathogens generally do not live in snow and ice. Water quality is divided into three levels of safety with disinfecting as the most desired level, then purified, followed by potable.

a. Disinfecting. Water disinfecting removes or destroys harmful microorganisms. Giardia cysts are an ever-present danger in clear appearing mountain water throughout the world. By drinking non-potable water you may contract diseases or swallow organisms that could harm you. Examples of such diseases or organisms are Dysentery, Cholera, Typhoid, Flukes, and Leeches.

b. Remember impure water, no matter how overpowering the thirst, is one of the worst hazards in a survival situation.

c. The first step in disinfecting is to select a treatment method. The two methods we will discuss are as follows: (FMST.07.08j)

(1) Heat. The Manual of Naval Preventive Medicine (P-5010) states that you must bring the water to a rolling boil before it is considered safe for human consumption. This is the most preferred method.

(a) Bringing water to the boiling point will kill 99.9% of all Giardia cysts. The Giardia cyst dies at 60°C and Cryptosporidium dies at 65C. Water will boil at 14,000' at 86°C and at 10,000' at 90C. With this in mind you should note that altitude does not make a difference unless you are extremely high.

(2) Chemicals. There are numerous types of chemicals that can disinfect water. Below are a few of the most common. In a survival situation, you will use whatever you have available.

(a) Iodine Tablets.

(b) Chlorine Bleach.

(c) Iodine Solution.

(d) Betadine Solution.

(e) Military water purification tablets. These tablets are standard issue for all DOD agencies. These tablets have a shelf life of four years from the date of manufacture, unless opened. Once the seal is broken, they have a shelf life of one year, not to exceed the initial expiration date of four years.

49703
Month / Year / Batch Number

(3) Water Disinfecting Techniques and Halogen Doses.

Iodination techniques Added to 1 liter or quart of water	Amount for 4 PPM	Amount for 8 PPM
Iodine tablets Tetraglcine hydroperiodide EDWGT Potable Aqua Globaline	½ tablet	1 tablet
2% iodine solution (tincture)	0.2 ml 5 gtts	0.4 ml 10 gtts
10% povidone-iodine solution*	0.35 ml 8 gtts	0.70 ml 16 gtts
Chlorination techniques	Amount for 5 PPM	Amount for 10 PPM
Household bleach 5% Sodium hypochlorie	0.1 Ml 2 gtts	0.2 ml 4 gtts
AquaClear Sodium dichloroisocyanurate		1 tablet
AquaCure, AquaPure, Chlor-floc Chlorine plus flocculating agent		8 PPM 1 tablet

*Providone-iodine solutions release free iodine in levels adequate for disinfecting, but scant data is available.

Measure with dropper (1 drop=0.05 ml) or tuberculin syringe

PPM-part per million gtts-drops ml-milliliter

Concentration of **Contact time in minutes at various water temperatures**

Halogen

	5 C / 40 F	15 C / 60 F	30 C / 85 F
2 PPM	240	180	60
4 PPM	180	50	45
8 PPM	60	30	15

NOTE: These contact times have been extended from the usual recommendations to account for recent data that prolonged contact time is needed in very cold water to kill *Giardia* cysts.

NOTE: Chemicals may not destroy Cryptosporidium.

d. Purification. Water purification is the removal of organic and inorganic chemicals and particulate matter, including radioactive particles. While purification can eliminate offensive color, taste, and odor, it may not remove or kill microorganisms.

(1) Filtration. Filtration purifying is a process by which commercial manufacturers build water filters. The water filter is a three-tier system. The first layer, or grass layer, removes the larger impurities. The second layer, or sand layer, removes the smaller impurities. The final layer, or charcoal layer (**not the ash but charcoal from a fire**), bonds and holds the toxins. All layers are placed on some type of straining device and the charcoal layer should be at least 5-6 inches thick. Layers should be changed frequently and straining material should be boiled. Remember that this is not a disinfecting method, cysts can possibly move through this system.

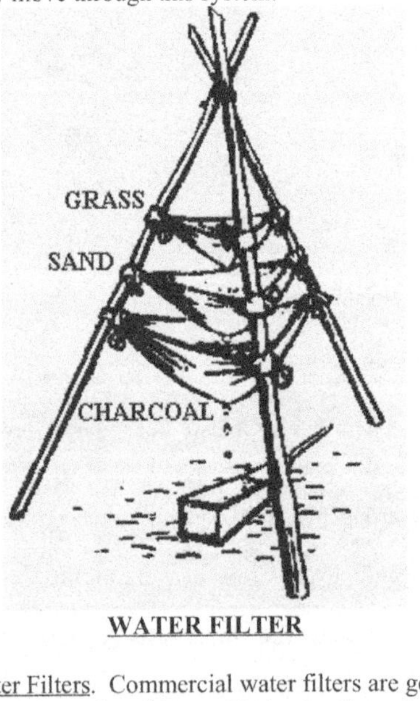

WATER FILTER

(2) Commercial Water Filters. Commercial water filters are generally available in most retail stores and may be with you. Understanding what the filter can do is the first step in safeguarding against future illnesses.

(a) A filter that has a .3 micron opening or larger will not stop Cryptosporidium.

(b) A filter system that does not release a chemical (i.e., iodine) may not kill all pathogens.

(c) A filter that has been overused may be clogged. Usage may result in excessive pumping pressure that can move harmful pathogens through the opening.

e. Potable. Potable indicates only that a water source, on average over a period of time, contains a "minimal microbial hazard," so the statistical likelihood of illness is acceptable.

(1) Sedimentation. Sedimentation is the separation of suspended particles large enough to settle rapidly by gravity. The time required depends on the size of the particle. Generally, 1 hour is adequate if the water is allowed to sit without agitation. After sediment has formed on the bottom of the container, the clear water is decanted or filtered from the top. Microorganisms, especially cysts, eventually settle, but this takes longer and the organisms are easily disturbed during pouring or filtering. Sedimentation should not be considered a means of disinfecting and should be used only as a last resort or in an extreme tactical situation.

5. **SOLAR STILLS.** (FMST.07.08k)

a. Solar stills are designed to supplement water reserves. Contrary to belief, they will not provide enough water to meet the daily requirement for water.

b. AboveGround Solar Still. This device allows the survivor to make water from vegetation. To make the aboveground solar still, locate a sunny slope on which to place the still, a clear plastic bag, green leafy vegetation, and a small rock.

(1) Construction.

(a) Fill the bag with air by turning the opening into the breeze or by "scooping" air into the bag.

(b) Fill the bag half to three-quarters full of green leafy vegetation. Be sure to remove all hard sticks or sharp spines that might puncture the bag.

CAUTION
Do not use poisonous vegetation. It will provide poisonous liquid.

(c) Place a small rock or similar item in the bag.

(d) Close the bag and tie the mouth securely as close to the end of the bag as possible to keep the maximum amount of air space. If you have a small piece of tubing, small straw, or hollow reed, inserts one end in the mouth of the bag before tying it securely. Tie off or plug the tubing so that air will not escape. This tubing will allow you to drain out condensed water without untying the bag.

(e) Place the bag, mouth downhill, on a slope in full sunlight. Position the mouth of the bag slightly higher than the low point in the bag.

(f) Settle the bag in place so that the rock works itself into the low point in the bag.

(g) To get the condensed water from the still, loosen the tie and tip the bag so that the collected water will drain out. Retie the mouth and reposition the still to allow further condensation.

(h) Change vegetation in the bag after extracting most of the water from it.

(i) Using 1-gallon zip-loc. bag instead of trash bags is a more efficient means of construction.

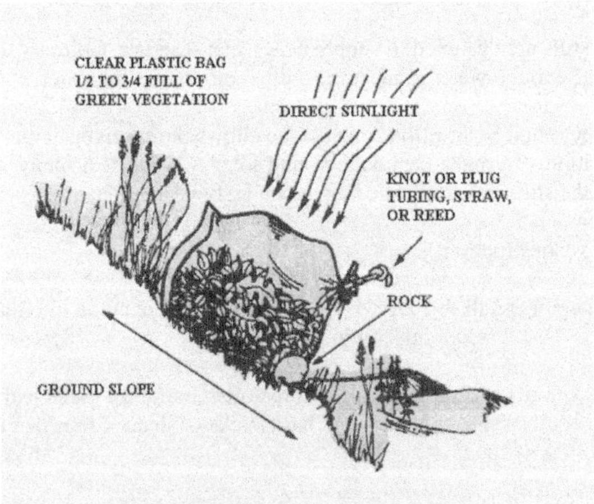

ABOVE GROUND SOLAR STILL

c. BelowGround Solar Still. Materials consist of a digging stick, clear plastic sheet, container, rock, and a drinking tube.

(1) Construction.

(a) Select a site where you believe the soil will contain moisture (such as a dry streambed or a low spot where rainwater has collected). The soil should be easy to dig, and will be exposed to sunlight.

(b) Dig a bowl-shaped hole about 1 meter across and 24 inches deep.

(c) Dig a sump in the center of the hole. The sump depth and perimeter will depend on the size of the container you have to place in it. The bottom of the sump should allow the container to stand upright.

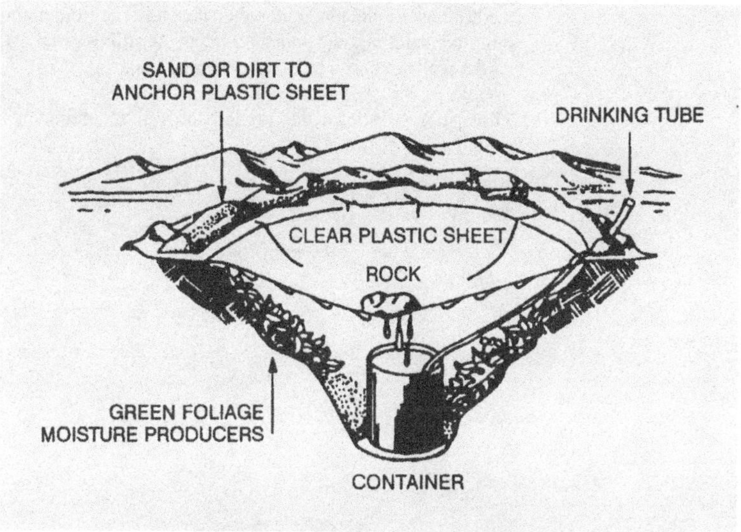

BELOW GROUND SOLAR STILL

(d) Anchor the tubing to the container's bottom by forming a loose overhand knot in the tubing. Extend the unanchored end of the tubing up, over, and beyond the lip of the hole.

(e) Place the plastic sheet over the hole, covering its edges with soil to hold in place. Place a rock in the center of the plastic sheet.

(f) Lower the plastic sheet into the hole until it is about 18 inches below ground level. Make sure the cone's apex is directly over the container. Ensure the plastic does not touch the sides of the hole because the earth will absorb the moisture.

(g) Put more soil on the edges of the plastic to hold it securely and prevent the loss of moisture.

(h) Plug the tube when not in use so that moisture will not evaporate.

(i) Plants can be placed in the hole as a moisture source. If so, dig out additional soil from the sides.

(j) If polluted water is the only moisture source, dig a small trough outside the hole about 10 inches from the still's lip. Dig the trough about 10 inches deep and 3 inches wide. Pour the polluted water in the trough. Ensure you do not spill any polluted water around the rim of the hole where the plastic touches the soil. The trough holds the polluted water and the soil filters it as the still draws it. This process works well when the only water source is salt water.

(k) Three stills will be needed to meet the individual daily water intake needs.

UNITED STATES MARINE CORPS
Mountain Warfare Training Center
Bridgeport, California 93517-5001

WMC
04/07/02

STUDENT HANDOUT

FORAGING PLANTS AND INSECTS FOR SURVIVAL USES

PURPOSE The purpose of this period of instruction is to familiarize the Wilderness Medicine Student with the potential resources for survival sustenance.

OUTLINE

1. GENERAL CONSIDERATIONS. There are few places without some type of edible vegetation. Plants contain vitamins, minerals, protein, carbohydrates, and dietary fiber. Some plants also contain fats. The following are general considerations:

a. Do not assume that because birds or animals have eaten a plant, it is edible by humans.

b. Poor plant recognition skills will seriously limit your ability to survive.

c. Plant dormancy and snowfall make foraging plants difficult during the winter months.

d. Plants generally poison by:

(1) Ingestion: When a person eats a part of a poisonous plant.

(2) Contact: When a person makes contact with a poisonous plant that cause any type of skin irritation or dermatitis.

(3) Absorption: When a person absorbs the poison through the skin, which can interrupt a bodily function.

(4) Inhalation: Poisoning can occur through the inhalation of smoke that contains poisonous plant residue.

A7

e. Plant properties can change throughout the growing season. Plants can be edible during certain periods while poisonous in others.

2. PLANT RECOGNITION. Plant recognition can take years of study to learn. The following is used as a general guideline.

a. Types of plants that should be avoided are:

(1) All mushrooms.

(2) Plants with a milky sap.

(3) White and yellow berries should be avoided, as they are almost always poisonous.

(4) Plants with shiny leaves should be considered poisonous.

(5) Plants that are irritants to the skin should not be eaten.

(6) All beans and peas should be avoided.

b. Edible Plants: There are several edible plants found at the Marine Corps Mountain Warfare Training Center. All are easy to recognize and many of the same plants are located throughout the United States and abroad.

(1) Wild Onions - Complete plant.

(2) Dandelion - Roots and leaves.

(3) Mallow - Leaves, stem and flowers.

(4) Watercress -Complete plant.

(5) Cattail - Root, stalk and stem.

(6) Bull Thistle - The flower.

(7) Juniper Tree - Berries and cambium layer.

(8) All Conifers - Cambium layer, needles, and nuts within the cones.

(9) Currants - Berries.

c. Preparing an edible plant. All edible plants should be boiled twice with a change of water to remove any bitterness, acids, or bacteria that the plant may contain.

3. PLANT TESTING PROCEDURE. Always adopt the following procedure when trying unknown plants as food.

a. Wait 8 hours without eating.

b. Select a plant that grows in sufficient quantity in the local area. Separate the part of the plant you wish to test root, stem, leaf, or flower. Certain parts of plants are poisonous while the, other parts may be edible.

c. Rub a portion of the plant you have selected on your inner forearm. Wait 15 minutes and look for any swelling, rash, or irritation.

d. Boil the plant or plant part in changes of water. The toxic properties of many plants are water-soluble or can be destroyed by heat. Cooking and discarding two changes of water can lessen the amount of poisonous material or remove it completely. These boiling periods should last at least 5 minutes each.

e. Place 1 teaspoon of the prepared plant food in the mouth and chew for 5 minutes, but do not swallow. If unpleasant effects occur (burning, bitter, or nauseating taste), remove the plant from the mouth at once and discard it as a food source. If no unpleasant effects occur, swallow the plant material and wait 8 hours.

f. If after 8 hours no unpleasant effects have occurred (nausea, cramps, and diarrhea), eat two tablespoonfuls and wait 8 hours.

g. In no unpleasant effects have occurred at the end of this 8-hour period, the plant may be considered edible.

h. Completely document and sketch the plant in a logbook to refer to for future use. This will aid in future procurement of this plant. If plant properties have changed, you will have to repeat the plant testing procedure.

NOTE: Have charcoal ready to eat. If any side effects occur, attempt to induce vomiting and immediately eat charcoal. Charcoal will absorb toxins and reduce the chance of Death.

4. MEDICINAL PLANTS One of the many things that hinder a survivor's ability to survive is medical problems. Injuries incurred will reduce survival expectancy and the ability to evade.

a. Coyote Willow: Coyote Willow is a thick forming shrub with clustered stems and very narrow leaves. It habits wet soils, especially riverbanks, sandbars, and silt flats.

(1) Drinking tea made from the inner bark can reduce fever.

(2) The dried and powdered inner bark can be used to stop bleeding.

(3) Administering a half-cup of willow charcoal dissolved in water can treat diarrhea.

(4) The Coyote Willow roots can be mashed and applied to tooth aches.

b. Yarrow: Yarrow is a plant (tall, usually not branched, with many white, slat topped group flowers). Do not confuse Yarrow with Water Hemlock.

(1) Rubbing a handful of crushed Yarrow flowers and leaves on any exposed can make insect repellent skin.

(2) Placing a Yarrow leaf poultice on the wound can stop bleeding.

(3) Relief from many rashes, including poison oak and ivy, can be achieved by applying a Yarrow leaf compress to the effected area.

(4) The Yarrow root can be chewed to relieve the pain of a toothache or break a fever.

(5) A potent anesthetic can be made by scrubbing fresh Yarrow roots in water to clean them. Once the roots are clean, crush them into a spongy mass and apply gently to the wounded area.

c. Tea To make a tea, place the plants in a container and pour in two cups of boiling water. Let the mixture steep for 20 minutes and drink.

d. Poultice: Place plant (Coyote Willow, Yarrow, etc.) in gauze or other similar material and fold it so the gauze will hold the plant in place. Put gauze in a larger cloth, about 6"x8", and roll the sides inward. Fold the cloth over without losing the plant. While boiling water, dip the bottom portion of the cloth containing the plant into the hot water by holding the edges. Keep the plant submerged in the boiling water until it becomes saturated. Bring the cloth straight up and with a twisting motion, wring the excess water back into the pot. Apply the poultice to the affected area as soon as it has cooled enough to place on the wound. To be effective, the poultice should be as hot as you can tolerate it.

e. Compress: A compress is made just like a poultice; except it is cold when applied to the wound. NOTE: Poultice or Compress should be applied for 1 to 24 hours, as needed. When applying a poultice, you may experience a throbbing pain as it draws out the infection and neutralizes toxins. When the pain subsides, the poultice has accomplished its task and should be removed. Apply a fresh poultice as needed until the desired level of healing has been reached.

5. POISONOUS PLANTS. Rarely will a survivor have an ideal means of killing large game, however there are certain plants that can aid the survivor. The two plants that we will talk about are the Water Hemlock and Monkshood.

a. Preparing Poisonous Plants: Once a poisonous plant has been located, dig up the root of the plant. The roots of the Water Hemlock and Monkshood generally grow 8 to 10 inches deep. Extreme caution must be used when handling the root. Do not handle the root without a barrier between your hands and the root. A barrier can be gloves, socks, T-shirt, or even moss. Split the root lengthwise to expose the inside of the root where the toxin is located. With the root split, rub the tip of your spear/arrow inside of the root opening. In a slow controlled manner, work from the bottom of the tip to the top. Once the tip is thoroughly coated, allow the toxin to dry and apply another coat to the tip. Continue to apply coat after coat of toxin, until the root is completely drained of its' toxin. You are now finished and the tip is ready for use.

NOTE: Do not eat the meat in the vicinity of the poisoned wound.

CAUTION: Do not apply poison to the actual point and edges of tip. This avoids accidental puncture to you. Wash and boil all gloves and clothing used as a barrier before wearing

6. INSECTS. Insects are the most abundant life form on earth and are an excellent survival food. They are easy to catch and provide 65-80% protein; compared to 20% for beef. They aren't too appetizing, but personal bias has no place in a survival situation. The focus must remain on maintaining your health.

a. Insects to avoid.

(1) All adults that sting or bite.

(2) Hairy or brightly colored insects.

(3) All Caterpillars.

(4) Insects that have a pungent odor.

(5) All spiders.

(6) Disease carriers like ticks, flies, or mosquitoes.

b. Edible Insects.

(1) Insect Larvae.

(2) Grasshoppers.

(3) Beetles.

(4) Grubs.

(5) Ants.

(6) Termites.

(7) Worms.

c. Foraging for Insects. One must be careful not to expend more energy harvesting food than can be replaced. For example, catching insects such as grasshoppers can become exasperating and tiring.

(1) At night grasshoppers climb tall plants and clings to the stalks near the top. They can be picked from the plants in the early morning while they are chilled and dormant.

(2) Dig for worms in damp humus soil, under rocks/logs or look for them on the ground after it has rained.

(3) Carpenter ants are found in dead trees and stumps, which can be gathered by hand.

(4) Most other insects can be found in rotten logs, under rocks, and in open grassy areas.

d. Preparing Edible Insects.

(1) Insects with a hard outer shell have parasites. Remove the wings and barbed legs before cooking.

(2) Drop worms into potable water for at least a half-hour. They will naturally purge themselves. You can either cook or eat them raw.

(3) Most other insects can be eaten raw. Cooking insects will improve their taste. If the thought of eating insects is unbearable, grind them into a paste and mix with other foods.

UNITED STATES MARINE CORPS
Mountain Warfare Training Center
Bridgeport, California 93517-5001

WMC.A8.0
05/02/01

STUDENT HANDOUT

SURVIVAL TRAPS AND SNARES

PURPOSE: The purpose of this class is to familiarize the student with some of the techniques used to take game. Game can be useful for both food and clothing.

OUTLINE

1. GENERAL TIPS AND TECHNIQUES FOR TRAPPING

a. General Tips. Knowing a few general hints and tips will make the trapping of animals much easier and considerably more effective. The eight general tips for trapping are:

(1) Know your game: Knowing the habits of the animal you want to trap will help lure it into your trap. Such things as when and where they move, feed, and water will help you determine where the trap can be most effectively placed.

A8

(2) Keep things simple: You don't have time in a survival situation to construct elaborate traps and they do not necessarily do a better job.

(3) Set traps in the right place: Animals will travel and stop in certain locations. That is where to best employ traps.

(4) Cover up your scent: Animals will avoid an area, which smells strange to them. Smoke from your fire is the best cover to use.

(5) Use the right type of trap: Some traps work better than others do (i.e. a deadfall works better on squirrels than a spring pole).

(6) Use the right size trap: Adjust the size of your traps to the size of the animal.

(7) Check traps: Check your traps twice daily: morning and evening. Checking your traps less than twice a day can allow your game to escape, rot, or be taken by other predators.

(8) Bait your traps: Bait of any type will add to your chances of success.

b. General Techniques. A general technique is the method in which the trap is intended to kill or hold the animal.

(1) Strangulation: This method strangles the animal, such as a snare.

(2) Mangling: This method crushes the animal, such as a deadfall.

(3) Entanglement: This method entangles the animal, such as a net.

(4) Live: This method holds the animal, such as a box trap.

c. Types of Triggers. There are three basic triggers used for all traps and path guards. Depending on the situation, variations of these triggers can be used.

(1) Puite Figure 4.

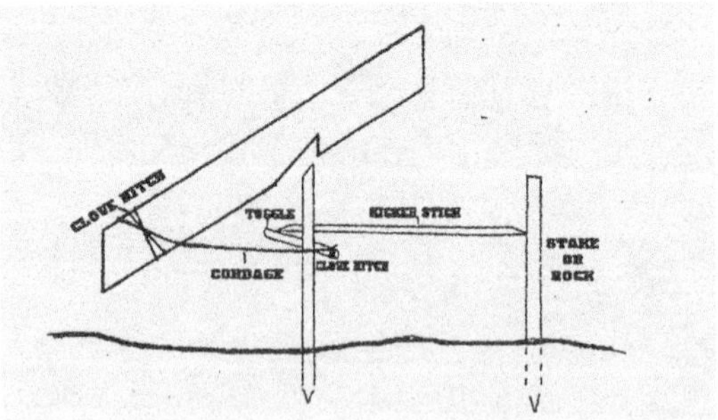

(2) Toggle.

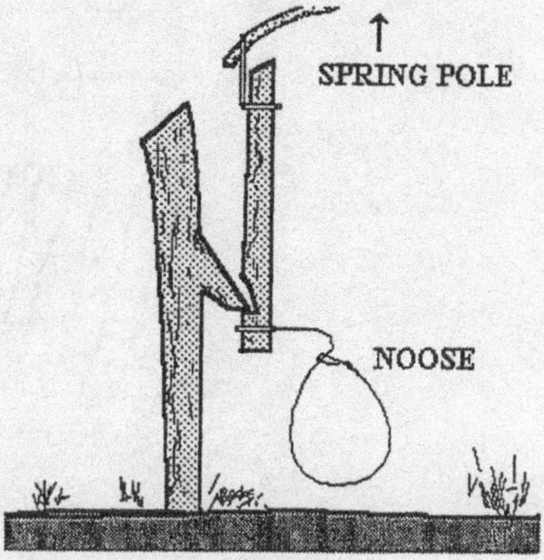

TOGGLE

(c) Universal.

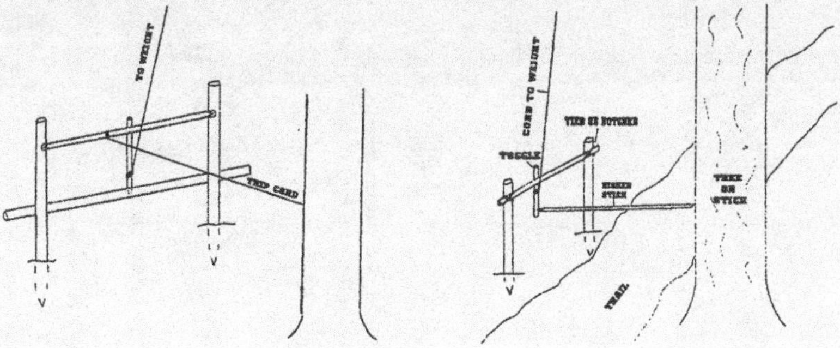

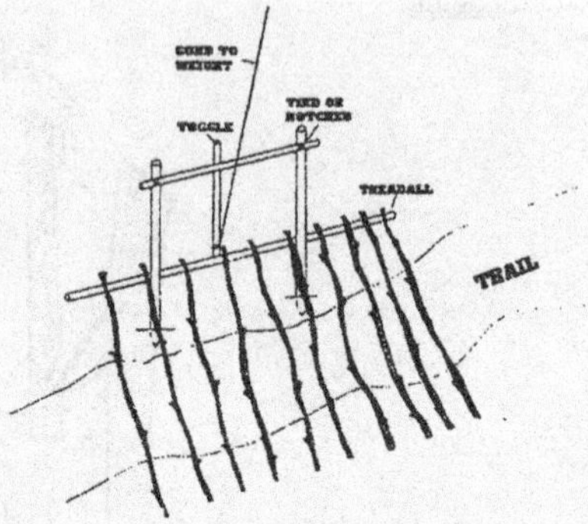

2. TRAPS. Traps are c
   action is generally caused by either a weight or spring loaded device.

a. Puite Deadfall: The Puite deadfall requires a knife and piece of cord to construct. It is designed to mangle small rodents. The trigger for the deadfall is the Puite figure 4.

b. Spring Pole: The spring pole requires a small sapling and cordage to construct. The trigger for the spring pole is the toggle. It is designed to lift the animal off the ground; not allowing predatory animals to take your game. Remember, the trigger can not be so tight that the intended game can not set it off.

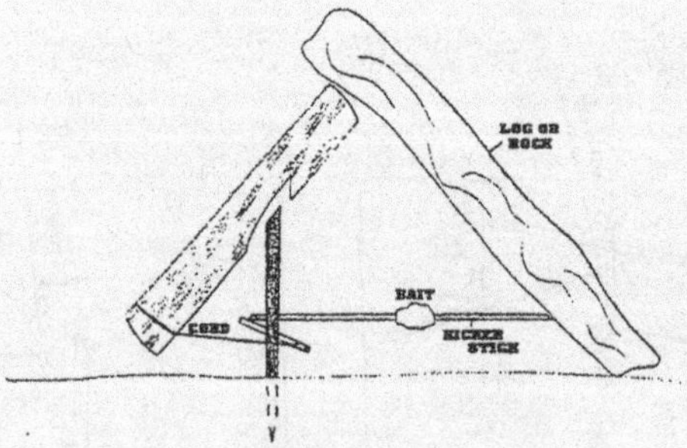

SPRING POLE

c. Box Trap: The box trap requires limited cordage to construct. It is designed to hold live small rodents and birds. The box trap trigger is the Puite figure 4.

3. PATH GUARDS Path guards is designed to protect and provide security for your shelter area against the enemy and predatory animals. They are classified into noise and casualty producing path guards.

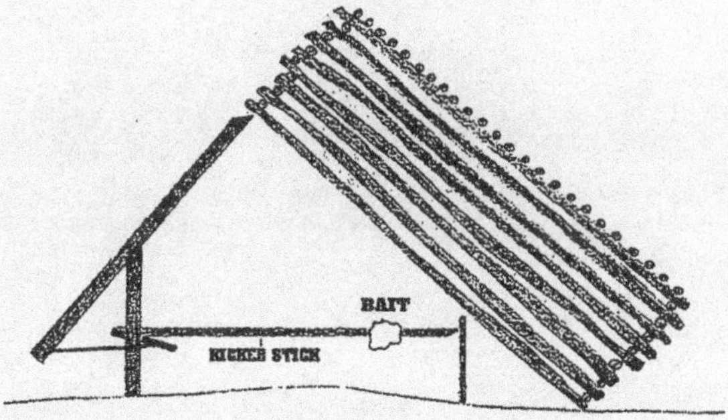

BOX TRAP

a. Noise producing path wards. Noise producing pathguards serve as an alarm for your shelter area. When triggered, it should produce some type of loud noise. Although construction can vary, depending on materials available, one example is as follows:

(1) Secure a young sapling to a universal trigger.

(2) At the end of the sapling, tie several pieces of metal to the sapling. Use what ever is available for metal.

(3) Camouflage the metal on the ground.

(4) When triggered, the sapling should swing back and forth, causing the metal rattle.

b. Casualty producing path wards. Casualty producing path guards, when triggered, should cause death or injury to the enemy or predatory animal. Tips should be poisoned as discussed in Survival Plant Uses class. Triggers for this type of path guard should be the universal triggers.

(1) Log Jerk

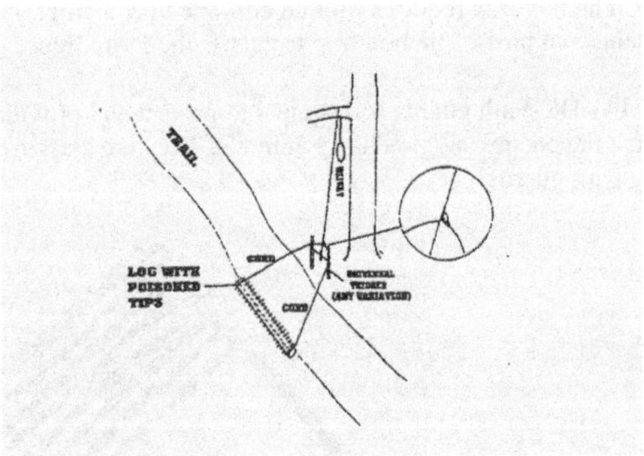

(2) Fish Hook Nightmare

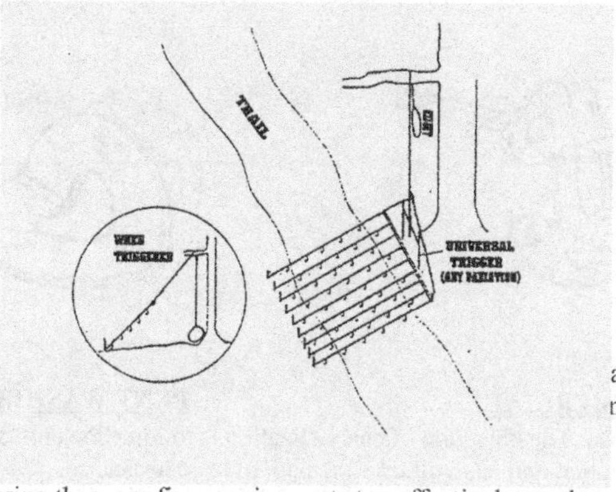

4. SNARING. A snare … a loop at one end, which tightens, … nd less time consuming to constr…

a. Requirements for snaring there are five requirements too effectively employ snares. They are as follows:

(1) Wire: Although snares can be used with rope or cord, they are less effective than wire. Wire should have memory and resist kinking. Aircraft cable type in size 1/32 to 3/16 inch should be used. This type of wire prevents animal chew out and resists breakage. Remember, you want the smallest diameter cable capable of holding the animal.

(2) Locking device: A locking device is imperative for a snare to work properly. Locking devices secure the snare around the animal's neck. There are several methods available for a locking device.

COMMERCIAL

BENT WASHER

(3) Channelization: The objective of channelization is to force the animal to move through the snare. Remember, animals will take the path of least resistance.

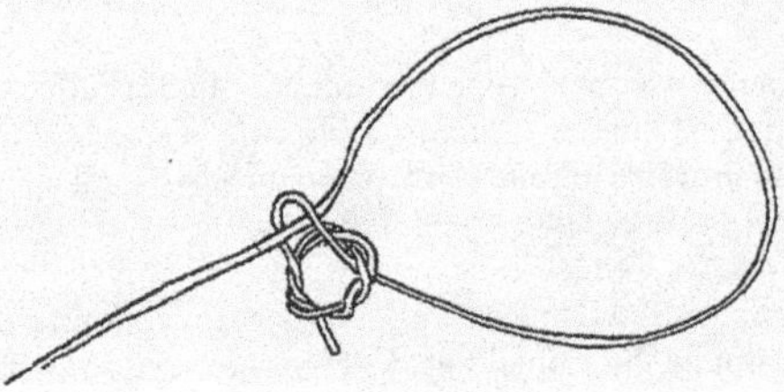

KNOT

(4) Loop size: Probably the single most important requirement in snaring. A correctly employed snare will have the snare holding the animal around the neck. Loop size is placed on the snare according to the intended animal. Too large will result in a body or leg catch, resulting in possible chew out or breakage. Too small will enable the animal to force the snare to the side, resulting in a miss. Additionally, the loop must be placed with specific ground clearance. The following chart will give a general guideline.

ANIMAL	NOOSE SIZE	GROUND
CLEARANCE		
SQUIRREL	2 1/2 TO 3 INCHES	1/2 TO 1 1/2 INCHES
RABBIT	4 TO 5 1/2 INCHES	1 1/2 TO 3 INCHES
RACCOON	6 INCHES	3 TO 4 INCHES
FOX	7 TO 10 INCHES	8 TO 10 INCHES
COYOTES	12 TO 14 INCHES	12 INCHES
BOBCAT	9 INCHES	8 INCHES

Note: (1) Noose size is the diameter of the snare loop.

(2) Ground clearance is measured from the bottom of the loop to the ground.

(5) Bait. In a survival situation, you will not be able to employ numerous snares. Baiting all snares increases your chances of success.

b. Snares although there are numerous ideas to employ snares, here are a few ideas.

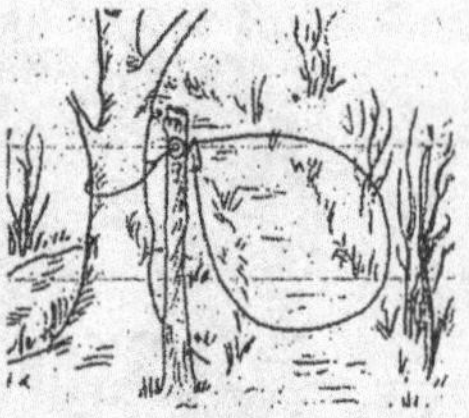

TRAIL SET

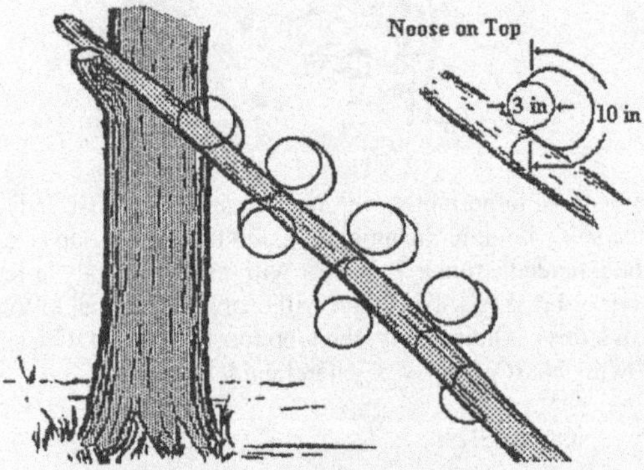

SQUIRREL POLE

DEN SET

CUBBIE SET

5. FISH. In a mountainous region, fish are normally an abundant resource. Not only are they a food source, the "left-overs" provide an excellent bait for traps and snares.

a. Fishing locations: Fishing in mountain streams is generally best done with a hand line. When fishing these streams, always look for these places to fish.

FISHING LOCATIONS

b. Expedient hooks: Although hooks should be carried in a survival kit, the survivor should be able to construct additional hooks if the situation arises. Expedient hooks are made to become lodged in the throat of the fish. Below are a few examples.

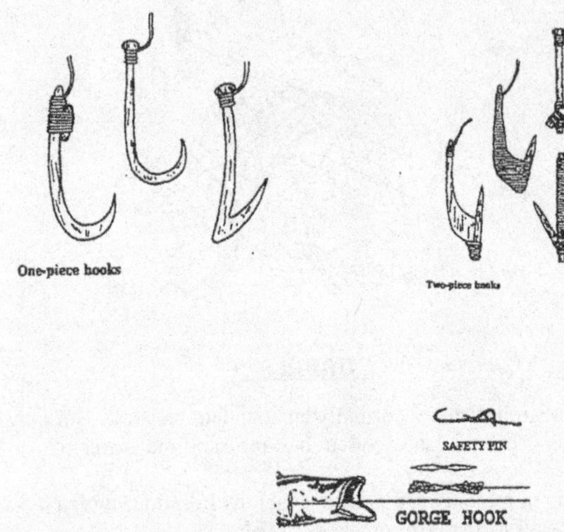

EXPEDIENT HOOKS

c. Fish traps: A fish trap can be effective if you have a shallow stream and time to construct it. A basic fish trap in nothing more than a barricade of rocks or sticks across a stream with another barricade using a funnel-type entrance which fish can be driven into but have a difficult time finding their way out. Once fish are trapped between these two barricades they may be speared, clubbed, or grabbed. This can be very effective when certain types of fish are moving in large groups to spawn. This type of trap is very effective in catching fish. A door can be constructed at the mouth of the trap so that excess fish can be kept live until needed.

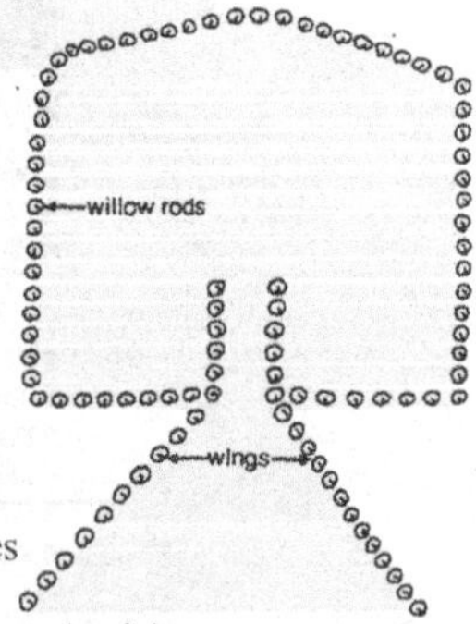

FISH TRAP

d. Set Lines: Set lines are an effective method of fishing which conserves energy. Put them out over night with several boated hooks attached. Place them with the hooks either on the bottom or suspended off the bottom, until you have determined where the fish are feeding.

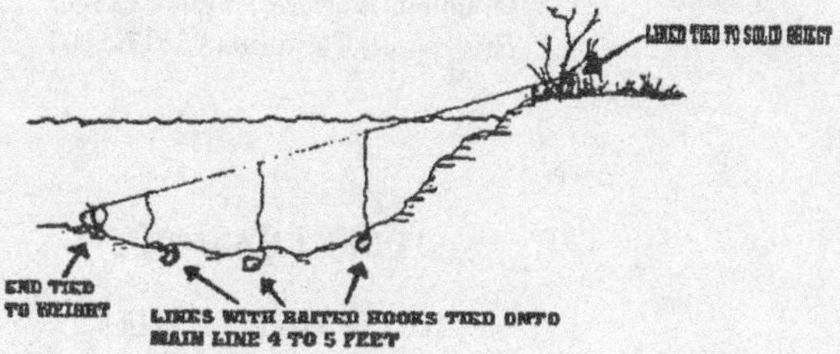

SET LINES

e. Netting: A gill net is most effective in still water, e.g., a lake (near the inlet and outlet are good locations) or back water in a large stream (for survival doesn't hesitate to block the stream). Nets can be constructed using the inner cords of parachute shroud lines. Place floats on top and weights on the bottom to keep the net vertical in the water. When the ice is on a lake, the fish are inclined to stay deeper. The smaller the mesh, the smaller the fish you can catch, but a small mest will still entangle a large fish. A mesh of 3 inches is a good standard. A gill net is very time consuming and requires allot of material to construct.

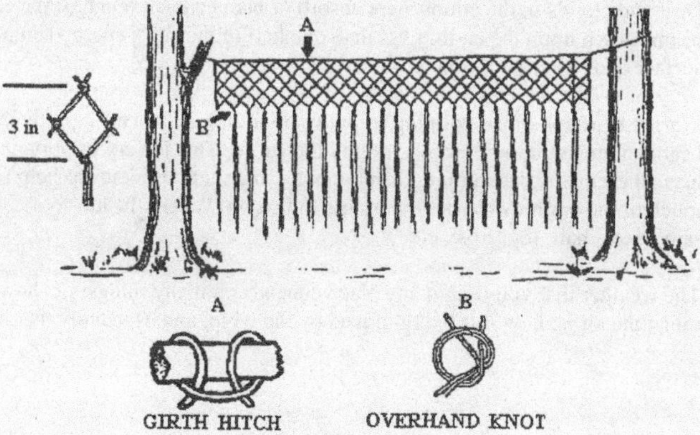

GILL NET

UNITED STATES MARINE CORPS
Mountain Warfare Training Center
Bridgeport, California 93517-5001

WMC
04/07/02

STUDENT HANDOUT

MOUNTAIN WEATHER

LESSON PURPOSE: The purpose of this period of instruction is to introduce you to weather patterns, the invisible aspects of weather, and some visible clues to help you forecast the weather. This lesson relates to all operations in the mountains at MWTC.

OUTLINE

1. GENERAL

a. The earth is surrounded by the atmosphere, which is divided, into several layers. The world's weather systems are in the troposphere, the lower of these layers. This layer reaches as high as 40,000 feet.

b. Dust and clouds in the atmosphere absorb or bounce back much of the energy that the sun beams down upon the earth. Less than one half of the sun's energy actually warms the earth's surface and lower atmosphere.

c. Warmed air, combined with the spinning (rotation) of the earth, produces winds that spread heat and moisture more evenly around the world. This is very important because the sun heats the Equator much more than the poles and without winds to help restore the balance, much of the earth would be impossible to live on. Where the air-cools, you can get clouds, rain, snow, hail, fog, frost, etc.

d. The weather that you find in any place depends on many things, i.e. how hot the air is, how moist the air is, how it is being moved by the wind, and especially, is it being lifted or not?

A10

2. PRESSURE

a. All of these factors are related to air pressure, which is the weight of the atmosphere at any given place. The lower the pressure, the more likely are rain and strong winds.

b. In order to understand this we can say that the air in our atmosphere acts very much like a liquid.

c. Areas with a high level of this liquid would exert more pressure on the Earth and be called a "high pressure area".

d. Areas with a lower level would be called a "low pressure area".

e. In order to equalize the areas of high pressure it would have to push out to the areas of low pressure.

f. The characteristics of these two pressure areas are as follows:

(1) High-pressure area. Flows out to equalize pressure.

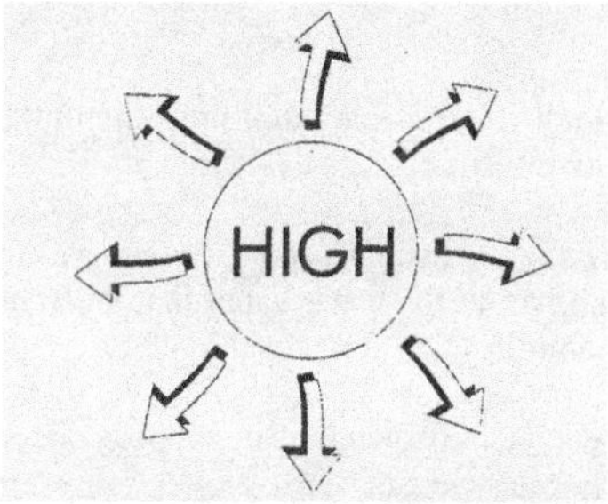

(2) Low pressure area. Flows in to equalize pressure.

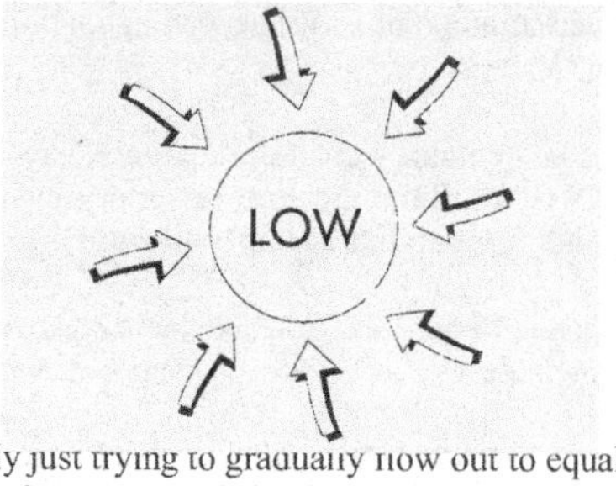

g. The air from the high-pressure area is basically just trying to gradually flow out to equalize its pressure with the surrounding air, while the low pressure is beginning to build vertically. Once the low has achieved equal pressure, it can't stop and continues to build vertically; causing turbulence, which results in bad weather.

NOTE: When looking on the weather map, you will notice that these differences in pressure resemble contour lines. These contour lines are called isobars and are translated to mean equal pressure area.

h. Isobars: Pressure is measured in millibars or another more common measurement is inch mercury.

i. Fitting enough, areas of high pressure are called ridges and areas of low pressure are called troughs.

NOTE: The average air pressure at sea level is:

29.92 inches mercury.
1,013 millibars.

j. As we go up in altitude, the pressure (or weight) of the atmosphere decreases.

EXAMPLE: At 18,000 feet in elevation it would be 500 millibars vice 1,013 millibars at sea level.

3. HUMIDITY Humidity is the amount of moisture in the air. All air holds water vapor, although it is quite invisible.

a. Air can hold only so much water vapors, but the warmer the air, the more moisture it can hold. When the air has all the water vapor that it can hold, the air is said to be saturated (100% relative humidity).

b. If the air is then cooled, any excess water vapor condenses; that is, it's molecules join to build the water droplets we can see.

c. The temperature at which this happens is called the "condensation point". The condensation point varies depending on the amount of water vapor and the temperature of the air.

d. If the air contains a great deal of water vapor, condensation will form at a temperature of 20°C (68°F). But if the air is rather dry and does not hold much moisture, condensation may not form until the temperature drops to 0°C (32°F) or even below freezing.

e. Adiabatic Lapse Rate: The adiabatic lapse rate is the rate that air will cool on ascent and warm on descent. The rate also varies depending on the moisture content of the air.

(1) Saturated Air = 2.2°F per 1,000 feet.

(2) Dry Air = 5.5°F per 1,000 feet.

4. WINDS. As we stated earlier, the uneven heating of the air by the sun and rotation of the earth causes winds. Much of the world's weather depends on a system of winds that blow in a set direction. This pattern depends on the different amounts of sun (heat) that the different regions get and also on the rotation of the earth.

a. Above hot surfaces, air expends (air molecules spread out), and move to colder areas where it cools and becomes denser, and sinks to the earth's surface. This forms a circulation of air from the poles along the surface or the earth to the Equator, where it rises and moves towards the poles again.

b. Once the rotation of the earth is added to this, the pattern of the circulation becomes confusing.

c. Because of the heating and cooling, along with the rotation of the earth, we have these surface winds. All winds are named from the direction they originated from:

(1) Polar Easterlies: These are winds from the polar region moving from the east. This is air that has cooled and settled at the poles.

(2) Prevailing Westerlies: These winds originate from approximately 30 degrees North Latitude from the west. This is an area where prematurely cooled air, due to the earth's rotation, has settled back to the surface.

(3) Northeast Tradewinds: These are winds that originate from approximately 30 degrees North from the Northeast. Also prematurely cooled air.

d. Jet Stream: A jet stream can be defined as a long, meandering current of high speed winds near the tropopause (transition zone between the troposphere and the stratosphere) blowing from generally a westerly direction and often exceeding 250 miles per hour. The jet stream results from:

(1) Circulation of air around the poles and Equator.

(2) The direction of airflow above the mid latitudes.

(3) The actual path of the jet stream comes from the west, dipping down and picking up air masses from the tropical regions and going north and bringing down air masses from the Polar Regions.

NOTE: The average number of long waves in the jet stream is between three and five depending on the season. Temperature differences between polar and tropical regions influence this. The long waves influence day to week changes in the weather; there are also short waves that influence hourly changes in the weather.

5. AIR MASSES. As we know, all -of these patterns move air. This air comes in parcels known as "air masses". These air masses can vary in size from as small as a town too as large as a country. These air masses are named for where they originate:

a. Maritime: Over water.

b. Continental: Over land.

c. Polar: Above 60 degrees North.

d. Tropical: Below 60 degrees North.

e. Combining these give us the names and description of the four types of air masses:

(1) Continental Polar: Cold, dry air mass.

(2) Maritime Polar: Cold wet air mass.

(3) Continental Tropical: Dry, warm air mass.

(4) Maritime Tropical: Wet, warm air mass.

ABBREVIATED MEANINGS
CP = Continental Polar.
MP = Maritime Polar.
CT = Continental
Tropical. MT =
Maritime Tropical.

f. The thing to understand about air masses; they will not mix with another air mass of a different temperature and moisture content. When two different air masses collide, we have a front, which will be covered in more detail later in this period of instruction.

g. Here are some other types of winds that are peculiar to mountain environments but don't necessarily affect the weather:

(1) Anabolic wind: These are winds that blow up mountain valleys to replace warm rising air and are usually light winds.

(2) Catabolic wind: These are winds that blow down mountain valley slopes caused by the cooling of air and are occasionally strong winds.

6. LIFTING/COOLING. As we know, air can only hold so much moisture depending on its temperature. If we cool this air beyond its saturation point, it must release this moisture in one form or another, i.e. rain, snow, fog, dews, etc. There are three ways that air can be lifted and cooled beyond its saturation point.

a. Orographic uplift: This happens when an air mass is pushed up and over a mass of higher ground such as a mountain. Due to the adiabatic lapse rate, the air is cooled with altitude and if it reaches its saturation point we will receive precipitation.

OROGRAPHIC UPLIFT

b. Convection effects: This is normally a summer effect due to the sun's heat radiating off of the surface and causing the air currents to push straight up and lift air to a point of saturation.

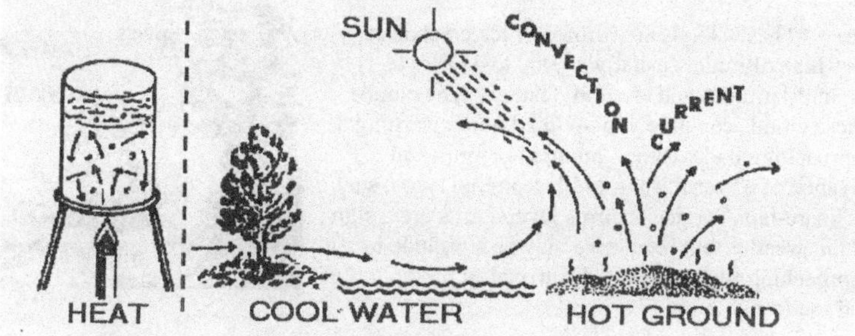

CONVECTION EFFECTS

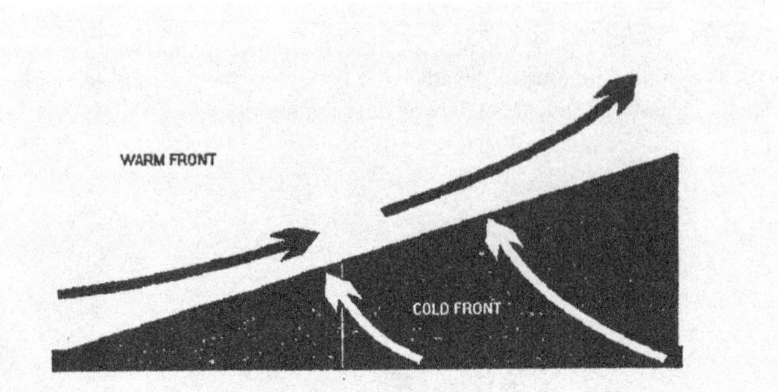

FRONTAL LIFTING

c. Frontal ... rature content ... is forced aloft; fi ... where we receive ... ifting is not uncom

7. CLOUDS. Anytime air is lifted or cooled beyond its saturation point (100% relative humidity), clouds are formed. Clouds are one of our signposts to what is happening. Clouds can be described in many different ways; they can also be classified by height or appearance, or even by the amount of area covered, vertically or horizontally.

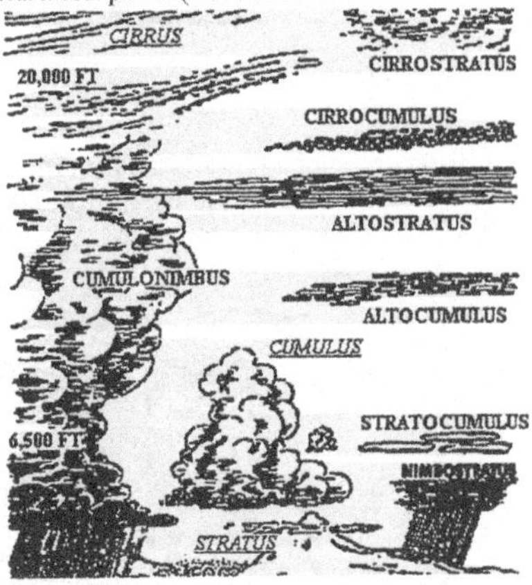

a. Cirrus: These clouds are formed of ice crystals at very high altitudes (usually 20,000 to 35,000 feet) in the mid-latitudes and are thin, feathery type clouds. These clouds can give you up to 24 hours warning of approaching bad weather, hundreds of miles in advance of a 'warm front. Frail, scattered types, such as "mare-tails" or dense cirrus layers, tufts are a sign of fair weather but predictive may be a prelude to approaching lower clouds, the arrival of precipitation and the front.

b. Cumulus: These clouds are formed due to rising air currents and are prevalent in unstable air that favors vertical development. These currents of air create cumiliform clouds that give them a piled or bunched up appearance, looking similar to cotton balls. Within the cumulus family there are three different types to help us to forecast the weather:

(1) Cotton puffs of cumulus are Fair Weather Clouds but should be observed for possible growth into towering cumulus and cumulonimbus.

(2) Towering cumulus are characterized by vertical development. Their vertical lifting is caused by some type of lifting action, such as connective currents found on hot summer afternoons or when wind is forced to rise up the slope of a mountain or possibly the lifting action that

may be present in a frontal system. The towering cumulus has a puffy and "cauliflower-shaped" appearance.

(3) Cumulonimbus clouds are characterized in the same manner as the towering cumulus, form the familiar "thunderhead" and produce thunderstorm activity. These clouds are characterized by violent updrafts, which carry the tops of the clouds to extreme elevations. Tornadoes, hail and severe rainstorms are all products of this type of cloud. At the top of the cloud, a flat anvil shaped form appears as the thunderstorm begins to dissipate.

c. Stratus: Stratus clouds are formed when a layer of moist air is cooled below its saturation point. Stratiform, clouds lie mostly in horizontal layers or sheets, resisting vertical development. The word "stratus" is derived from the Latin word "layer". The stratus cloud is quite uniform and resembles fog. It has a fairly uniform base and a dull, gray appearance. Stratus clouds make the sky appear heavy and will occasionally produce fine drizzle or Very light snow with fog. However, because there is little or no vertical movement in the stratus clouds, they usually do not produce precipitation in the form of heavy rain or snow.

d. As previously stated, clouds are formed when air is lifted to a point where it cools to its saturation point. We also know that frontal lifting affects our fronts, which produce the largest portion of our precipitation.

8. FRONTS. As we know, fronts often happen when two air masses of different moisture and temperature content interact. One of the ways we can identify that this is happening is by the progression of the clouds.

a. Warm Front: A warm front occurs when warm air moves into and over a slower (or stationary) cold air mass. Since warm air is less dense, it will rise naturally so that it will push the cooler air down and rise above it. The cloud you will see at this stage is cirrus. From the point where it actually starts rising, you will see stratus. As it continues to rise, this warm air-cools by the cold air and, this, receiving moisture at the same time. As it builds in moisture, it darkens becoming "nimbus-stratus", which means rain of thunderclouds. At that point some type of moisture will generally fall.

b. Cold Front. A cold front occurs when a cold air mass (colder than the ground that it is traveling over) overtakes a warm air mass that is stationary or moving slowly. This cold air, being denser, will go underneath the warm air, pushing it higher. Of course, no one can see this "air", but they can see clouds and the clouds themselves can tell us what is happening. The cloud progression to look for is cirrus to cirrocumulus to cumulus and, finally, to cumulonimbus.

c. Occluded Front. Cold fronts move faster than warm ones so that eventually a cold front overtakes a warm one and the warm air becomes progressively lifted from the surface. The zone of division between cold air ahead and cold air behind is called a "cold occlusion". If the air behind the front is warmer than ahead, it is a warm occlusion. Most land areas experience more occlusions than other types of fronts. In the progression of clouds leading to fronts, orographic uplift can play part in deceiving you of the actual type of front, i.e. progression of clouds leading to a warm front with orographic cumulus clouds added to these. The progression of clouds in an occlusion is a combination of both progressions from a warm and cold front.

9. USING PRESSURE AS AN INDICATOR. A very important factor of telling us what might happen is the pressure. As we know, low pressure or dropping pressure normally indicates deteriorating weather whereas high pressure usually gives us more good weather or clearing of bad weather. There are a couple of ways to monitor our pressure and are as follows:

a. Barometer: A barometer could be described as a pan of mercury with a tube leading out of the pan. Pressure from the atmosphere causes the mercury to rise in the tube.

(1) The tube is marked in millibars and the station that's reading these millibars will know how much it should rise for that location. Once again, if it rises more than normal, it would be considered a high-pressure reading.

b. Altimeter: Another means that is used to measure pressure is an altimeter, which is commonly used by mountaineers. It works like this:

(1) As you rise in elevation the pressure becomes less, thus allowing the needle in the altimeter to rise. If the needle rises without you rising with it, there is less pressure in the atmosphere than before and thus, a low-pressure area.

c. Contrail Lines: A basic way of identifying a low-pressure area is to note the contrail lines from jet aircraft. If they don't dissipate within two hours, that indicates a low-pressure area in your area. This usually occurs about 24 hours prior to an oncoming front.

d. Lenticulars: These are optical, lens-shaped cumulus clouds that have been sculpted by the winds. This indicates moisture in the air and high winds aloft. When preceding a cold front, winds and clouds will begin to lower.

10. USING SIGNS FROM NATURE. These signs will give you a general prediction of the incoming weather conditions. Try to utilize as many signs together as possible, which will improve your prediction. All of these signs have been tested with relative accuracy, but shouldn't be depended on 100%. But in any case you will be right more times than wrong in predicting the weather. From this we can gather as much information as needed and compile it along with our own experience of the area we are working in to help us form a prediction of incoming weather. The signs are as follows:

a. A spider's habits are very good indicators of what weather conditions will be within the next few hours. When the day is to be fair and relatively windless, they will spin long filaments over which they scout persistently. When precipitation is imminent, they shorten and tighten their snares and drowse dully in their centers.

b. Insects are especially annoying two to four hours before a storm.

c. If bees are swarming, fair weather will continue for at least the next half day.

d. Large game such as deer, elk, etc., will be feeding unusually heavy four to six hours before a storm.

e. When the smoke from a campfire, after lifting a short distance with the heated air, beats downward, a storm is approaching. Steadily rising smoke indicates fair weather.

f. A red sun or sky in the morning indicates nearing rain due to the excessive amount of moisture in the air.

g. A red evening sky indicates that the air contains so little moisture that rain within the next 24 hours is highly improbable.

h. A gray, overcast evening sky indicates that moisture carrying dust particles in the atmosphere have become overloaded with water; this condition favors rain.

i. A gray morning sky indicates dry air above the haze caused by the collecting of moisture on the dust in the lower atmosphere; you can reasonably a fair day.

j. When the setting sun shows a green tint at the top as it sinks behind clear horizon, fair weather is probable for most of the next 24 hours.

k. A rainbow in the late afternoon indicates fair weather ahead. However, a rainbow in the morning is a sign of prolonged bad weather.

l. A corona is the circle that appears around the sun or the moon. When this circle grows larger and larger, it indicates that the drops of water in the atmosphere are evaporating and that the weather will probably be clear. When this circle shrinks by the hour, it indicates that the water drops in the atmosphere are becoming larger, forming into clouds, rain is almost sure to fall.

m. In fair weather, air currents flow down streams and hillsides in the early morning and start drifting back up towards sunset. Any reversal of these directions warns of a nearing storm.

n. When the breeze is such that the leaves show their undersides, a storm is likely on the way.

o. It is so quiet before a storm, that distant noises can be heard more clearly. This is due to the inactivity of wildlife a couple of hours before a storm.

p. When in the mountains, the sight of morning mist rising from ravines is a good sign of clear weather the rest of the day.

q. A heavy dew or frost in the morning is a sign of fair weather for the rest of the day. This is due to the moisture in the atmosphere settling on the ground vice in the form of precipitation such as rain, snow, etc.

UNITED STATES MARINE CORPS
Mountain Warfare Training Center
Bridgeport, California 93517-5001

FMST.07.42
03/14/02

STUDENT HANDOUT

RESCUE PHILOSOPHY FOR SWIFTWATER RESCUE TEAMS

TERMIANL LEARNING OBJECTIVE In a swift water rescue scenario, organize a swift water rescue, in accordance with the reference. (FMST.07.42)

ENABLING LEARNING OBJECTIVES.

(1) With out the aid of reference, from a given list select the correct components of rescue preparation, in accordance with the reference. (FMST.07.42a)

(2) With out the aid of reference, from a given list select the correct role description of the rescue team members, in accordance with the reference. (FMST.07.42b)

(3) With out the aid of reference, from a given list select the correct methods of rescue, in accordance with the reference. (FMST.07.42c)

OUTLINE.

1. **COMPONENTS OF RESCUE PREPARATION**. (FMST.07.42a) These components comprise the backbone of both technical skills and the proper mental attitude required of a rescue professional. Remember that this course only provides you with the first component.

a. Training. Provides the students with the basic technical skills required to perform a safe rescue, these can be gained by attending an approved SWIFTWATER RESCUE COURSE.

b. Practice. Use of the skills learned in training.

c. Experience. Gained through advanced training, practical exercises and real-world situations.

d. Judgment. The combination and application of skills for a given situation. Remember that every rescue situation is different. Don't get into a SOP mindset.

2. **PRIORITIES OF SAFETY**. Most of you have heard the saying that "It is better to be judged by twelve then carried by six". With that in mind you must realize that, at times, there is no way to affect a safe rescue. If you place yourself or your team in a hostile environment, from which you cannot escape, you are adding to the problem, not assisting in the solution. For this reason we set priorities of safety.

a. Self-aid. Be able to rescue yourself.

b. Buddy-aid. Ensure all safety measures are in force and that team members can rescue each other in an emergency.

c. Victim-aid. Only after self and buddy-aid are ensured should you attempt to rescue the victim. This will enable you to concentrate and focus your entire effort on the rescue.

3. **ROLES OF THE TEAM MEMBERS**. (FMST.07.24b) Often when dealing with rescues, you will find yourself working with outside agencies and organizations. In order to successfully coordinate the entire effort the Incident Command System (ICS) was developed. In swiftwater rescue situations there are four specific roles that must be filled.

a. Incident Commander. Is in overall command of the rescue efforts. Does not necessarily have the technical knowledge to conduct the rescue, but must be able to coordinate the rescue effort.

b. Rigger. Personnel responsible for equipment and rope systems, i.e. establishing the tension diagonal or the tyrolean high line.

c. Gofer. This group of individuals fills all other supporting roles, i.e. Communications, logistics, medevac, accounting, equipment accountability etc.

d. Rescuer. The individuals who are actually conducting the rescue.

4. **COMPONENTS OF SEARCH AND RESCUE**. In order to successfully complete a rescue there is four objectives that must be met. In order to remember them we use the acronym LAST.

a. Locate. This is the search in the search and rescue business.

b. Access. The determination of the assets, equipment, technical skills, and personnel required safely reaching and extracting the victim.

c. Stabilization. Providing the on scene medical attention required in order to extract the victim without causing further injury or aggravating an existing condition.

d. Transport. Removing the victim from the hostile environment to a safe one where further treatment can be provided.

5. **METHODS OF RESCUE**. (FMST.07.42c) In any rescue, we must access the situation to determine the best techniques to apply. Bearing in mind the self-aid, buddy-aid, victim-aid concept as well as time, weather, victim's condition, etc. In all cases we will attempt the lowest risk methods first, while preparing for a higher risk option. In all cases remember the KISS principle.

a. Reach. Use a ladder, fire hose, stick, paddle, rifle or any other sturdy object to extend your reach in order to reach the victim.

b. Throw. Get some type of floatation device to the victim. Or throw a line to the victim in order to pendulum or pull him/her to the shore.

c. Row. Putting a boat in the water to reach the victim.

d. Go and Tow. These are the contact rescues that, while look good on film, put you at the extreme risk of becoming a second victim.

e. Helo. Most civilian helicopter crews are not trained in SAR evolutions. Military aircraft and crews are a better bet. Several uses for helos are in the support role. Locating the victim, inserting and extracting teams and moving equipment are some of these roles. If using the helo in a direct role it is better to use a qualified SAR swimmer/diver to hook-up or pick-up victims.

UNITED STATES MARINE CORPS
Mountain Warfare Training Center
Bridgeport, California 93517-5001

FMST.07.25
03/14/02

STUDENT HANDOUT

PERSONAL AND TEAM EQUIPMENT

TERMINAL LEARNING OBJECTIVE As a member of a swiftwater rescue team, maintain swiftwater rescue equipment, in accordance with the reference. (FMST.07.25)

ENABLING LEARNING OBJECTIVES.

(1) Without the aid of references, from a given identify the correct items of personal equipment, in accordance with the reference. (FMST.07.25a)

(2) Without the aid of references, from a given identify the correct items of team equipment, in accordance with the reference. (FMST.07.25b)

OUTLINE.

1. **PERSONAL EQUIPMENT**. (FMST.07.25a) The following items of personal equipment should be in each man's possession for a swiftwater rescue.

   a. Dry Suit. Warmest, most versatile, full body protection available. Extra thermal layers can be added for varying water temperatures. Provides wind protection out of the water and provides protection against contaminated water.

   b. Wet Suit. Excellent protection from flotsam. Variety of thick nesses for thermal protection. Insulating quality is degraded in moving water due to the flushing effect.

   c. Footwear. Primary is full sole scuba booties. You can use tennis shoes or jungle boots with heavy wool socks.

   d. Gloves. Neoprene is recommended.

   e. Hoods. Wetsuit hood or pile hood for insulation. Remember 30-50% of your body heat losses can occur through the head and neck region of your body if left unprotected.

B2

f. Helmet. Protects you from flotsam and impact injuries. Should be padded and ventilated to allow water to flow through.

g. Knife. Fixed or locking blade attached to the vest, handle down.

h. Whistle. Required for communication. Best without "pea".

i. Eye Protection. Protects subject from debris.

j. Personal Floatation Device. This is the single most important piece of equipment. Jackets should provide at least 15 lbs. of floatation. Proper fit is a priority. Rescue vests with a "Blow-Out" belt should only be used by trained professionals.

2. **TEAM EQUIPMENT**. (FMST.07.25b) The following is a list of some of the team equipment for a typical four-man team. Of course the situation will dictate which equipment will be required for the mission on hand, but all hands should be familiar with this equipment. This is only a list. The uses and application of this gear, along with practical application exercises later in this course will provide you with the hands-on training and working knowledge to properly and safely use this equipment.

QTY	ITEM
1	300 ft, l Static Rope
4	150 ft, 11 mm Static Rope
25	85 Locking Stubai Carabiners
4	2" Pulleys
2	Rescue Figure 8
6	3 ft, 7mm Prussic
20	15 ft, 1" Tubular Nylon Runners
2	Ascenders
3	Throw Bags
2 pr	Fins
1	Rescue PFD
1	Self Bailing Raft (13 or 14 foot)
1	Oar Frame w/ Oars, or Seven Paddles
3	Radios (waterproofed)
1	Medical Kit
1	Litter

NOTE: This is not a complete list of rescue equipment. However, it is a good basic load.

UNITED STATES MARINE CORPS
Mountain Warfare Training Center
Bridgeport, California 93517-5001

FMST.07.41
03/14/02

STUDENT HANDOUT

SWIFTWATER TERMINOLOGY AND DYNAMICS

TERMINAL LEARNING OBJECTIVES In a swift water rescue scenario, evaluate swift water river conditions in accordance with the reference. (FMST.07.41)

ENABLING LEARNING OBJECTIVES.

1) Without the aid of reference, from a given list correctly identify the swiftwater terminology, in accordance with the reference. (FMST.07.41a)

2) Without the aid of reference, from a given list correctly identify the classifications of rivers, in accordance with the reference. (FMST.07.41b)

OUTLINE.

1. **SWIFTWATER TERMINOLOGY**.
(FMST.07.41a)

a. Hydraulic or Hydraulic effect: Movement of water caused by pressure.

b. Eddy: Horizontal reversal. Current is blocked by an obstacle creating a vacuum which, when filled, causes water to flow upstream.

c. Eddy Fence: Reversal line between downstream current and the eddy. This fence can be very strong and the water flow may vary in height by several inches.

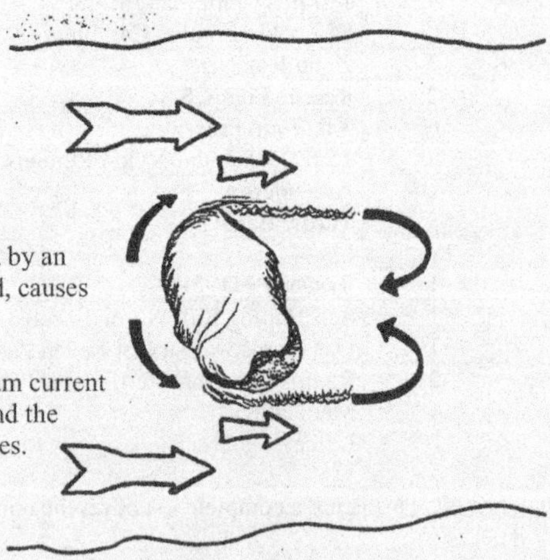

EDDY/EDDY FENCE

B3

d. Hole, Stopper, Keeper: Vertical reversal as water flows over an obstacle the main current follows the base of the river causing a surface vacuum. Downstream water flows back upstream to fill this void. This sets up a recirculating effect.

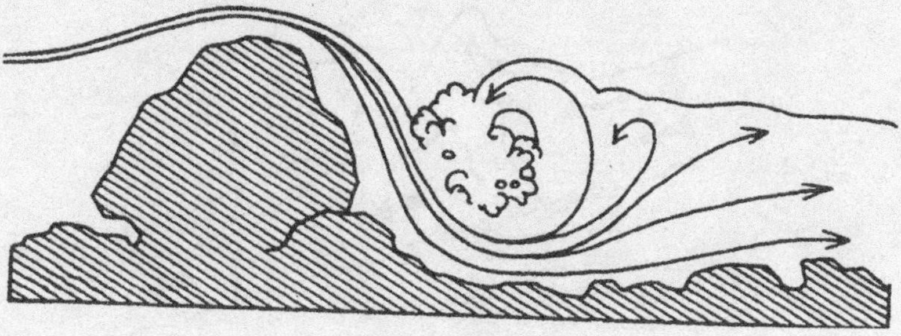

HOLE/STOPPER/KEEPER

e. Standing Wave. Haystack: Series of waves in the main channel caused by obstacles, ledges, or increases in the flow rate, gradient, or volume.

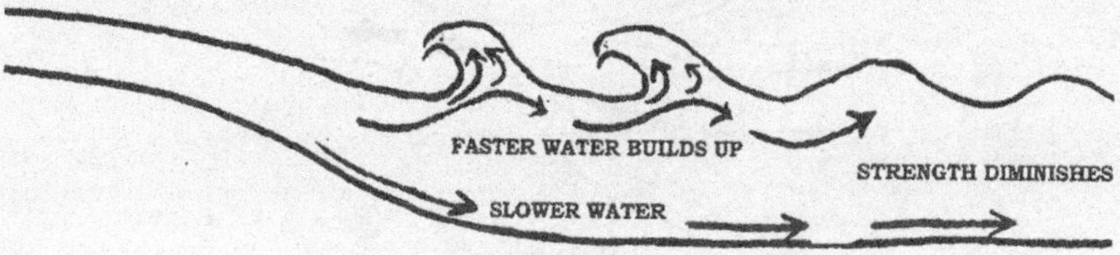

STANDING WAVES/HAYSTACKS

f. Downstream V. Tonal: Represent largest volume of water or main channel.

g. Upstream V: Caused by obstacles submerged just below the surface.

h. Strainer: Debris, logs, small trees on flooded banks, which allow water to flow through but will not allow a body to pass.

STRAINER / GUARD RAIL

STRAINER / TREE

i. Cushion, Pillow: Build-up of water on the upstream side of obstacles or the outside of curves.

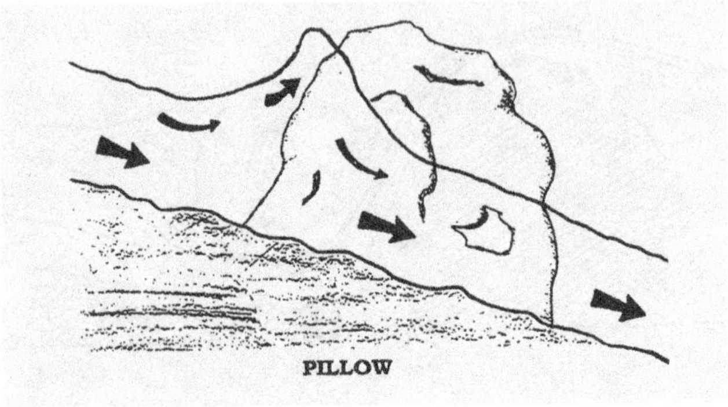

j. Confluence: Merging of streams.

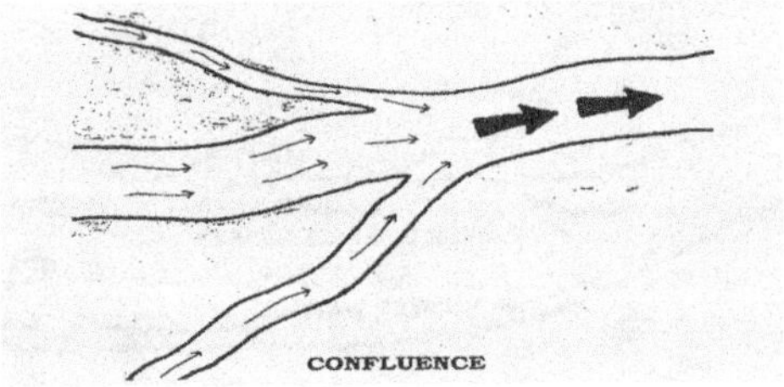

k. River Right. On the right side of the river looking downstream.

l. River Left. On the left side of the river looking downstream.

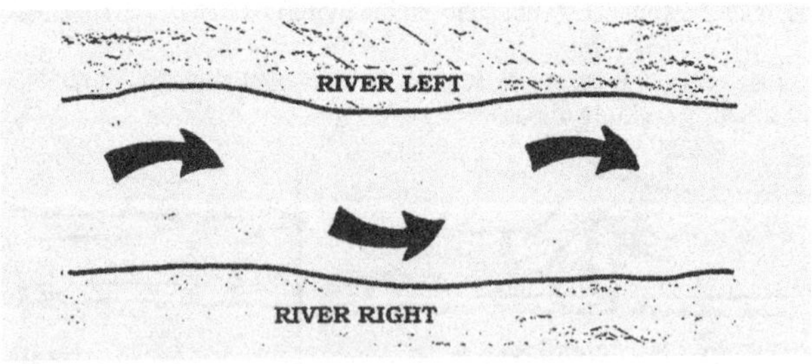

2. **CLASSIFICATIONS OF RIVERS**: (FMST.07.41b)

a. Class 1. Moving water, few obstacles, small ripples or waves.

b. Class 2. Moving water, clear channels, wide waves to 3', no scouting required.

c. Class 3. Rapids, narrow channels that require complex maneuvering, waves capable of flipping rafts or swamping canoes, scouting required.

d. Class 4. Long hazardous rapids, precise boat handling in rough water, scouting required, rescues hazardous.

e. Class 5. Severe, extended, extremely violent rapids, professional boat handlers and scouting required. Rescues extremely hazardous, situations are life threatening.

f. Class 6. Only Expert/Professional River Runners. Rescue may be impossible.

g. CLASSES U DON'T EVEN THINK ABOUT IT! Waters are not navigable.

3. **RIVER STRENGTH**.

a. Volume of Water. The volume of water in a given stretch of the river between two points is based on a simple formula and is measured in Cubic Feet Per Second (CFS). Simply take the river depth x river width x river velocity = volume of water in CFS.

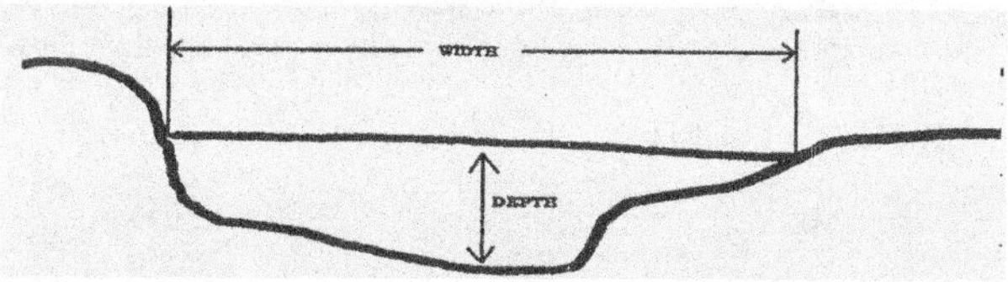

For example: 6 feet deep x 100 feet wide x 2 feet per second = 1,200 CFS

b. River Velocity. In order to determine the velocity or current speed measure a 100-foot section of the river and drop a float at the upstream end. Then, time its movement to the lower end and use the chart below.

TIME IN SECONDS	VELOCITY FT PER SEC	VELOCITY MPH
5	20	13
10	10	7
15	6.7	5
17	5.9	4
20	5	3.5
25	4	3
29	3.5	2.5
37	2.7	2
50	2	1.5
80	1.3	1
110	0.7	0.5

NOTE: Should the shape of the river change by becoming wider or narrower, once you have the flow in CFS, you can determine the new velocity by using the following formula:

$$\text{Velocity} = \frac{\text{Volume in CFS}}{\text{Width x Depth}}$$

c. Force Exerted. The velocity of the water will determine the force exerted X's the amount of exposed area under pressure. For simplicity we will use the following chart:

CURRENT VELOCITY	FORCE EXERTED ON OBJECT		
MPH	LEGS RAFT	BODY	WRAPPED
3	17	34	168
6	67	134	672
9	151	302	1,512
12	269	538	2,688

4. **LOW HEAD DAMS**. Low head dams are used throughout the world for irrigation, flood control, at hydroelectric plants and for many other uses. Basically they are pouring over obstacles that create hydraulics, keeper holes, and recirculating currents at their bases. Unlike natural obstacles, that rarely span the entire river and that allow you to be "flushed" out of the bottom or sides, low head dams create a hydraulic that extends from shore to shore.

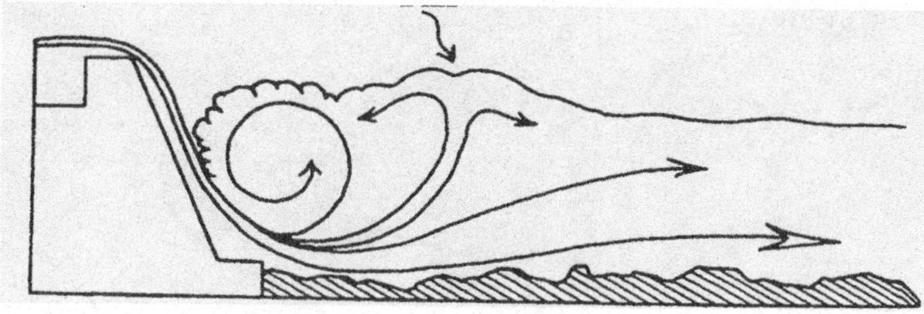

a. The size of the hydraulic created is based on several denominators:

(1) Height of drop.

(2) Angle of drop.

(3) Volume of water CFS.

(4) Velocity of water MPH.

(5) Bottom composition of runout.

b. The degree of difficulty for the rescuer is also compounded by:

(1) Water depth below dam.

(2) Debris build-up.

(3) Dam construction/conformation.

(4) Lack of access.

(5) Lack of natural anchors.

(6) Distance between riverbanks.

c. The basic hydrology is the same as any other pour over obstacle. As the river flows over the dam, the water follows the face of the dam to the river bottom. When sufficient speed diminishes the water rebounds to the surface and continues downstream. This flow has caused a vacuum or pocket directly below the dam that must be filled. Therefore, when the water rebounds to the surface, some of it flows back upstream to form a vertical reversal or pour-over eddy. The line of demarcation for the upstream/downstream flow is called the boil line. The shape and angle of the dam face and the volume and velocity of the water flowing over the dam create the distance and width of the boil line.

5. **REVIEW**.

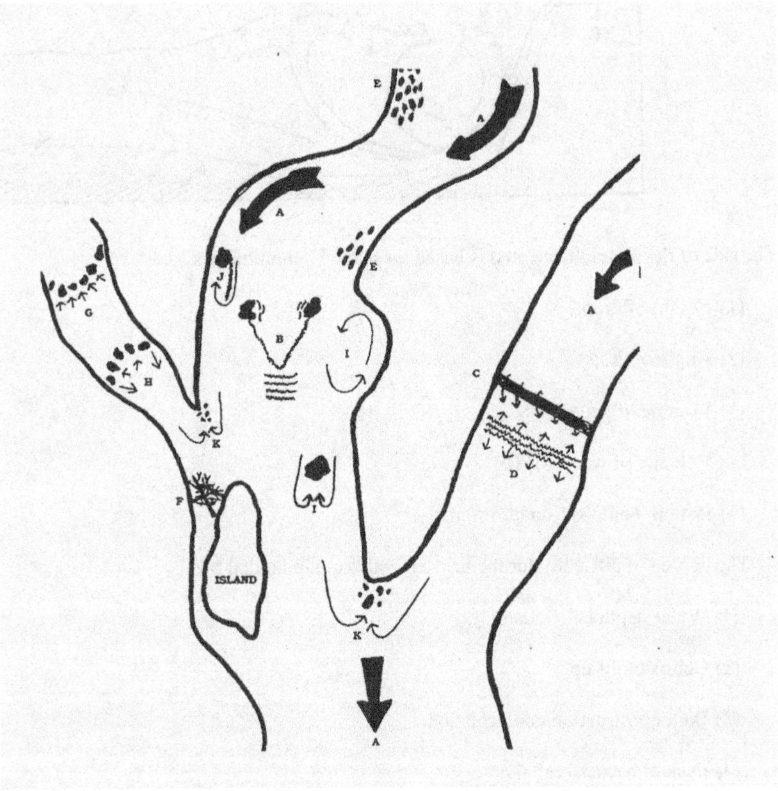

A. Main Current
B. Downstream V
C. Low head Dam
D. Boil Line
E. Rock Garden
F. Debris Pile/Strainer
G. Frowning Hole
H. Smiling Hole
I. Eddy
J. Eddy Fence
K. Confluence

UNITED STATES MARINE CORPS
Mountain Warfare Training Center
Bridgeport, California 93517-5001

FMST.07.24
04/14/02

STUDENT HANDOUT

COMMUNICATIONS

TERMINAL LEARNING OBJECTIVES Given a casualty in a swift water environment, assignment as a member of a rescue team, and a whistle, employ swift water rescue communication signals to support mission requirements per the reference. (FMST.07.24)

ENABLING LEARNING OBJECTIVES.

(1) In a summer mountainous environment, prior to swift water rescue operations, when the instructor verbally states a hand and arm signal, the student will demonstrate that signal, in accordance with the references. (FMST.07.24a)

(2) In a summer mountainous environment, standing at the edge of a swift water stream, when the instructor blasts a whistle a varied amount of times, the student will focus their attention in the proper direction, in accordance with the references. (FMST.07.24b)

OUTLINE.

1. **RELIABLE COMMUNICATION**. In many cases time is critical in a SWR scenario. Therefore immediate, understandable communication is a must. Radios, bull horns, and voice work well in some situations, but wet equipment, dead batteries, and ambient noise levels can render these methods ineffective. Therefore, rescue professionals have developed a series of simple hand and arm, and whistle signals. Any rescue team can develop their own signals. The only requirement is that all hands understand the signals and that they are briefed to any other agency you are working with.

B4

2. **HAND AND ARM SIGNALS**. (FMST.07.24a)

ACTION	MEANING
1 hand extended over head	Require Assistance
Both hands extended forming an "O" with one hand touching The head	Okay
Extend both hands above head, then point left/right	Move in direction indicated
Extend both hands above head, wave, then point left/right	Eddy out in direction indicated
Cross both arms over chest	Require medical assistance or Corpsman

3. **WHISTLE BLASTS**. (FMST.07.24b)

ACTION	MEANING
One Blast	Stop/Attention
Two Blasts	Upstream
Three Blasts	Downstream
Three Blasts Repeated	Emergency

UNITED STATES MARINE CORPS
Mountain Warfare Training Center
Bridgeport, California 93517-5001

FMST.0
7.27
04/14/02

STUDENT HANDOUT

THROW BAGS AND FOOT ENTANGLEMENTS

TERMINAL LEARNING OBJECTIVES In a swiftwater rescue scenario employ a throw bag, in accordance with the reference. (FMST.07.27)

ENABLING LEARNING OBJECTIVES.

1) Without the aid of reference, provided with a throw bag and swimming victim, rescue the victim, in accordance with the reference. (FMST.07.27a)

OUTLINE.

1. **THE THROW BAG**: The throw bag is very simply a cloth bag with a small float insert, packed with 50 to 70 feet of 1/4" to 3/8" floating polypropylene or spectra/poly line. The concept is to hold the exposed end of the line and throw the packed bag to a swimmer. As the bag flies, the rope deploys enabling you to reach a victim 50 to 70 feet from shore. It is used either from a standing position, or a moving position on open shores.

   a. Standing Procedures. (FMST.07.27a)

1. Assume a stable belay position and ensure your footing is solid. Avoid loose, sloped or wet rock.

2. Attract victim by blowing your whistle (1 blast). Show him/her the throw bag and yell, "grab the rope". NOTE: If the victim grabs the bag rather than the rope, the rope will continue to deploy to its full length.

3. Grasping the exposed end of the rope, throw the bag underhand in a smooth motion directly at the victim. You may need to lead them in order to score a direct hit.

4. Tell the victim to place the rope over his/her upstream shoulder. This will establish a good ferry angle.

B5

5. Do not wrap the rope around yourself or tie it off at anytime. You may have to release the victim to avoid an obstacle or strainer. Use a dynamic belay and pull the victim to shore or pendulum the victim to other rescuers downstream.
6. If available have other rescuers downstream to recover the victim or repeat the procedure if you miss!

b. Moving Procedures. The moving procedures are the same as the standing procedures except you are on a smooth, open shoreline, so you move slightly downstream of the victim as you pull them gently ashore.

2. **FOOT ENTRAPMENTS AND VERTICAL PINNINGS**. These scenarios are caused by a variety of actions. Swimmers put their feet down or fishermen, in waders, get their feet stuck between rocks and are pushed over by the current. Swimmers are stuck in a strainer. Rafters get wrapped on a rock with victims pinned between the raft and rock or kayakers drop down a vertical shoot and get lodged in the rocks at the bottom. Whatever way it occurs you must respond quickly in order to save the victim. Initially take the following precautions:

    1. Position upstream spotters, to provide warning of boats or debris.

    2. Plan alternate methods if initial try doesn't work.

    3. Have back-up rescuers downstream in order to retrieve the victim.

a. Once these actions are complete conduct the rescue using one of the five following methods.

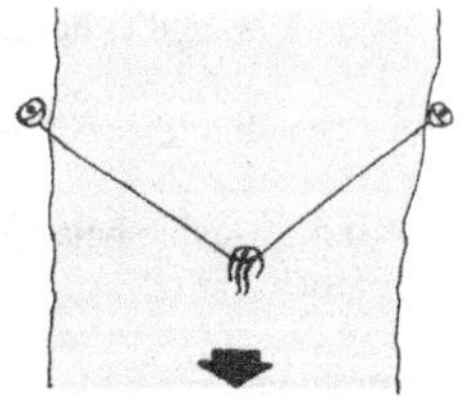

1. Bechel-Ray One Line Method. The simplest and safest method. Get a rescuer on both sides of the river. Establish a rope between the two rescuers and downstream of the victim. Walk upstream until the rope hooks the victim. If you can position the line near the victims ankles you may be able to pop him/her loose and the downstream safety can pick him/her up. If not you can haul the victim upstream until he/she is clear of the obstacle, then by one rescuer playing out line and the other retrieving it you can bring the victim ashore. Remember to bring the victim to the shore with the medevac capability.

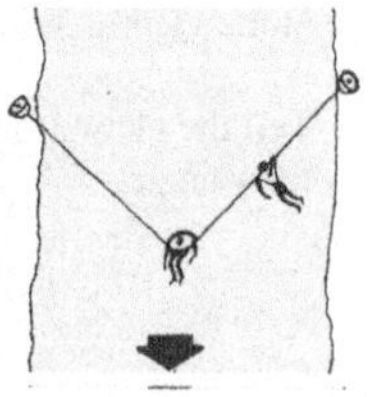

2. The Strong Swimmer-variation. If the previous method doesn't work, leave the line in place and anchor it to stabilize the victim above the surface. Then have a "strong swimmer" go down one side of the rope, as would a tension diagonal. This rescuer is not tied in. Upon arrival the rescuer can readjust the line on the victim, attempt to pull them free, give them a PFD and helmet, or if a debris pile or rock is near, get on it and assist in further efforts.

3. Two Line Loop. There are two variations on this method both of which are applied using the initial rope used in the Bechel-Ray attempt.

(1) Deploy a second rope across the river and hook it onto the first line. Allow it to go one-half way down the rope to the victim. Pull at a 90° angle to rope #1. This will establish a mechanical advantage and pull the victim loose.

(2) Once the victim is free walk the line downstream trapping the victim in a loop. Then by releasing slack on line one, you can bring him/her ashore.

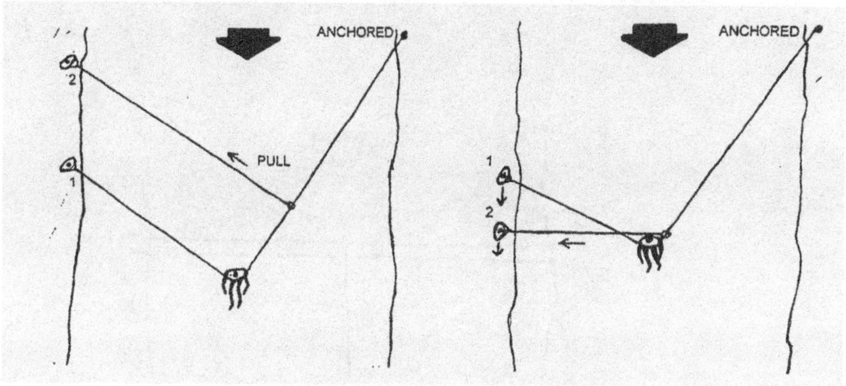

4. European Tethered Rescuer Lower. Using a rescue vest with a "blow-out" belt establish an upstream belay and lower the tethered rescuer to the victim. The rescuer's body will block the water forming an eddy around the victim enabling him to be pulled free. Ensure that the rescuer has a working rescue vest, as this will enable him to cut away if the situation dictates.

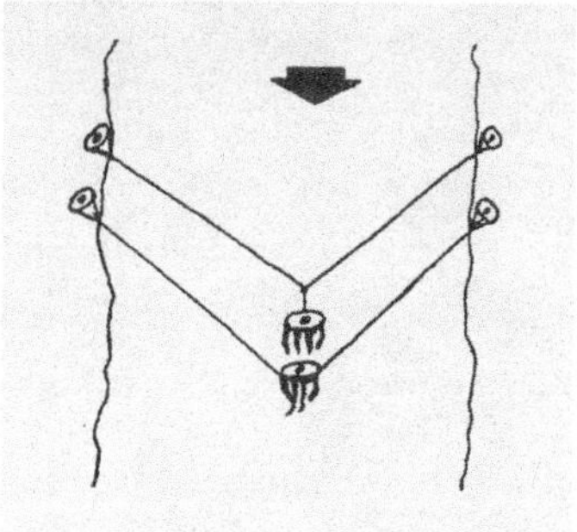

5. Mechanical Advantage Systems. As a last resort you can use a mechanical advantage system. This method is usually used only for body recoveries, as the amount of force exerted on the victim is tremendous and can cause severe injuries.

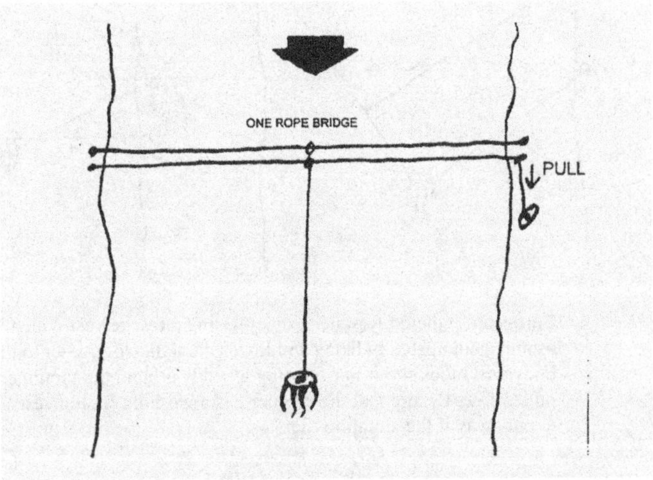
ONE ROPE BRIDGE
PULL

UNITED STATES MARINE CORPS
Mountain Warfare Training Center
Bridgeport, California 93517-5001

FMST.07.28
04/14/02

STUDENT OUTLINE

DEFENSIVE SWIMMING AND FERRY ANGLE

TERMINAL LERNING OBJECTIVES Given a swift water river and throw bags, perform aggressive swift water swimming to meet mission requirements per the reference. (FMST.07.28)

ENABLING LEARNING OBJECTIVES.

(1) Without the aid of reference, demonstrate the defensive swimming position, in accordance with the reference. (FMST.07.28a)

(2) Without the aid of reference, demonstrate use of the ferry angle to cross the river, in accordance with the reference. (FMST.07.28b)

OUTLINE

1. **THE DEFENSIVE SWIMMING POSITION (DSP).** (FMST.07.28a) The DSP is assumed by lying flat on your back, chin on chest (to read the river & select a route), with your feet up, heels just below the level of your buttocks. Holding your feet in this position will enable you to fend off of shallow rocks and will reduce the likelihood of a foot entrapment or pinning. Your arms and hands should be kept close to the body but may be used like oars on a boat to maneuver. Your head will remain upstream except to cross a strainer or to over take a victim in a swimming rescue.

2. **FERRY ANGLE.** (FMST.07.28b) A ferry angle is used to maneuver diagonally across a river. By placing your body at a 45° angle to the current, the river will push you across the river. This is because the force of the current will travel down the exposed, upstream, side of the body moving it laterally more rapidly then the current pushing straight. This is like traversing a slope with edged skis vice sideslipping down on flat skis. In order to cross faster or to reach an eddy you may roll to your belly and execute the rescue crawl stroke or breaststroke. Remember to maintain your ferry angle with your head upstream this will enable you to travel diagonally across the river.

3. **STRAINER CROSSING**. The first rule is to avoid the strainer or shallow submerged obstacle. Read the river and maneuver to the main channel by using downstream "V's" or tongues. If the obstacle cannot be avoided, roll on your belly, head downstream and use the rescue crawl stroke. You must get your body moving faster than the current. Upon arrival at the obstacle use both hands to "vault" yourself over. A well-timed dolphin kick will assist in keeping your legs clear of entanglements. Once clear of the obstacle, resume the defensive swimming position. Bear in mind that it is usually safer to remain in the main current or to eddy hop downstream than it is to attempt to get out on a flooded riverbank, the outside of a bend or on slippery rocks. Also, use the dolphin kick rather than placing your feet on the bottom and kicking off. This will reduce the likelihood of foot entrapments.

UNITED STATES MARINE CORPS
Mountain Warfare Training Center
Bridgeport, California 93517-5001

FMST.07.25
03/14/02

STUDENT HANDOUT

FEET-WET RESCUES AND C-SPINE ROLLS

TERMINAL LEARNING OBJECTIVES In a swift water rescue scenario, conduct a swimming rescue, in accordance with the reference. (FMST.07.25)

ENABLING LEARNING OBJECTIVES.

(1) Without the aid of reference, while in a river and provided with a victim, perform a contact rescue, in accordance with the reference. (FMST.07.25a)

(2) Without the aid of reference, while in a river and provided with a simulated unconscious victim, perform a C-spine roll, in accordance with the references. (FMST.07.25b)

OUTLINE.

1. **FEET WET RESCUE GENERAL CONSIDERATIONS**. (FMST.07.25a)

   a. Always wear the proper personal equipment, i.e. PFD, helmet, drysuit, etc.

b. Don't make physical contact if it can be avoided. Make him/her "chase" you ashore. Use a life ring, can, or boogie board, and tow it along with the victim. Toss him a throw bag and swim/tow him/her until you reach the shore, then pendulum or haul the victim in. Remember, to a panicked victim YOU are a floatation device.

2. **CONTACT RESCUE**.

a. Approach the victim from upstream. Use the rescue crawl stroke to catch-up to him/her. Talk to the victim and attempt to get him/her to swim to shore or "chase" you to the shore.

b. Explain to the victim the defensive swimming position and attempt to get them in this position. Assume this position yourself, in contact rescues this is called the "Reverse and Ready" position. Tell the victim what you are going to do prior to grabbing them.

c. Gain contact by grabbing the shoulders of the life jacket or utilize a cross-chest carry. If the victim is unconscious and face down use a C-Spine roll down and maintain stability as much as possible.

d. Assume a good ferry angle with the victim downstream of you and use his/her body to protect you from obstacles.

e. If the victim panics and attempts to climb on top of you, break away or lock the victim up and go for an alligator roll. Reestablish control and calm the victim as soon as possible.

f. Once contact is made remember that you may be held accountable under the "abandonment" laws if you quit.

3. **TOWED VICTIM SWIM**. When towing a victim it is essential that you establish a ferry angle. If the victim is on a boogie board or float, explain to him/her what you are doing and have them help by kicking their feet. If you are towing by the PFD grasp the shoulder towards the far bank and remain in the DSP (defensive swimming position) in case your victim panics.

4. **LIVE BAIT RESCUES**. The final go and tow method is the "live bait" rescue. Using a rescue vest with a "blow out" belt and a mule team on the shore to belay you, swim to the victim as explained in contact rescues. Once you establish contact you are, in effect, a throw bag with hands. The belayer will then either pendulum you ashore or hauls you in. In an emergency you can cut away and conduct a normal contact rescue.

5. **C-SPINE ROLLS AND STABILITY**.

a. Basic Rules: (FMST.07.25b)

(1) Suspect C-Spine involvement in any head injury.

(2) Stabilize and immobilize, use C-Collar, Sam Splints, or a field expedient method.

(3) Keep the victim in the water until a spine board or equivalent is available, or there are enough rescuers to move the victim safely.

(4) Observe for other signs, i.e. cranial bleeding, bleeding from ears, eyes, nose (Do not attempt to stop this bleeding) or unequal/unresponsive pupil response.

b. Methods:

(1) First, approach the victim from behind. Reach under the arms and grasp the victim's head with both hands, then roll the victim. This is the least desirable method as you cannot swim properly or evaluate the ABC's of the victim.

(2) Second, approach the victim from the side, extend his/her arms above and alongside his/her head, reach across and roll the victim onto your arm maintaining the victims arms in place to ensure stabilization. Evaluate the ABC's, provide rescue breathing and use your free arm to swim.

(3) Finally, approach a victim wearing a PFD from the head. If face down, cross your arms thumbs inboard and use your forearms to immobilize the neck. Then roll the victim over. If face up, simply approach as if gaining contact on a contact rescue and maintain control of the C-Spine.

UNITED STATES MARINE CORPS
Mountain Warfare Training Center
Bridgeport, California 93517-5001

WMC

04/14/02

STUDENT HANDOUT

BOATS AND BOAT HANDLING

LESSON PURPOSE. The purpose of this period of instruction is to introduce the student to the boats available at MWTC for river rescue evolutions.

OUTLINE

1. **GENERAL**. There are a wide variety of boats and hybrid river craft available to rescue crews. Whitewater rafts, rigid raiders, zodiacs, airboats, and regular pleasure boats can all be used in the proper situations. All have been used both safely and have caused deaths and injuries when used improperly. Wherever you are, learn the equipment available, both its assets and its limitations.

2. **WHITEWATER RAFTS**. The whitewater raft of today is an inflatable, self-bailing, multi-chambered vessel with several tie rings and foot cups. Additionally, the materials used in construction resist rips, tears, and punctures much better than their predecessors. The size most often found on local rivers is 12 to 14 feet long due to the size of the rivers. In the MWTC SAR locker, we have a 13-foot self-bailer. This will be our primary rescue boat.

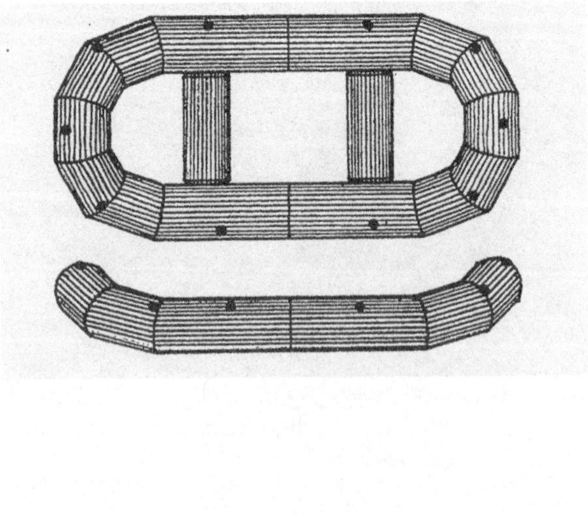

3. **RAFTING COMMANDS**. In order to control the direction of a peddler powered boat the coxswain or guide must use a series of simple, concise, and understandable commands.

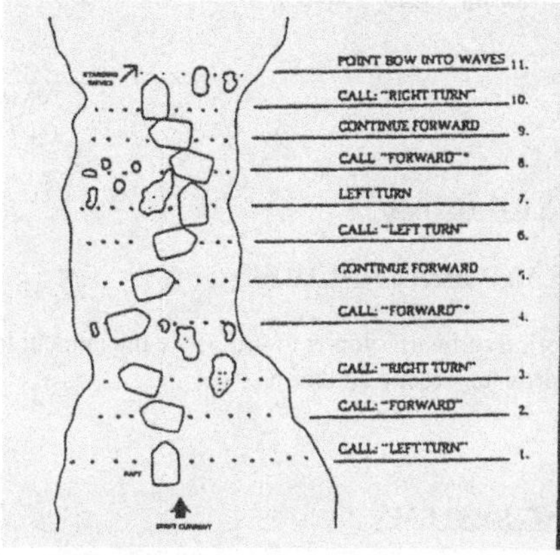

NORMAL TACTICS APPROACH

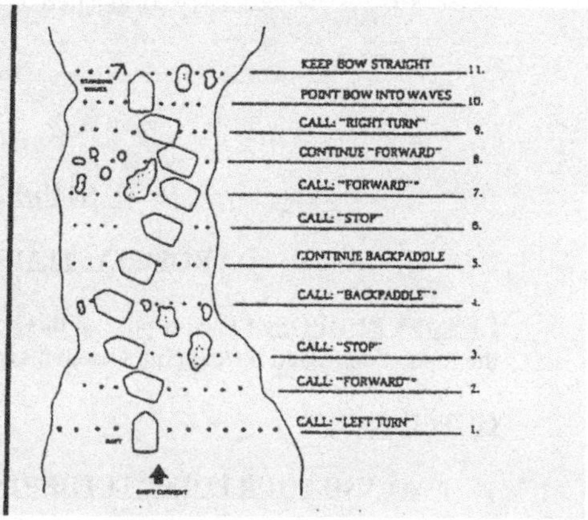

CONSERVATIVE/FAST CURRENT

a. Forward - All hands paddle forward together.

b. Back Paddle - All hands back paddle together.

c. Stop - Stop paddling - Remove blades from water.

d. Left Turn - Left side back paddles, right side forward.

e. Right Turn - Right side back paddles, left side forward.

f. High Side - Move quickly to the down streamside of the boat.

4. **PADDLE STROKES**. In order to control the raft from the coxswain position, either with a crew or alone on a tether system, you will need to apply the following strokes:

a. Sweep - Stroke in a wide arc around the fantail of the boat.

b. Draw - Reach out 90° away from the boat and "draw" it straight back to you.

c. Pry - Put paddle blade beside raft and "pry" the water away from the boat.

d. "J" - Used by lone guide to keep boat straight. Regular straight stroke finished with an outboard stroke or tail.

UNITED STATES MARINE CORPS
Mountain Warfare Training Center
Bridgeport, California 93517-5001

WMC
04/14/02

STUDENT HANDOUT

ROPE SYSTEMS AND BOAT ANCHORS

LESSON PURPOSE: The purpose of this period of instruction is to introduce the student to the rope systems and boat anchors used in swiftwater rescue situations.

OUTLINE

1. **TWO AND FOUR POINT TETHERING SYSTEMS**.

a. With all rope systems the first order of business is to get a rope across the river. Best case is to throw a rope bag across to someone already on the other side. Next best is to use another boat or raft to ferry the rope across. Last we use a swimmer to drag the rope across. Remember to use a smaller line on the swimmer and not to tie it to him/her. When swimming, start high and pendulum across using a good ferry angle.

b. Clip in on the side "D" rings forward for a two point or on the two forward and two aft "D" rings for a four-point tether system. For a two point you now need to simply load in a rescuer (if used) and move the boat into the river by taking slack on the far side and giving slack from the near side. For a four point man the two nearside lines and load a far side aft lineman and a rescuer (if used) into the boat and ferry the boat across. The far side aft belayer will get out and the boat is on a four point system ready for action.

c. This system is good for Class II Rivers with clear banks, few mid stream obstacles, and little floating debris. Use more belayers as required by river speed and boatload.

d. In all rope systems the whistle and hand and arm signals used to communicate are those taught in the communications class earlier in this course.

e. To break down the system simply reverse the procedures. The only difference is that the last man rides in the boat instead of swimming.

B9

2. **THREE POINT AND BOATMEN'S SELF-EQUALIZING ANCHORS.**

a. These anchors have in common the fact that they are self equalizing. This means that, regardless of the direction of pull, all anchor points ("D" rings, thwarts, etc) will remain under equal tension.

b. Using ropes the knots required are the double figure 8 on a bight, a directional figure 8, and a figure 8 on a bight.

c. Using webbing (one inch tubular nylon) a water tape knot is required.

d. If you have an abundance of carabineers, simply clip into the anchor points and back into the small loop on the double figure 8 on a bight, or into the directional figure 8. If not run the rope through the small loop. Remember that this is not preferred as it causes rope on rope stress and rubbing that can melt the rope.

e. When using webbing the same concept applies. Ensure you put a twist in the webbing at the contact point to ensure that a single blow-out will not collapse the entire system.

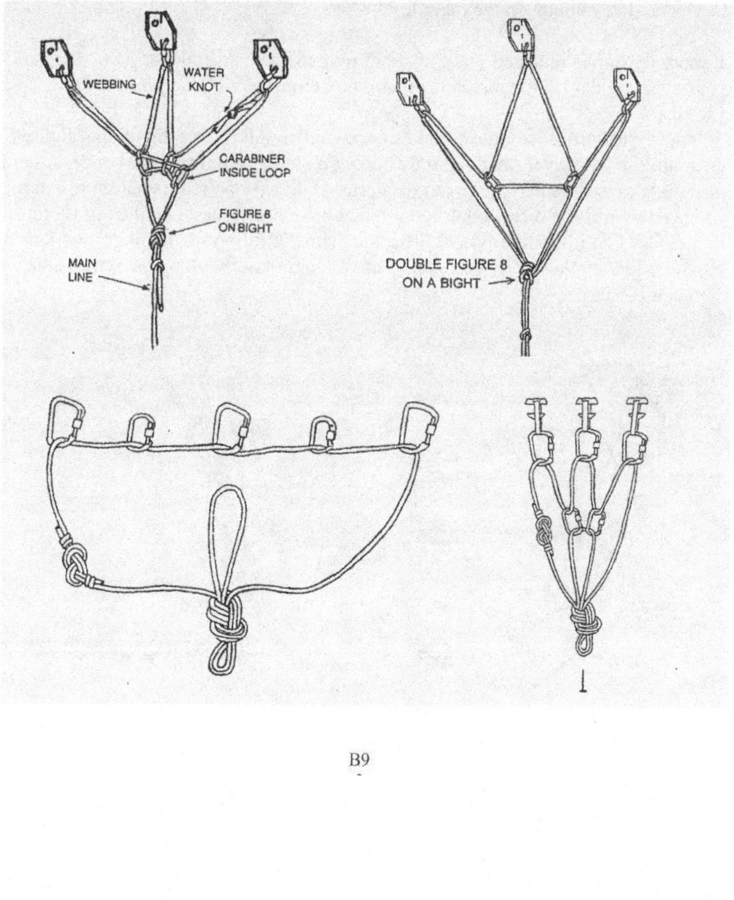

3. **PIG-RIG FOR MECHANICAL ADVANTAGE.** At times you will find yourself with ropes that are long enough to span the river, but not long enough to establish a mechanical advantage system. In these situations, you can use a piggy-back line or "pig-rig" by following the procedures below:

a. Tie off the mainline on either ends, or tie off the far side and set up a belay device on the near side using a munter hitch or rescue 8.

b. Establish an anchor for the pig-rig below the near side anchor.

c. Tie a prussic on the mainline and clip in a carabiner. Tie a figure 8 on a bight into the end of the pig-rig line and clip it into the prussic's carabineer.

d. Run the pig-rig line back to the anchor carabineer and then back through the prussic carabiner.

e. Pull on the running end towards the anchor. This will give you a 4:1 mechanical advantage on the main line.

f. If more tension is required you can either retie the nearside anchor (if tied), or use your belaying device to tension the mainline before moving the prussic.

g. If no prussic cord is available, you can accomplish this with a directional figure 8 on the mainline. However, this will only allow you to tighten the system once. A better method is to use the pig-rig line to tie a prussic directly on to the main line and run your system off of a directional figure 8 tied below the prussic on the pig-rig line. Remember that if the lines are of the same diameter this system will not work as the prussic will slip, therefore, you should use a kragur knot with ropes of the same diameter.

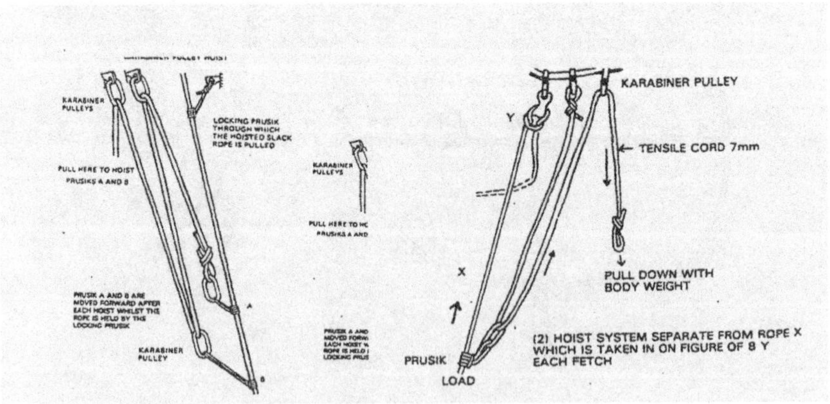

4. **TYROLEAN HIGHLINE AND TELFER LOWERING SYSTEMS** Both of these systems are used to maintain a boat as a stable rescue platform. The lateral movement (side to side/shore to shore) in both systems is controlled by mule teams on the shore. The main difference is that in the tyrolean highline the movement up and down-stream is also controlled from the shore, while in the telfer lowering system, the rescuer in the boat controls that aspect of movement. Both systems have good and bad points and both should be mastered. The situation will dictate which system you will use. Follow the steps below to establish these systems:

a. First you must establish TWO lines across the river. You can use any method taught to accomplish this.

b. Establish one line as the anchor line and tighten it as taught in the "one rope bridge" class or use a pig-rig, if required. If you have uneven mule teams set it at a slight down-stream angle towards the weak team's shore (no more then 10 to 15 degrees) to assist the weak team.

c. Rig the boat with a self-equalizing anchor system.

d. Rig the lowering systems as shown in the diagrams. Either tyrolean or telfer.

e. Load a rescuer or rescue team into the boat and execute. Remember the rescuer can act as coxswain and assist in maneuvering the boat by establishing a proper ferry angle.

f. Communications will be as taught earlier. Ensure you have a shore commander who maintains eye contact with the rescuers and mule teams on both the near and far shores.

g. When using the telfer system, a counter hitch, stitch plate or rescue 8 can be used as a belay device. This enables you to tie-off when you reach the victim so that you can affect the rescue without worrying about controlling the boat.

h. When using either of these systems it is a good idea to set-up a mechanical advantage system between the control (central) pulleys aid the raft/rescue platform.

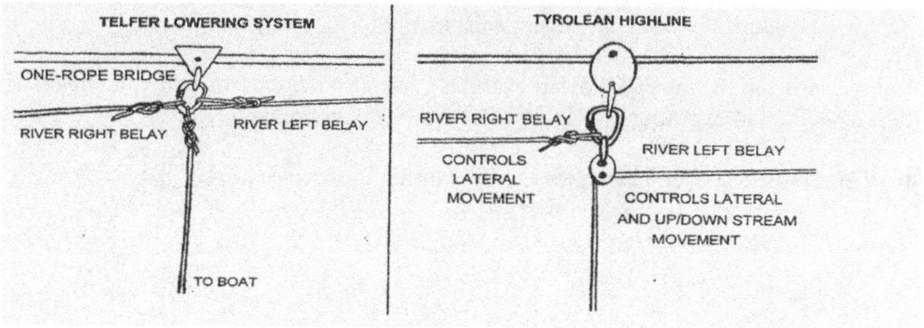

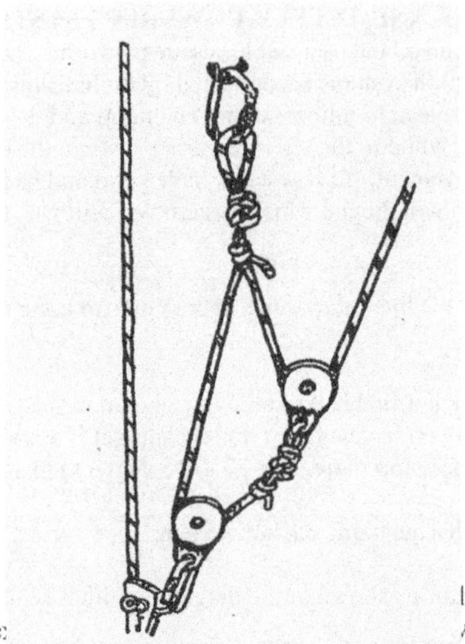

5. **BOAT WRAPS**. As logy and Dynamics" class, the river can put an e: /rapped around an obstacle in the river. You first job is to ensure that NO ONE is trapped between the boat and the obstacle. If you do not have full accountability of personnel, don't hesitate! CUT THE BOAT! The easiest method to do this is to cut the floor out of the older "bucket" boats, or to cut the laces holding in the floor of a "self bailing" raft. Depending on how tightly wrapped the boat is you have several options available:

   a. First ensure the crew is accounted for and treated for any injuries.

   b. Establish a tag-line and attempt to bounce and pull the raft free.

   c. Set a Z-drag inside the boat to dump water.

   d. Use an oar or tree to lever the boat off the obstacle.

e. Set up a mechanical advantage system and attempt to "peal" the boat free.

f. Release air from the tubes. Be aware that this can cause more harm than good by reducing tube strength and exposing more surface area to the water pressure.

g. If all else fails, consider cutting out the floor as discussed earlier.

UNITED STATES MARINE CORPS
Mountain Warfare Training Center
Bridgeport, California 93517-5001

WMC
04/14/02

STUDENT HANDOUT

VEHICLE RESCUES

LESSON PURPOSE. The purpose of this period of instruction is to train the students in the procedures used to rescue personnel from vehicles trapped in swift water.

1. **VEHICLE BEHAVIOR**. There are three types of hazards in a river. Floating objects such as trees and branches, suspended objects or tumbling objects such as sawyers, rocks, and other debris, and stationary objects such as live trees on flooded banks and boulders. A vehicle will become all of these "loads". For the first minute, or so, they will float, as they become partially filled with water, they will be suspended. This is normally when they tumble or roll, and finally they will become stationary as they lodge against another stationary object or fill completely. Once they become stationary, depending on the water depth, they may form an eddy, a pour through or pour over, creating a hydraulic or a standing wave if fully covered. Also, the straighter a vehicle is (head-on) to the water flow the more stable it is. Finally, bear in mind that changes in water flow or displacement (weight) of the vehicle (taking a victim off it) may cause the vehicle to once again become a suspended load and tumble.

2. **RESCUE PROCEDURES**

a. The initial action is to secure the vehicle. Ensure that you have adequate anchors. Trees, guardrails, and bridge abutments are good. Avoid using another vehicle if possible.

b. Approach the vehicle on the downstream side by using the eddy created by the vehicle. Use a boat, either tethered or under power, a strong swimmer or boogie board. Another method is to use the line securing the vehicle as a tension diagonal to reach the vehicle. Then establish a second tension diagonal or endless rope to move the victims ashore. Ensure your upstream spotters are in place to warn of other suspended debris coming your way.

B10

c. If the vehicle is sealed tight, crack one window slightly to allow water to get in the vehicle slowly and equalize the pressure. Warning: If you pop or shatter a window you may cause a sympathetic detonation of the remaining windows.

d. Equip the victims with PFD's and remove to the boat or onto the rope system. Consider the following points:

(1) The victims weight might be the only thing stopping it from rolling. You might need to transfer a rescuer in a one for one exchange.

(2) On hard pack or asphalt the vehicle will shift more easily.

(3) If the vehicle is sideways keep the victims on the upstream side. The high side rule is reversed.

(4) Vehicles roll slowly.

(5) Helo's are a last resort.

UNITED STATES MARINE CORPS
Mountain Warfare Training Center
Bridgeport, California 93517-5001

FMST.07.3
0
04/14/02

STUDENT HANDOUT

STREAM CROSSING

TERMINAL LEARNING OBJECTIVE: Given a mountain swift water environment, and a staff, conduct a mountain stream crossing to meet mission requirements per the reference. (FMST.07.30)

ENABLING LEARNING OBJECTIVES:

(1) Without the aid of references, choose from a given list the considerations in site selection for a stream crossing, in accordance with the references. (FMST.07.30a)

(2) Without the aid of the reference, choose from a given list the safety precautions taken while conducting a stream crossing, in accordance with the reference. (FMST.07.30b)

(3) In a summer mountainous environment, execute the individual preparations taken prior to crossing a stream, in accordance with the reference. (FMST.07.30c)

(4) In a summer mountainous environment and given a staff, cross a mountain stream using the staff method, in accordance with the reference. (FMST.07.30d)

(5) In a summer mountainous environment, cross a mountain stream using team methods, in accordance with the reference. (FMST.07.30e)

1. **SITE SELECTION** (FMST.07.30a)

   a. Mountain streams and rivers are military obstacles and therefore are danger areas for units crossing them. In order to reduce the time in the vicinity of the danger area, a recon team should precede the main body and select the best crossing site. The site selection for a stream crossing should include these eight considerations:

B11

(1) Look for logjams, rocks or fallen trees that will provide a dry crossing if possible.

(2) If a dry crossing is not possible, select a crossing point at a wide and shallow point where the current is slower.

(3) Avoid sharp bends. They can be deep with a strong current on the outside of the bend.

(4) Look for a firm, smooth bottom. This is because large rocks and boulders provide poor footing and cause a great deal of turbulence in the water.

(5) It may be easier to cross several small channels of water rather than one large one.

(6) Do not cross just above rapids, falls, or logjams, taking a fall or slipping could have serious consequences.

(7) Cross in the early morning. The water level will be lower since there has been less daylight for the snow to melt. Also, on sunny days, you will have more time to dry clothing and equipment.

(8) There should be a suitable spot downstream for safety swimmers.

2. **SAFETY PRECAUTIONS** (FMST.07.30b). The following two safety precautions must be taken while conducting stream crossings.

   a. There must be a safety line at a 45-degree angle downstream across the stream. This is for anyone who slips and is swept downstream to grab so that he is capable to stopping himself.

   b. There must be safety swimmers downstream. These are strong swimmers who are positioned downstream to help anyone who is swept downstream. The safety swimmers will use throw bags.

NOTE: The safety line, as well as any other lines that must be taken across the stream, will be taken across by using the lead swimmer method taught in ONE-ROPE BRIDGE.

3. **INDIVIDUAL CROSSING PREPARATIONS.** (FMST.07.30c) Prior to beginning to cross a stream, there are certain preparations that each individual should take. The five preparations are as follows:

   a. Wear pack with shoulder straps fastened snugly. Waterproof the pack for buoyancy if possible.

   b. Weapons will be slung diagonally over the shoulder with the weapon itself being between the pack and the individual's back.

c. Button all pockets and remove blousing garters. This prevents the water from creating added drag against the individual, which it would do if it could flow into open pockets.

d. Wear boots to protect the feet, but remove socks and insoles to keep them dry:

e. Wear the minimum amount of clothing. This reduces the amount of clothing that must be dried after the crossing.

f. Wear helmet in slow moving, shallow water.

NOTE: If you are in a tactical situation, the actual situation will dictate which of the above precautions will be taken.

4. **INDIVIDUAL CROSSING METHODS**. There are three individual methods that may be used.

a. Staff Method. (FMST.07.30d) A strong staff or pole about 6 feet long is used as a crossing aid. It should be strong enough to support the Marine's weight and trimmed clean of any branches. Placing both hands on the pole, the Marine should place the staff just upstream of his intended path. He should use the staff as the third leg of a tripod and should move only one leg or the staff at a time. He should face upstream using the staff to retain his balance. The staff is also used as a probe to discover bottom irregularities that could trip the Marine. The Marine should drag his feet instead of picking them up.

STAFF CROSSING METHOD

b. Swimming. This is an obvious method, if your Marines are good swimmers. This is not always the case, so usually this method is not a preferred one.

(1) In fast, shallow water, the Marine should angle across on his back with his feet downstream and his head up. He should use his hands to tread water and his feet to fend off obstructions.

(2) In fast, deep water, the Marine should angle across the stream on his stomach with his head upstream, to establish a proper ferry angle.

c. Belayed Method. In chest deep or water with a strong current, a rope can assist greatly. A rope will be secured from bank to bank, with the far anchor slightly downstream from the near anchor. The rope will be anchored off so that it lays at a minimum of 45 degrees. The Marine attaches himself to the rope by using a sling rope as safety line, tying a bowline around his waist and a figure-of-eight loop with a steel locking carabiner inserted. He then attaches the carabiner to the crossing line and crosses using the current to assist him.

5. **TEAM CROSSING METHODS** (FMST.07.30e)

a. Line Abreast. Small units (squad to platoon) can cross in moderate currents up to chest deep, by lining arms in a line abreast or chain method. The largest man of the chain is placed on the upstream side of the group. The group will enter the stream parallel to the flow of the stream. The middleman of the chain will control the group's movement and give the command when to step.

LINE ABREAST

b. Line Astern Method. Three or more men can line up facing the current. The upstream man, who should be the largest man in the group, breaks the current while the downstream men hold him steady. The upstream man may use a staff, similar to the individual staff method, to steady himself. All men side step at the same time with one man calling the cadence.

LINE ASTERN

c. Huddle Method. Between three and eight men can face inboard as in a football huddle. They will wrap their arms around each other's shoulders and cross the stream in this formation. The upstream man should change position as they cross because the entire formation will rotate. This prevents one man from becoming exhausted in the upstream position.

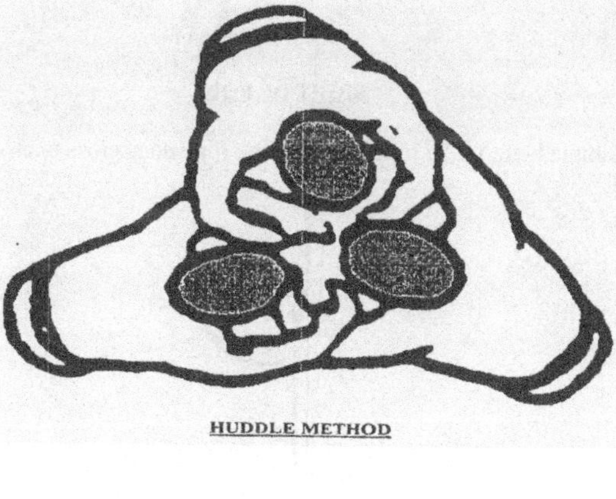

HUDDLE METHOD

UNITED STATES MARINE CORPS
Mountain Warfare Training Center
Bridgeport, California 93517-5001

WMC
04/09/02

STUDENT OUTLINE

ROPE MANAGEMENT

LESSON PURPOSE: To familiarize the student with some of the terms and procedures for working with a rope.

BODY

1. **TERMS USED IN ROPE WORK**

a. Bight. A simple bend in the rope in which the rope does not cross itself.

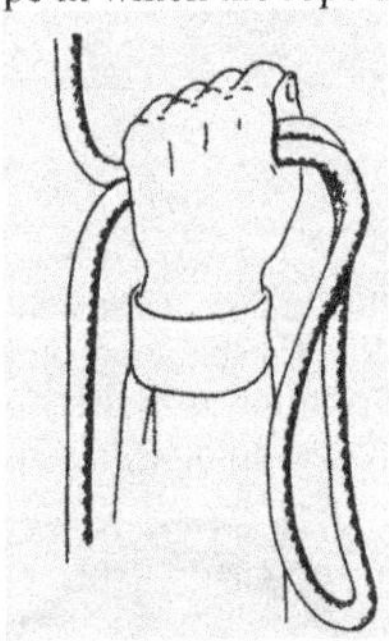

BIGHT OF ROPE

b. Loop. A simple bend in the rope in which the rope does cross itself.

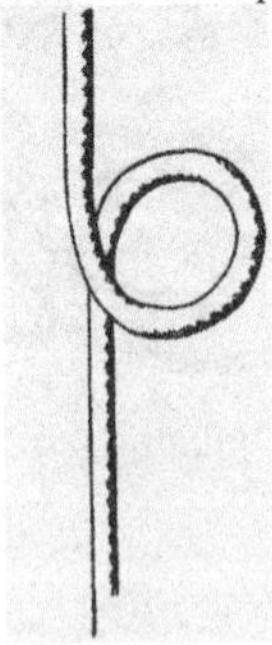

LOOP

c. Half Hitch. A loop which runs around an object in such a manner as to bind on itself.

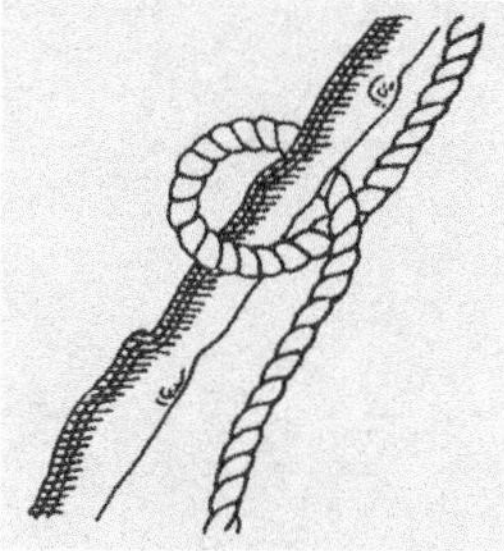

HALF HITCH

d. Standing End. The part of the rope which is anchored and cannot be used, also called the static end.

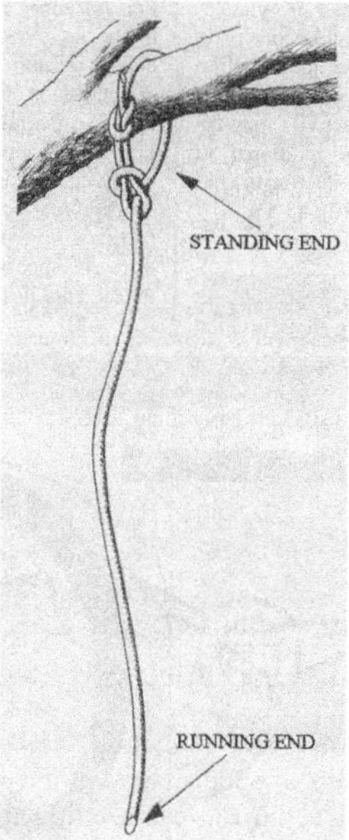

e. Running End. The free end of the rope which can be used.

f. Lay. The same as the twist of the rope. (Applies only to hawser laid ropes, such as manila.)

g. Pigtail. The short length left at the end of a rope after tying a knot or coiling a rope. It may or my not be tied off with a secondary knot, depending on the circumstance.

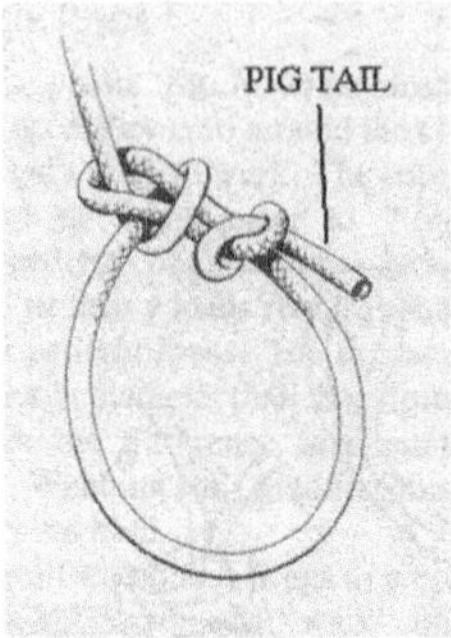

PIG TAIL

h. Stacking (or Flaking). Taking off one wrap at a time from a coil, and letting it fall naturally to the ground.

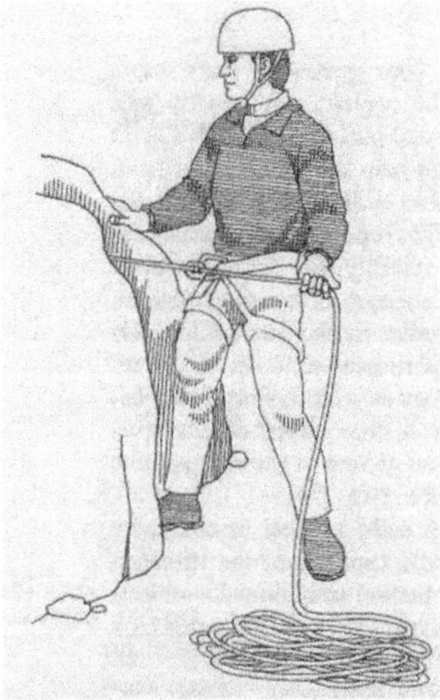

FLAKED OUT ROPEON THE GROUND

i. Dressing the knot. This involves the orientation of all of the knot parts so that they are properly aligned, straightened, or bundled, and so the parts of the knot look like the accompanying pictures. Neglecting this can result in an additional 50% reduction in knot strength.

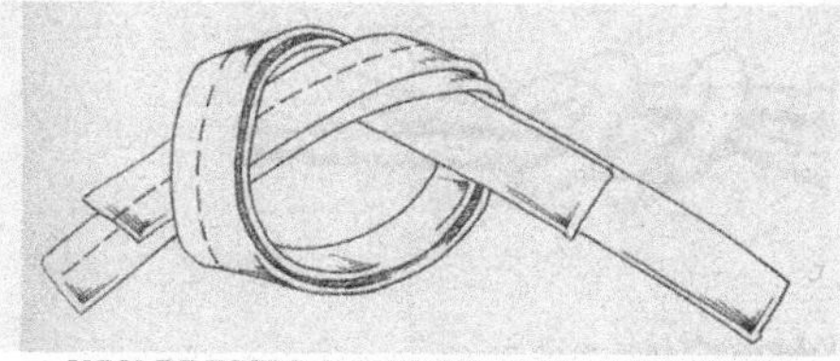

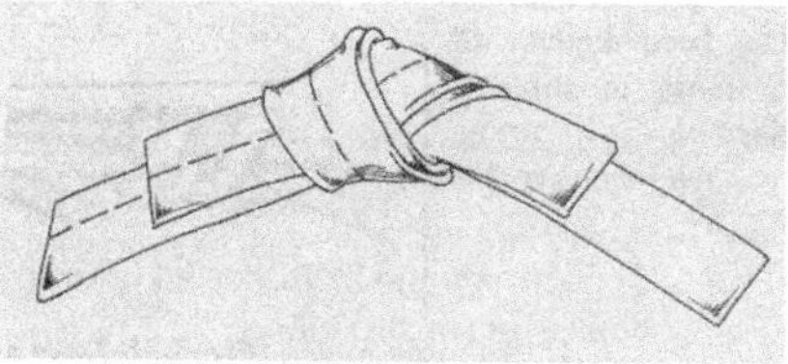

NON-DRESSED DOWN WATER TAPE KNOT **DRESSED DOWN WATER TAPE KNOT**

j. Setting the knot. This involves tightening all parts of the knot so that all of the rope parts bind upon other parts of the knot so as to render it operational. A loosely tied knot can easily deform under strain and change character.

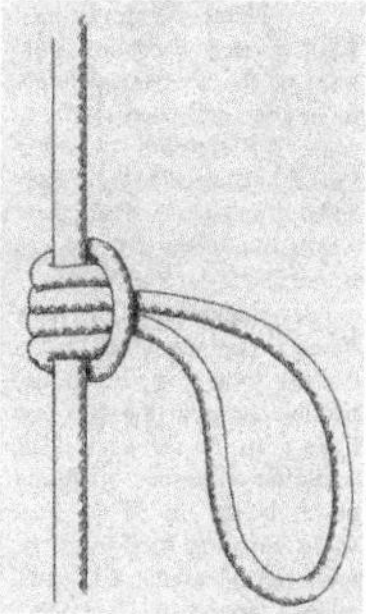

A FULLY SET PRUSIK KNOT

2. CONSIDERATIONS FOR THE CARE OF ROPE

a. The rope should not be stepped on or dragged on the ground unnecessarily. Small particles of dirt will get into and through the sheath causing unnecessary wear to the rope within.

b. The rope should never come in contact with sharp edges of any type. Nylon rope is easily cut, particularly when under tension. If a rope must be used around an edge which could cut it, then that edge must be padded or buffed using fire hose if available, or several small sticks.

c. Keep the rope as dry as possible. If it should become wet, hang it in large loops, above the ground, and allow it to dry. A rope should never be dried out by an open flame, or be hung to dry on metal pegs, as this will cause rust to get in the rope thus rendering it unserviceable.

d. Never leave a rope knotted or tightly stretched longer than necessary.

e. When using rope installations, never allow one rope to rub continually against another.

NOTE: With manila ropes this will cause the rope to fray, whereas nylon ropes can melt under the friction that this causes.

f. The rope should be inspected prior to each use for frayed or cut spots, mildew, rot or defects in construction.

g. Mark all climbing ropes at their midpoints to facilitate establishing the midpoint for a procedure requiring you to use the middle of the rope. The rope should be marked with a bright colored adhesive tape.

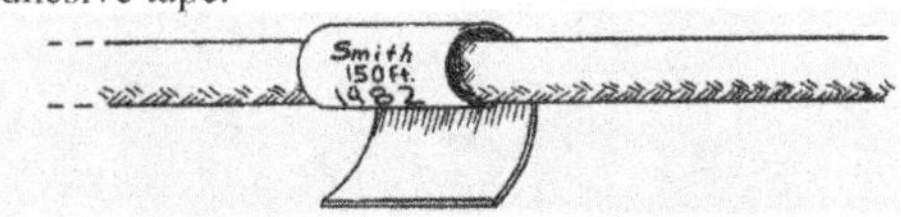

MARKING THE MIDDLE OF THE ROPE

h. The rope should not be marked with paints or allowed to come in contact with oils or petroleum products for these products will weaken it.

i. A climbing rope should NEVER be used for any other purpose except for mountaineering, i.e., towing vehicles.

j. The ends of a new rope or ends caused by a cut should be cut with the rope cutter contained in the MACK and marked with a serial number.

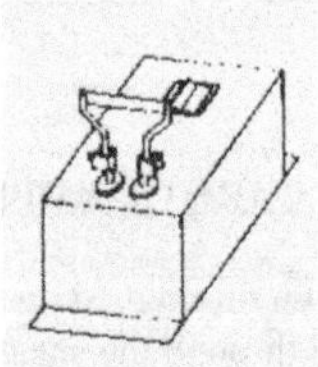

ROPE CUTTER

k. The rope should never be subjected to high heat or flame as this can significantly weaken it.

l. To clean rope use mild soap and rinse thoroughly with water. A rope washer can be used to clean or rinse the rope.

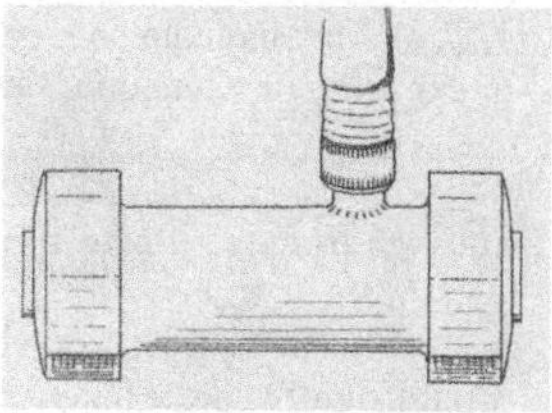

ROPE WASHER

m. When not in use, ropes should be coiled and hung on wooden pegs rather than on nails or any other metal object. They should be stored in a cool place out of the direct rays of the sun.

n. When in areas of loose rock, the rope must be inspected frequently for cuts and abrasions.

o. Always maintain an accurate Rope Log whenever using a rope.

p. Ropes 300-600 foot in length should be Mountain Coiled.

3. **INSPECTION OF ROPE**

a. All ropes have to be inspected before, during, and after all operations. Kernmantle rope is harder to inspect than a laid rope. I.e. green line. The Assault Climber must know what to look and feel for when inspecting a rope. Any of the below listed deficiencies can warrant the retirement of a rope.

(1) Excessive Fraying. Indicates broken sheath bundles or PIC breakage.

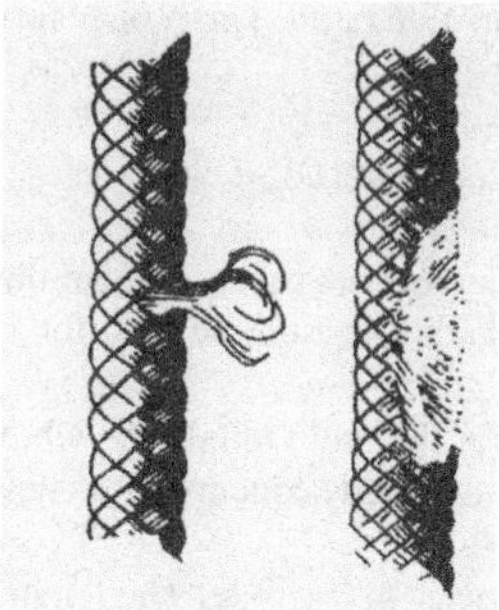

DAMAGED AND EXPOSED ROPE

(2) Exposed Core Fibers. Indicates severe sheath damage. (When you can see the inner core fibers)

(3) Glossy Marks. Signify heat fusion damage, also called a booger.

(4) Uniformity of Diameter / Size. May indicate core damage, noted by an obvious depression (hour glass) or exposure of white core fibers protruding from the sheath (puff).

(5) Discoloration. A drastic change from the ropes original color may indicate chemical change or damage.

(6) Stiffness or Soft Spots. Could signify core damage.

NOTE: Dynamic ropes measuring between 10mm and 12mm are marked at each end of its pigtails with a number "1" indicating that the rope is UIAA approved for single rope lead climbing. Dynamic ropes measuring between 8mm and 9mm are marked at each end of its pigtails with a "1/2" number indicating that two of these ropes are required to conduct a lead climb.

4. **ROPE LOG.** The purpose of the Rope Log is to maintain an accurate record for the use of each rope contained within the Marine Assault Climbers Kit (MACK). Due to the turnover of personnel, and the fact that no one person may have the same rope twice, the rope log is used to ensure the safe use, serviceability, and account of each rope.

a. Serial number. By assigning each rope a serial number, responsible units can determine information about a rope. As soon as the ropes are cut to the desired length for their intended purpose, each rope will be assigned a serial number by the responsible unit. That rope should then be labeled with that serial number in some permanent manner. The best method for this is as soon as the ends of the new rope have been whipped and fused, mark both ends of the rope with the serial number and then dip the ends of the rope, in effect, laminating the serial number to the rope ends.

(1) A rope serial number has five parts. The proper format for a rope serial number is as follows:

1. Type of rope: (S) - Static or (D) - Dynamic.

2. Last two digits of the year the rope was manufactured. Each rope has a shelf life of two years, after that it must not be used for any mountaineering purpose.

3. Four-digit number for that individual rope which is assigned by the responsible unit, and should be assigned sequentially as new ropes are issued out.

4. The length and diameter of the rope. The length may be recorded in feet, and the diameter in millimeters.

5. Responsible unit code. Example: S-96-0001-150/11mm-F2/8

Example meaning: This rope is a 150 foot, 11mm, static rope, manufactured in 1996. It is the first rope issued by Fox Co 2/8.

b. Recording Information in a Rope Log. Once a new rope has been serialized, the rope log for that rope should be started up. At a minimum, it should contain the following information:

(1) The rope serial number.

(2) Manufacturer. Depending on the manufacturer, ropes of the same type and diameter may vary in tensile strength, stretch factor, and durability.

(3) Date of manufacture. Five years from this shelf life date, ropes are considered to have reached their normal expiration date, and should be destroyed.

(4) Date in service. This is to be recorded for tracking purposes to establish how long a rope is in service. Two years is considered maximum rope life in service.

(5) Each time a rope is used, the using unit is responsible to record how that rope was used and how much use the rope received. Additionally, before checking out a rope and prior to turning it back in, the rope must be inspected by qualified personnel, and initial in the "Inspected By" block of the rope log.

NOTE: ANY TIME A DYNAMIC ROPE IS SUBJECTED TO A FALL FACTOR 2, THAT ROPE SHOULD NOT BE USED AGAIN FOR MOUNTAINEERING.

DYNAMIC					
SERIAL #					
MANUFACTURER					
DATE OF MANUFACTURE					
DATE IN SERVICE					
INDICATE THE DATE AND NUMBER OF EACH SPECIFIC USE					
CLIMBS	**FALLS (FACTOR)**	**TOP ROPES**	**RAPPELS**	**REMARKS**	**INSPECTED BY**

ROPE LOG

5. **COILING A ROPE**. There are two types of rope coils frequently used at MWTC. The Mountain Coil and the Butterfly Coil.

a. Mountain Coil. This coil is useful for carrying the rope over a pack or over a climber's shoulder and neck. It can be used for short time storage. The mountain coil can be tied in the following manner:

(1) Sit down with your leg bent at a 90-degree angle, heel on the deck. Starting at one end, the rope is looped around the leg in a clockwise fashion, going over the knee and under the boot sole until the entire rope is coiled.

(2) If coiling a 150-foot rope, use only one leg and offset the other, when coiling a 300-foot rope, use two legs and keep them together.

(3) With the starting end of the rope, form a 12-inch bight on the top of the coils.

(4) Uncoil the last loop and along the top of the coils, wrap 4-6 times towards the closed end of the bight.

(5) The end of the rope being wrapped is then placed through the closed end of the bight.

(6) The running end of the bight is then pulled snuggly to secure the coil.

(7) To prevent the coil from unraveling, the two pigtails are tied together with a square knot.

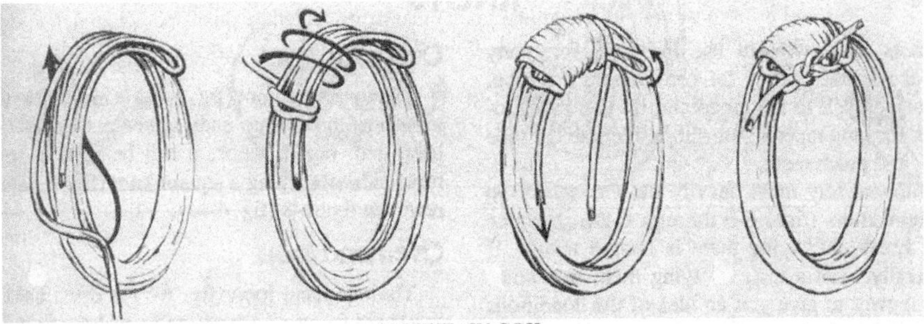

MOUNTAIN COIL

b. Butterfly Coil. This method is used for carrying a rope when the individual needs to have maximum use of his upper body, (i.e. while climbing), without the encumbrance of a large rope coil hanging across his chest.

(1) Coiling the Butterfly Coil

(a) Step 1: Find the middle of the rope, then form a three foot bight laying both ropes in the upraised palm at the two foot point.

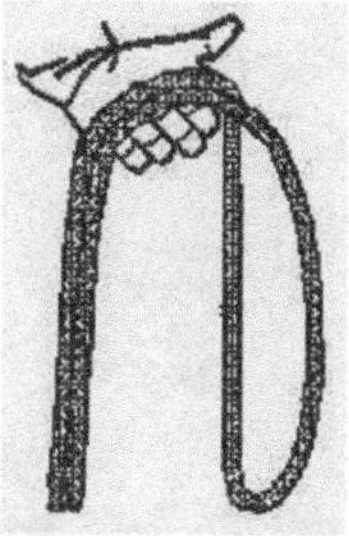

STARTING THE BUTTERLFY COIL

(b) Step 2: Form another two-foot bight with the running end. Place the rope at the two-foot bight along side on top of the original bight ensuring the running end is on the same side as the original bight.

(c) Step 3: Continue making two foot bights, laying them alternately into your palm until there is only six to eight feet remaining. At that point, begin wrapping the two pigtails horizontally four to six times at the mid way point of the ropes in a bight from bottom to top.

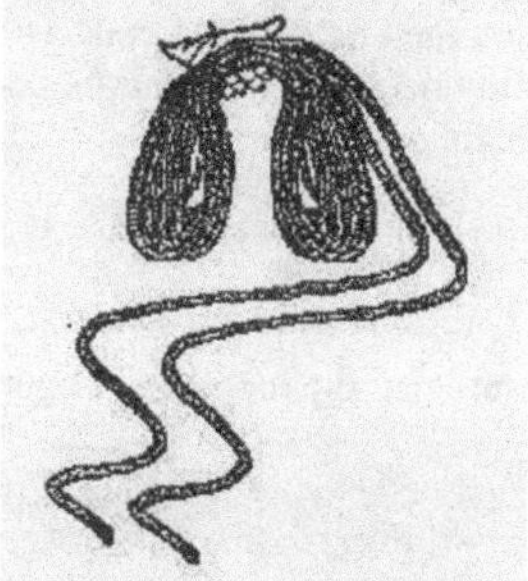

TWO FOOT BIGHTS ON BOTHS SIDES WITH 6 TO 8 FEET REMAINING

(d) Step 4: After completing your wraps, form a bight with the remaining pigtail and then thread it underneath your palm and upwards to one-foot above the coiled rope.

(e) Step 5: With the remaining pigtail, thread it through the one-foot bight in step four.

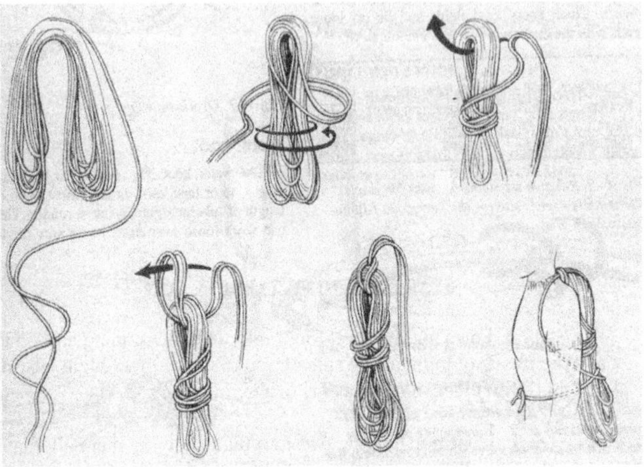

TYING AND CARRYING THE BUTTERFLY COIL

(2) Carrying the Butterfly coil. Separate the running ends, placing the coil in the center of the back of the carrier, then fun the two ends over his shoulders so as to form shoulder straps. The running ends are then brought under the arms, crossed in the back over the coil, brought around the body of the carrier and tied off with a square knot at his stomach.

6. **ROPE THROWING**. To insure that the rope will not get tangled when deployed, certain steps must be taken.

a. With a stacked rope, anchor off the standing end.

b. Take the opposite end of the rope and make 6-8 coils and place them in your strong arm. These wraps will serve as a throwing weight that you can aim.

c. 10-15 feet from the strong-arm coils create a second set of 6-8 wraps and place them in your weak arm.

d. From the edge of the cliff, sound off with the command "STAND-BY FOR ROPE". Just before you release the coils, sound off with the command "ROPE". At that time drop the weak arm coils from the cliff.

e. While taking aim, throw your strong-arm coils overhand or sidearm hard enough to hit your intended target.

f. If the throw was misdirected due to wind, tree, etc., reorganize and attempt to re-deploy the rope.

THROWING THE ROPE

6. MOUNTAINEERING KNOTS.

a. **Class I** – *End of the Rope Knots*

(1) Square Knot. Used to tie ends of two ropes of equal diameter together. It should be secured by overhand knots on both sides of the square knot.

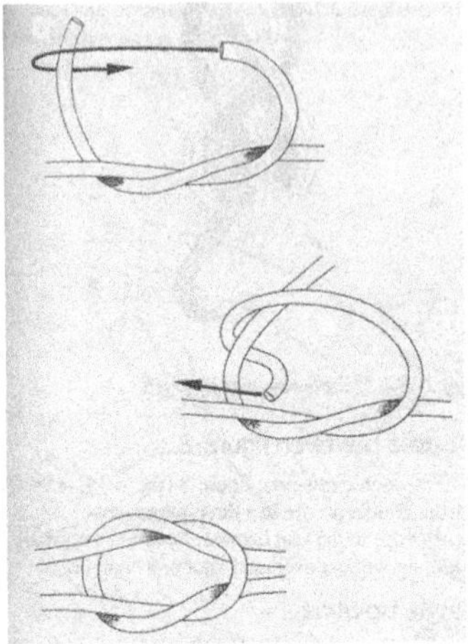

SQUARE KNOT

(2) Double Fisherman's Knot. It is a self-locking knot used for tying two ropes of equal diameter together. It can be tightened beyond untying.

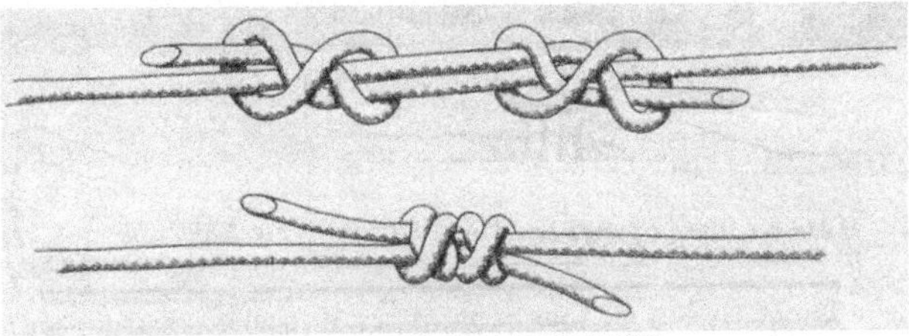

DOUBLE FISHERMAN'S KNOT

(3) Water/Tape Knot. Used to secure webbing or tape runners. It is constructed by tying an overhand knot (without twists) in one end of the tape, and threading the other end of the tape through the knot from the opposite direction. After the knot is dressed down, each pigtail should be a minimum of two inches long.

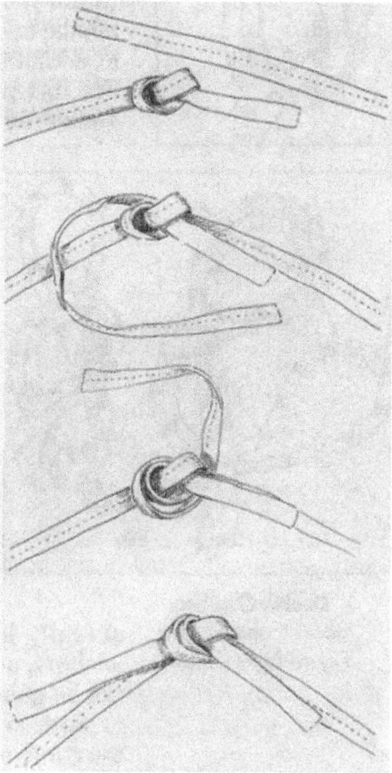

WATER TAPE KNOT

(4) Double Sheet Bend. Used to tie the ends of two or more ropes of equal or unequal diameter together.

SHEET BEND

b. **Class II** - *Anchor Knots*

(1) Bowline. Used to tie a fixed loop in the end of a rope. This knot is always tied with the pigtail on the inside and secured with an overhand knot.

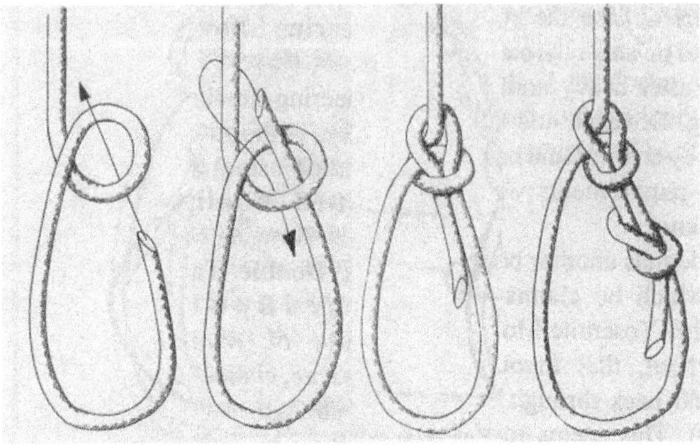

BOWLINE

(2) Round Turn with Two Half Hitches or a Bowline. A loop which runs around an object in such a manner as to provide 360 degree contact and may be used to distribute the load over a small diameter anchor. It will be secured with two half hitches or a bowline.

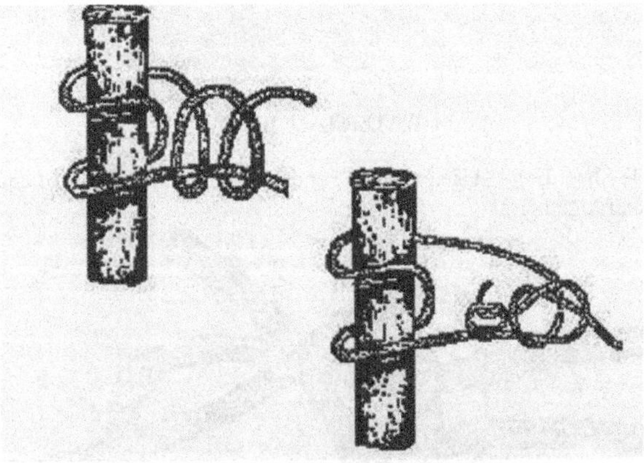

ROUND TURN AND TWO HALF HITCHES AND

ROUND TURN AND A BOWLINE

(3) Clove Hitch. This knot is an adjustable hitch. It could be considered a middle-of-the-rope anchor knot at the end-of-the-rope when used in conjunction with a bowline or round turn and two half hitches.

(1) Around-the-object Clove Hitch.

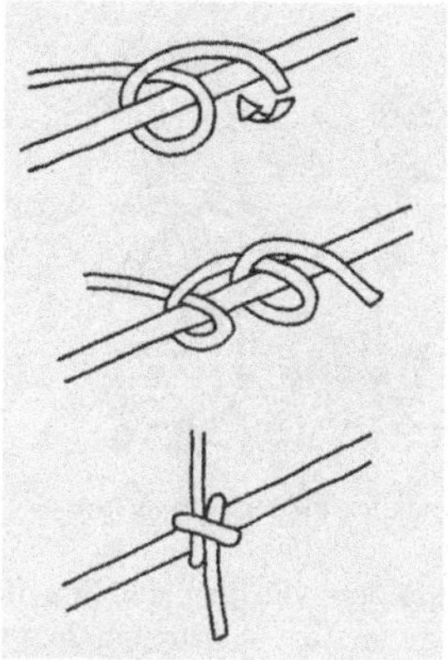

AROUND THE OBJECT CLOVE HITCH

(2) Over-the-object Clove Hitch.

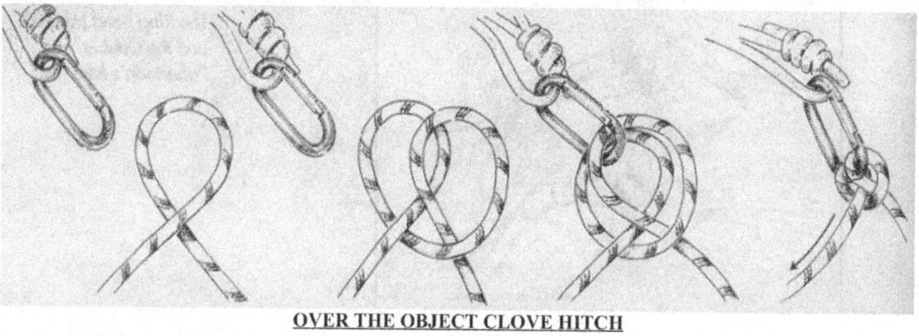

OVER THE OBJECT CLOVE HITCH

c. **Class III** - *Middle-of-the-rope*

(1) Figure-of-Eight-Loop. This is a strong knot that can be readily untied after being under load.

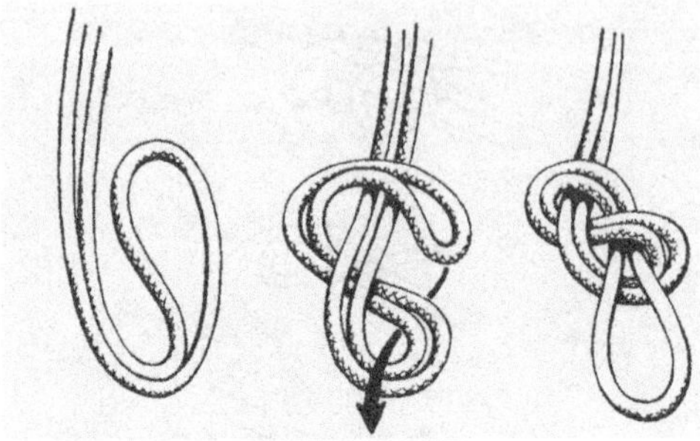

FIGURE 8 LOOP

(2) Double Figure of Eight Loop. The double Figure of Eight Loop is a strong knot and the double loop reduces the wear and strength loss from the rope bending around the carabiner by splitting the load between the two loops.

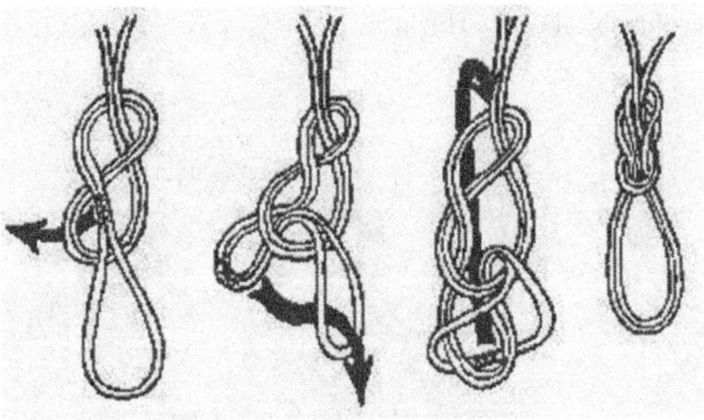

DOUBLE FIGURE 8 LOOP

(3) Two Loop Bowline. This knot can be used to construct a self-equalizing belay or to tie the middle person in on three people on a rope.

TWO LOOP BOWLINE

d. **Class IV** - *Special Knots*

(1) Prusik Knot. This knot functions by introducing friction that can be alternately set and released. For best results, tie the knot with a smaller diameter cord on a larger diameter cord. If slippage occurs, more wraps maybe used.

(a) Middle-of-the-Rope Prusik. This knot is created with an endless loop also known as a Prusik Cord. Do not tie this knot with tape due to less friction.

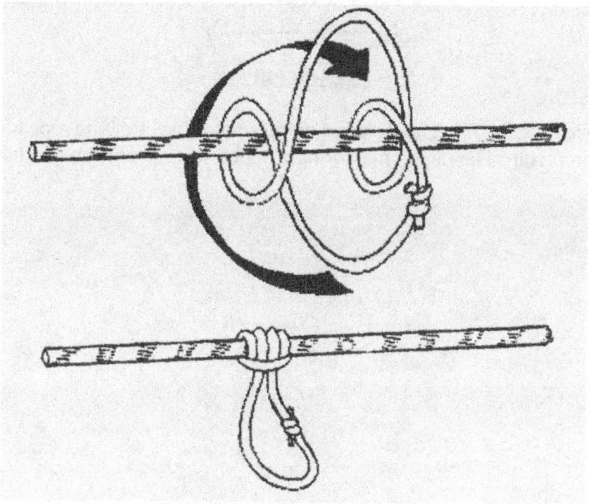

MIDDLE OF THE LINE PRUSIK

(b) End-of-the-Rope Prusik. This knot is always secured with a bowline. Do not tie this knot with tape due to less friction.

END OF THE LINE PRUSIK

(c) French Prusik. This can be constructed with a Prusik Cord or a single strand of cord with figure-of-eight loops on each end.

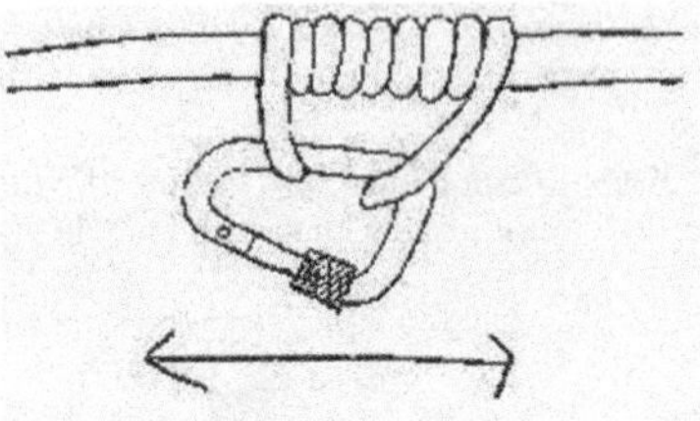

FRENCH PRUSIK

(2) Retraced Figure of Eight. Used to tie the end of the climbing rope into a harness or swammi wrap. The pigtail may or may not be secured with an overhand.

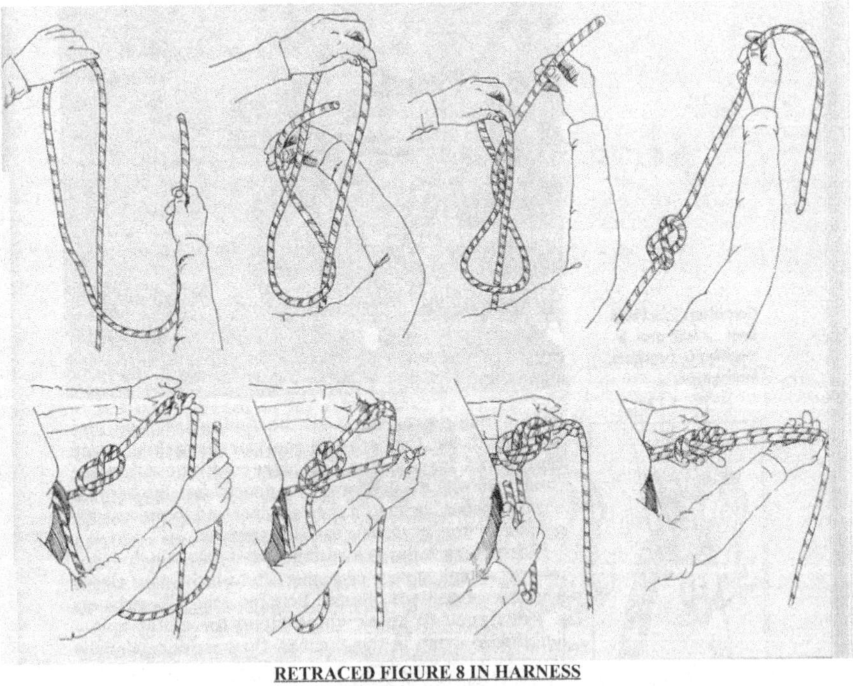

RETRACED FIGURE 8 IN HARNESS

(3) Directional Figure-Of-Eight. When tied and tensioned is applied to the standing and running ends of the rope, the knot will not pull apart. The loop will point toward the direction of pull.

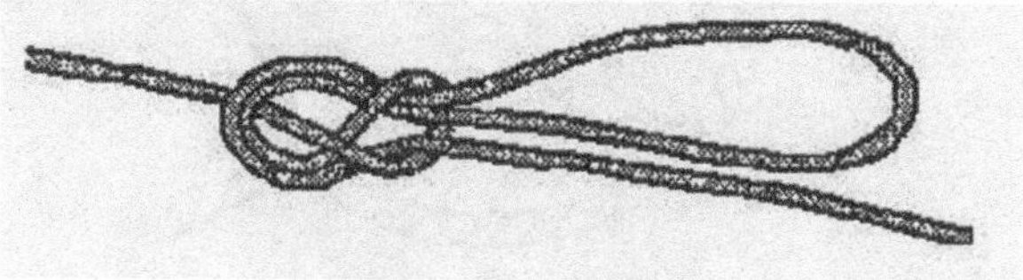

DIRECTIONAL FIGURE 8

(4) Slip Figure 8. This knot is used for retrievable anchors and fixed rope installations for the ease of untying the knot.

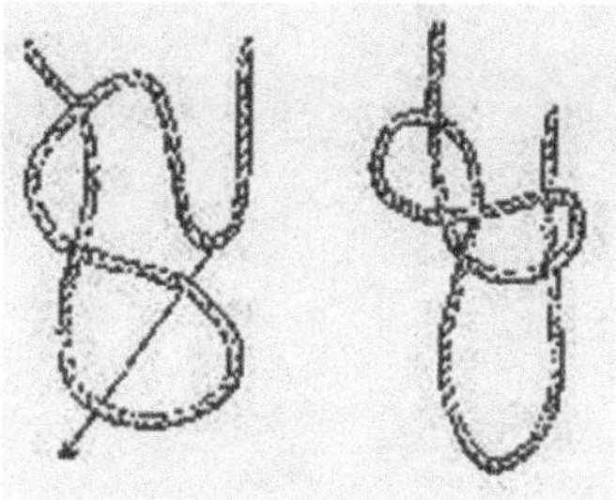

SLIP FIGURE 8

(5) Kliemheist. This is a friction knot.

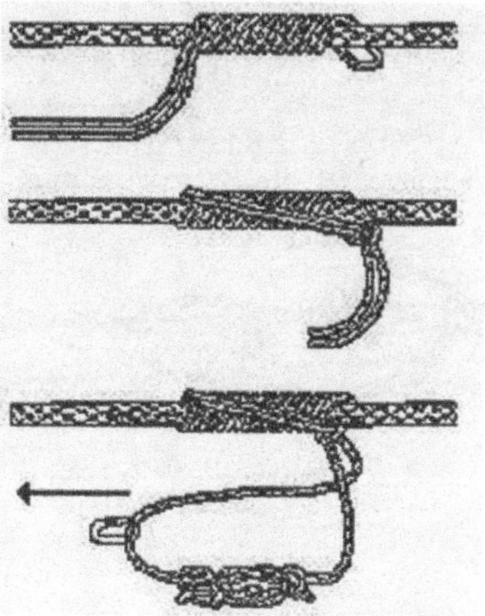

KLEIMHEIST

(6) Overhand Knot. Can be used to secure primary knots on itself.

OVERHAND

(7) M nter Hitch. This is a simple hitch in the rope that is clipped into a carabiner to put friction on the line.

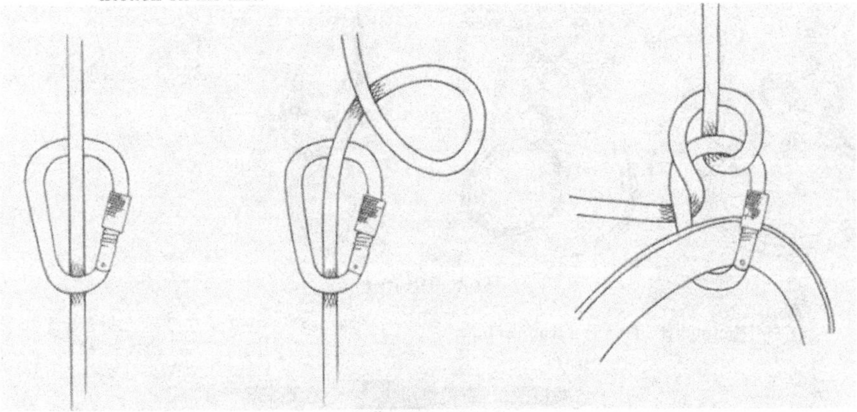

MÜNTER HITCH

(8) Timber Hitch. A Timber Hitch is used to fix a rope to a pole or equivalent for hoisting or towing purposes. It has the capability of casting off easily.

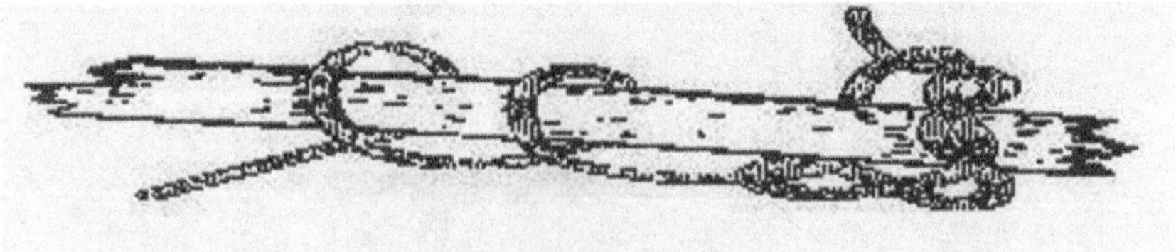

TIMBER HITCH

(9) Mariner's Hitch. This knot's advantage over the prusik is that it can be released under load.

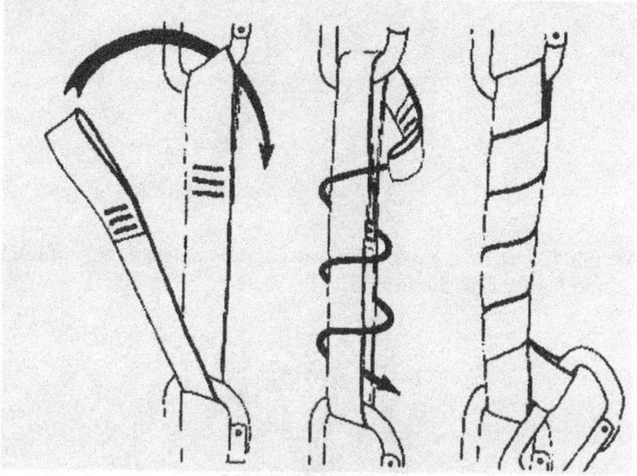

MARINER'S HITCH

(10) Krägur Knot. This friction knot is used to prevent unnecessary slippage when tied onto a rope of the same diameter.

KRÄGUR KNOT

(11) Rappel Seat. This is used as an expedient support harness for rappelling, crossing rope bridges, etc. It is constructed as follows:

(a) Center the sling rope on the left hip.

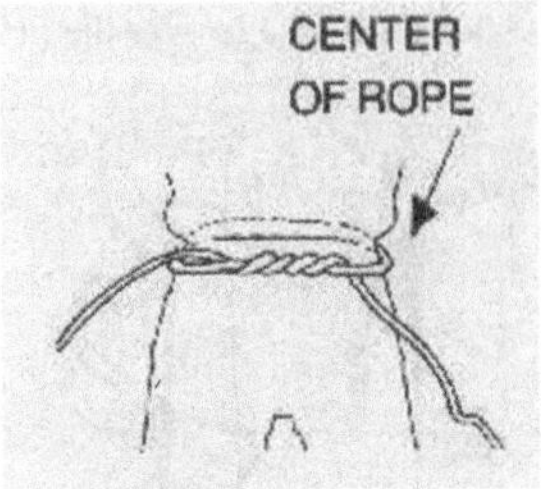

(b) Wrap the sling rope around the waist and tie at least one half hitch around itself (preferably two) in the front.

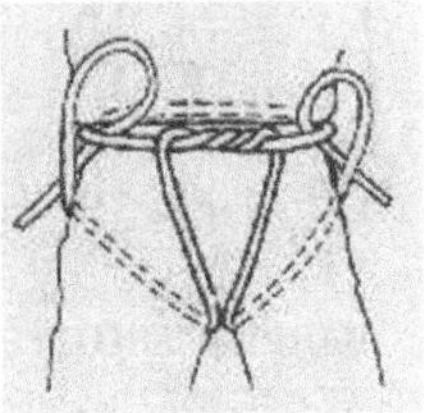

(c) Bring the running ends down through the legs, up over the buttocks, and over the original waist wrap, down between the waist wrap and the waist, and over itself, forming a bight. Cinch this up tightly.

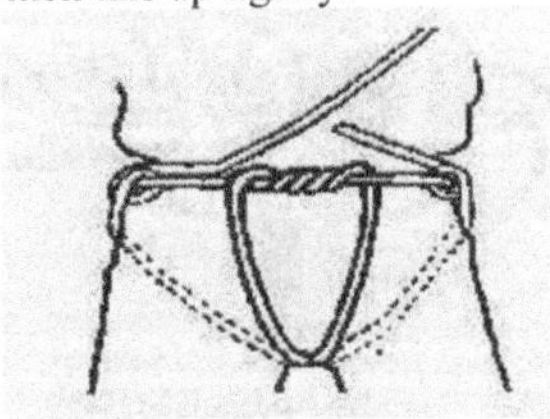

(d) Now tie a square knot with two overhand knots on the left hip ensuring that the overhand knots encompass all the wraps.

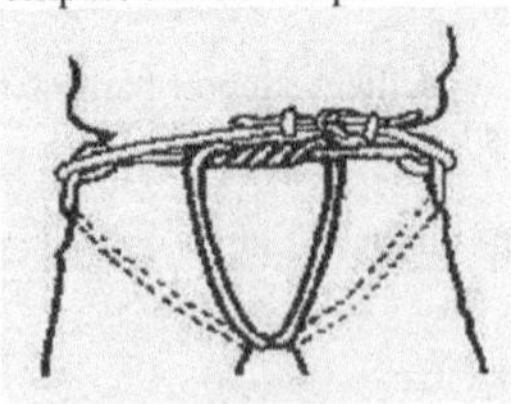

(e) Tuck any excess rope into a pocket.

(12) Bowline on a coil. Used by the first and last men on a climbing rope to tie into the rope. An overhand knot is used behind the knot. It distributes the force of a fall over a larger area of a climber's waist and is preferable to a single bowline around the waist. The bowline on a coil can also be used to take up excess rope. The bowline on a coil should have 4-6 wraps around the waist.

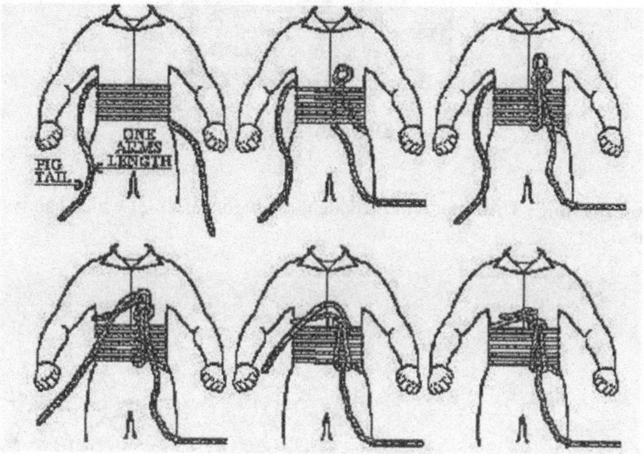

BOWLINE ON A COIL

(13) Swami wrap (Swami belt). The swami wrap (four to six wraps of rope or nylon webbing tied around the waist) popular with rock climbers who wanted to increase the working length of their climbing ropes. It also offers some improvement over the bowline on the coil in distributing the forces sustained in a fall over a larger area of the midsection of the body, but does not eliminate the problem of suffocation when hanging. The Swami wrap is constructed as follows:

(a) Take a sling rope and find the middle.

(b) Place the middle of the sling rope in the small of your back.

(c) Bring both pigtails to the front and continue to wrap around your body until both pigtails are about 15-20 inches long.

(d) Tie the pigtails on the left side with a square knot and two overhand knots.

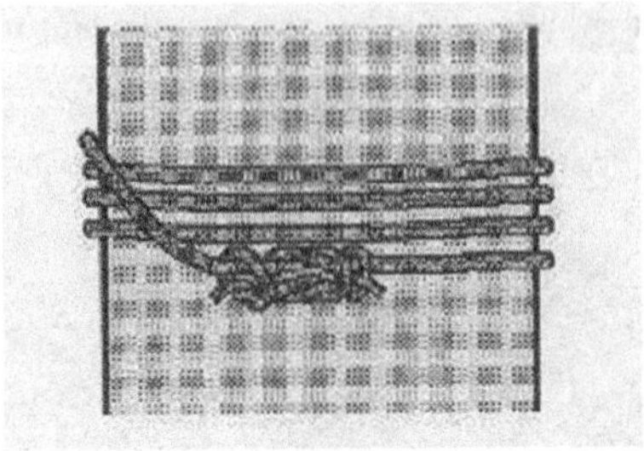

SWAMMI WRAP

(14) Three Loop Bowline. This knot will provide three bights, each of which can be adjusted against the others.

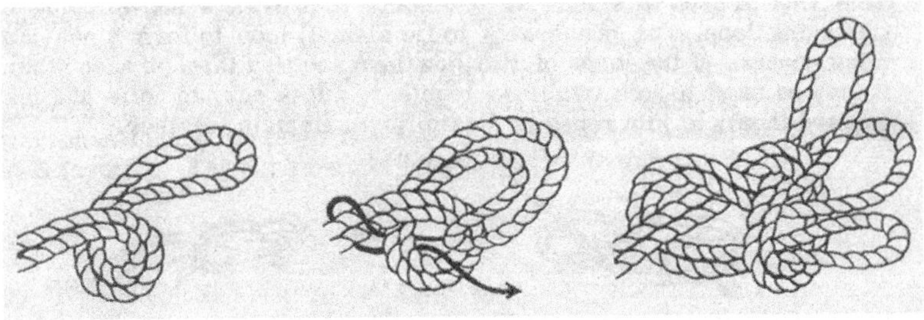

THREE LOOP BOWLINE

7 **KNOT STRENGTH.** Rope, cordage and webbing is strongest when loaded in a straight line. When bending a rope or web to create a knot, the strength of the rope is reduced. All knots should be dressed properly for maximum effective use.

KNOT	RELATIVE STRENGTH OF KNOT
No Knot	100%
Figure 8	75-80%
Bowline	70-75%
Double Bowline	70-75%

Double Fisherman	65-70%
Water Tape Knot	60-70%
Clove Hitch	60-65%
Overhand Knot	60-65%
Square Knot	45%

1. **KNOT TESTING TIME LIMIT.** The following times must be met to pass the knot tying portion of this course.

KNOT	**UNIT OPERATIONS TIME LIMIT**	**SML, ACC TIME LIMIT (BLINDFOLDED)**
Square knot	60 Seconds	30 Seconds
Double Fisherman's Knot	60 Seconds	30 Seconds
Water/Tape Knot	60 Seconds	30 Seconds
Round Turn and a Bowline	60 Seconds	30 Seconds
Round Turn and Two Half Hitches	60 Seconds	30 Seconds
Clove Hitch (around the object)	60 Seconds	30 Seconds
Munter Hitch	60 Seconds	30 Seconds
Slip Figure 8	60 Seconds	30 Seconds
Figure 8 Loop	60 Seconds	30 Seconds
Directional Figure 8	60 Seconds	30 Seconds
End of the Line Prusik	90 Seconds	45 Seconds
Retrace Figure 8	90 Seconds	45 Seconds
Rappel (Swiss) Seat	2 Minutes	60 Seconds

UNITED STATES MARINE CORPS
Mountain Warfare Training Center
Bridgeport, California 93517-5001

SMO
05/02/01

STUDENT OUTLINE

NOMENCLATURE AND CARE OF MOUNTAINEERING EQUIPMENT

LESSON PURPOSE. The purpose of this period of instruction is to introduce you to the types of equipment used here at MWTC and how to care for it so as to prevent its untimely failure. This lesson relates to all climbing and installation work that you perform here at MWTC.

OUTLINE

1. **ROPES.** All ropes used in the military must meet UIAA standards or U. S. Federal Test Standard 191A. Most ropes have a 5-year shelf life and maximum 2-year service life.

a. Static. Black in color.

(1) Construction. Kernmantle

(2) Minimum tensile strength. 7500 lbs.

(3) Maximum elongation. 1.5%

(4) Diameter. 11mm

(5) Sizes.

(a) 165 ft ± 5 ft

(b) 300 ft ± 10 ft

(6) Usage. Rescue operations and bridging where a low amount of elongation is desirable under a working load.

b. Dynamic. Olive Drab in color.

(1) Construction. Water-resistant treated Kernmantle to reduce friction.

(2) Minimum tensile strength. 6500 lbs.

(3) Maximum elongation. 6%

(4) Diameter. 10.5mm and 11mm

(5) Size. 165 ft ± 5 ft.

(6) Usage. For lead climbing/party climbing.

c. Maxim Dry Rope. Olive Drab or Multi-Colored.

(1) Construction. Water-repellent treated Kernmantle.

(2) Minimum tensile strength. 3472 lbs.

(3) Maximum elongation. 6%

(4) Diameter. 9mm

(5) Size. 150 ft ± 5 ft

(6) Usage. For glacier travel/ice climbing

d. Gold Line II.

(1) Construction. Eight strand braided nylon plymor.

(2) Minimum tensile strength. 4500 lbs.

(3) Maximum elongation. 20%

(4) Diameter. 11 mm

(5) Size. 300 ft or 600 ft spools

(6) Usage. Sling Ropes and litters only.

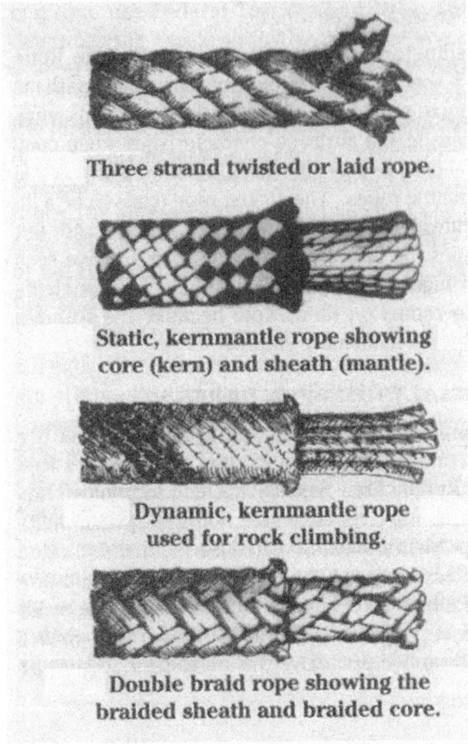

Three strand twisted or laid rope.

Static, kernmantle rope showing core (kern) and sheath (mantle).

Dynamic, kernmantle rope used for rock climbing.

Double braid rope showing the braided sheath and braided core.

<u>INSIDES AND OUTSIDE SHEATHS OF VARIOUS TYPES OF ROPES</u>

NOTE: Sling ropes are made from 15 foot lengths of plymor or dynamic rope ONLY. Twenty-five foot practice coils should be constructed with static rope, but dynamic rope can be used.

2. ADVANTAGES/DISADVANTAGES

a. <u>Advantages of Nylon Rope</u>

(1) High strength to weight ratio.

(2) Good energy absorption in dynamic ropes.

(3) Flexible.

(4) Rot resistant, not affected by frost.

b. <u>Disadvantages of Nylon Rope</u>

(1) Low melting point. Nylon fuses at 400°F and melts at 480°F.

(2) Susceptible to abrasions and cuts.

(3) Affected by chemicals and light.

c. Advantages of Manila Rope

(1) Easily gripped.

(2) Hard wearing.

(3) Does not deteriorate in heat.

d. Disadvantages of Manila Rode

(1) Heavy, kinks, especially when wet. Absorbs water and swells.

(2) Burns at +300°F.

(3) Edible by rodents.

3. GENERAL INFORMATION

a. Nylon rope stretches under tension and will rupture at between 30% and 70% elongation, depending on construction.

b. Nylon rope loses as much as 30% strength when wet.

c. Temperatures as low as 250°F will damage a nylon rope.

4. NYLON WEBBING

a. The type of nylon webbing available is tubular. Tubular nylon webbing is very strong and flexible. All rules that apply to nylon rope apply to tubular nylon webbing. The size of nylon webbing used is:

(1) 1 inch tubular nylon. Tensile strength approximately 4,000 - 4,500 lbs., depending on the manufacturer.

b. Pre-sewn Spectra Runners. Tensile strength approximately 5,500 lbs.

NYLON WEBBING

NOTE: These are minimum strengths. Some manufactures make even stronger webbing.

1. **CARABINERS**. Also commonly known as snap links. There are two different types of carabiners frequently used at MWTC, the steel locking carbine and the non-locking aluminum carabiners.

a. Steel Locking Carabiner. There are two different sizes of carabiners:

(1) Steel Locking Stubai 82.

(2) Large Locking "D" Stubai 85.

b. Aluminum Non-Locking Carabiner.

(1) The standard "D" non-locking carabiner with a tensile strength of 2000 lbs. or more.

2. **NOMENCLATURE OF A CARABINER**.

a. There are four parts to a non-locking carabiner. They are as follows:

(1) Gate.

(2) Gate pivot pin.

(3) Locking pin

(4) Body.

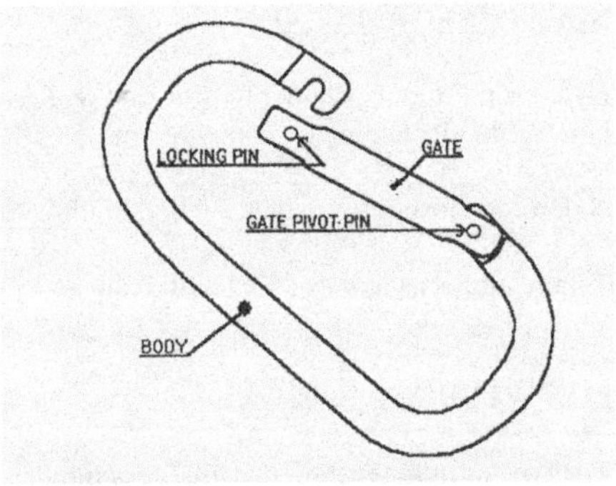

STEEL NON-LOCKING CARABINER

b. There are five parts to a locking carabiner. They are as follows:

(1) Gate

(2) Gate pivot pin

(3) Locking notch

(4) Locking nut

(5) Body

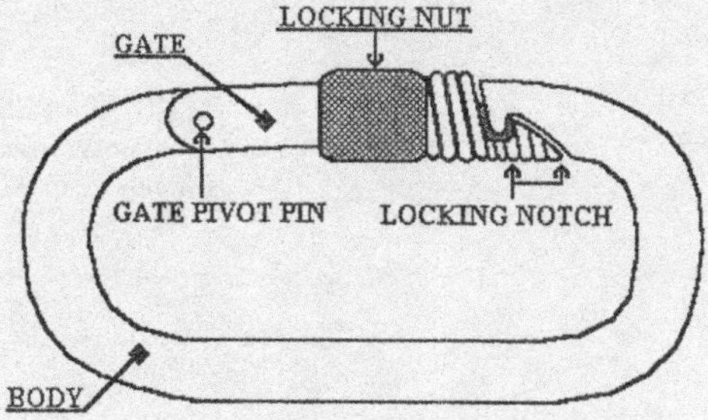

STEEL LOCKING CARABINER

3. **CARE, MAINTENANCE AND SERVICEABILITY OF A CARABINER**. The following inspections and maintenance should be performed on a routine basis:

a. Care.

(1) Avoid dropping of a carabiner onto a hard surface, hairline factures could occur.

(2) Keep the carabiners off of the bare ground because dirt and grit can get into the working parts.

(3) If the carabiner has fallen from a significant height, it should be retired.

b. Maintenance.

(1) Remove all dirt, moisture and grime after use.

(2) Lubricate with tri-flow graphite and wipe off with a clean cloth.

c. Serviceability Checks.

(1) The gate snaps shut with no friction and with no gap between the locking pin and locking notch.

(2) There is no excessive side to side movement of the gate

(3) The pivot pin is tight.

(4) That the spring-loaded gate functions properly.

(5) The locking pin is tight.

(6) The locking nut travels freely and locks securely.

(7) There are no cracks or flaws in the metal.

NOTE: The weakest part of a carabiner is the gate. If an engraver is used to mark a carabiner, it should be applied to the gate and not the load bearing side.

UNITED STATES MARINE CORPS
Mountain Warfare Training Center
Bridgeport, California 93517-5001

WMC
04/11/02

LESSON PLAN

BALANCE CLIMBING

1. LEARNING OBJECTIVES

a. TERMINAL LEARNING OBJECTIVE. In a summer mountainous environment, execute balance climbing, in accordance with the references. (WMC)

b. ENABLING LEARNING OBJECTIVES.
(1) With the aid of references, state orally the safety requirements for balance climbing, in accordance with the references. (WMC.C3a)

(2) With the aid of references, state orally the individual preparations for a balance climb, in accordance with the references. (WMC.C3b)

(3) In a summer mountainous environment, execute the duties of a spotter for a balance climb, in accordance with the references. (WMC.C3c)

(4) In a summer mountainous environment, execute the commands used between the climber and the spotter during a balance climb, in accordance with the references. (WMC.C3d)

(5) In a summer mountainous environment, execute each type of hold used in balance climbing, in accordance with the references. (WMC.C3e)

(6) Without the aid of references and given the acronym "CASHWORTH", describe orally the nine considerations for proper body position while climbing, in accordance with the references. (WMC.C3f)

OUTLINE

1. **SAFETY PRECAUTIONS**. There are two safety precautions that always apply to balance climbing. They are as follows:

a. Never climb more than 10 feet above the ground. By this it is meant that the climber's feet are never more than 10 feet above the ground.

b. A spotter is required for all balance climbs.

SPOTTER WATCHING THE CLIMBER

2. **INDIVIDUAL PREPARATIONS**. Prior to beginning a balance climb there are seven things that the climber must do to prepare himself. They are as follows:

a. Helmet with a serviceable chinstrap must be worn.

b. Sleeves rolled down_to give hand and arm freedom of movement. Blouse tucked in to your trousers. In case of fall, it may catch on a rock and cause you to flip over sideways.

c. All watches, rings, and jewelry must be removed before climbing.

d. Gloves will not be worn, as they can slip, and also give a false feel for the rock.

e. Unblouse trousers, if they restrict movement.

f. Soles of boots clean and dry.

g. Select route where vegetation is minimal. Never use vegetation for hand or foot holds.

3. **DUTIES OF THE SPOTTER**. The spotter is the balance climber's partner who, rather than climbing himself, acts as the safety man for the climber during the climb. The five duties of the spotter are as follows:

a. Positions himself directly behind the climber before the climb starts.

b. Maintains his position facing the cliff, directly below the climber and approximately 3-4 feet away from the base of the cliff, for the duration of the climb. He will move diagonally as necessary to remain below the climber.

c. The spotter will stand with his feet shoulder width apart and arms ready to stop the climber if he falls.

d. If the climber falls, the spotter will not "catch" him; he will prevent the climber from falling further down the hill. He will do this by pushing the climber towards the base of the cliff, thereby preventing him from tumbling backwards.

e. At no time will the spotter allow anyone to come between himself and the face of the cliff while a balance climb is taking place. He will require anyone who wants to pass by his position to go behind him.

4. **SPOTTING AND CLIMBING COMMANDS**. The following are the commands used by both the spotter and climber.

COMMAND	GIVEN BY	MEANING
"Last Name, climbing."	Climber	I am ready to climb.
"Climb Climber."	Spotter	I am ready to spot you.
"Last Name, off Climb."	Climber	I am off the climb.
"ROCK!"	Climber or Spotter.	A rock has been knocked off the rock face and is falling.
"FALLING!"	Climber	I am Falling.

a. If the command "ROCK" is given, all personnel in the vicinity will take the following action:

(1) If close to the cliff face, move against the cliff face with your face against the cliff face and your hands between you and the cliff face.

(2) If not close to the cliff face, look up to locate the rock and avoid it.

5. **ACTIONS IF FALLING**. If, while making a balance climb, the climber feels himself slipping and beginning to fall, he will take the following action:

a. Sound the command "falling".

b. Push himself away from the rock face.

c. Maintain proper body position as follows:

(1) Head up.

(2) Hands out toward the rock.

(3) Body relaxed.

(4) Feet kept below the body, slightly apart.

(5) Ensure you face the cliff face as you fall.

6. **TYPES OF HOLDS**. There are five basic holds that are used in balance climbing. They are as follows:

a. Push Holds.

(1) Most effective when hands are kept low.

(2) Often used in combination with a pull hold.

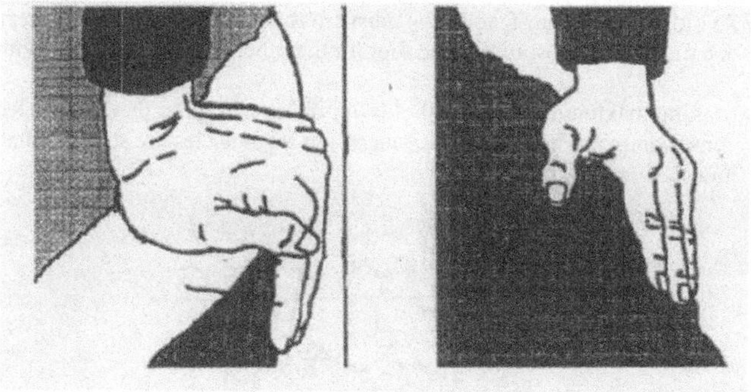

TYPES OF PUSH HOLDS

b. Pull Holds.

(1) The easiest hold to use and, consequently, often overused.

(2) Can be effective on small projections.

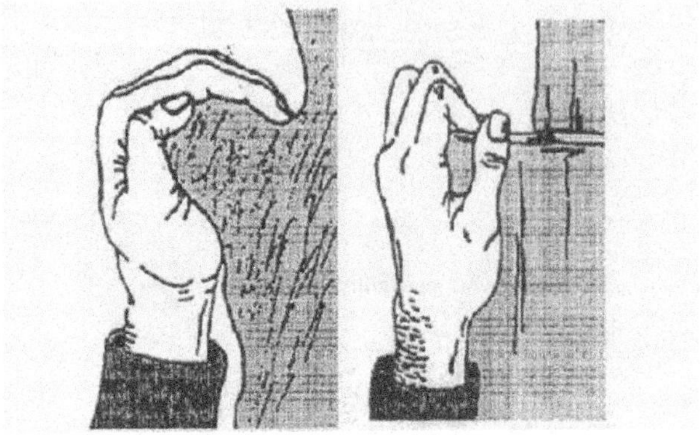

TYPES OF PULL HOLDS

c. Foot Holds

(1) Feet should be positioned with the inside of the foot to the rock.

(2) Use full sole contact as much as possible.

(3) Avoid crossing your feet. If you must cross your feet use a change step. A change step is a method of substituting one foot for the other foot on the same foothold.

(4) Making maximum use of footholds, climbing with your feet, is an effective means of conserving your body strength, since your leg muscles are stronger than your arm muscles.

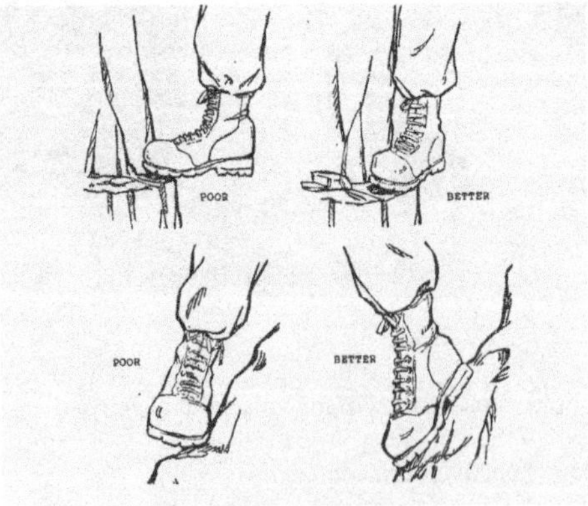

USE OF FOOT HOLDS

d. Friction Holds. A friction hold is anytime you are relying on the friction of your foot or hand against the face of the rock for traction, rather than pushing/pulling against a projection on the face of the rock.

(1) It is a type of hold that feels very insecure to an inexperienced climber.

(2) The effectiveness of this type of hold is dependent upon many things i.e., type, condition and angle of the rock face, type of boot soles, confidence, etc.

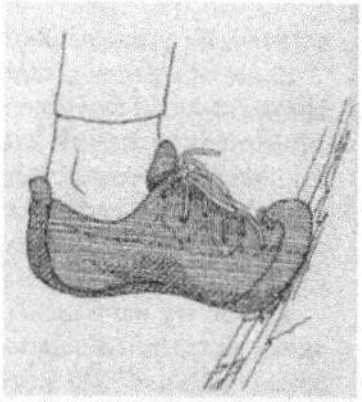

FRICTION USED IN SMEARING

e. Jam Holds. This type of hold involves jamming/wedging any part of your body or your entire body into a crack/opening in the rock.

(1) An important consideration is that you do not jam such that you cannot free that portion of your body after you complete the move. This sounds like a ridiculous statement; however, you must remember that after you complete your move, you may be withdrawing the portion of your body that you used from a different angle than you inserted it.

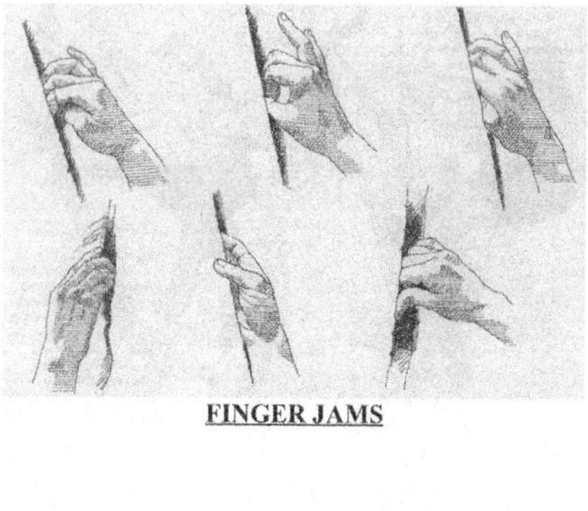

FINGER JAMS

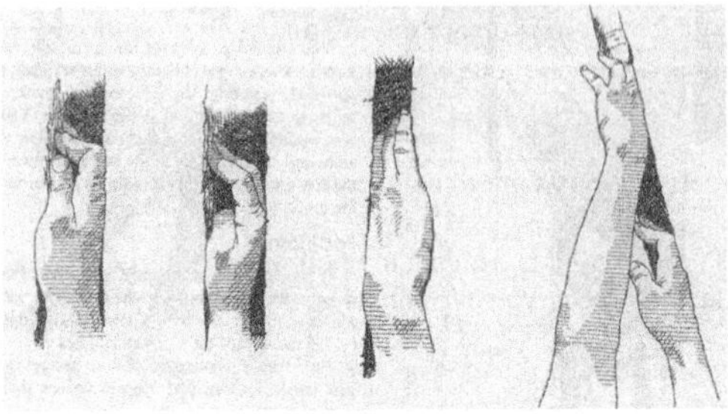

HAND JAMS

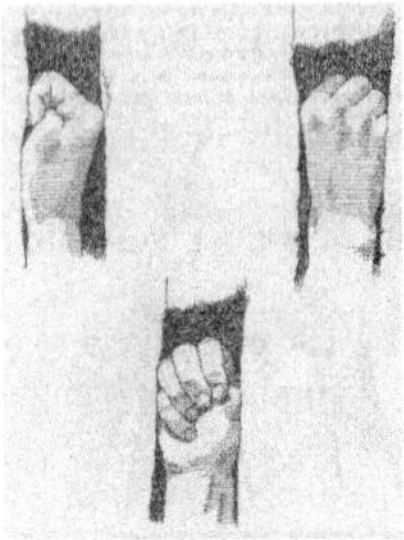

FIST JAMS

7.) **COMBINATION HOLDS.** The five types of holds just mentioned above are not just used individually. They are most often used in combinations with each other. Some examples are:

a. Chimney Climbing. This is when you insert your entire body into a crack in the rock and by using both sides of the opening, and possibly all five types of basic holds, move up' the crack.

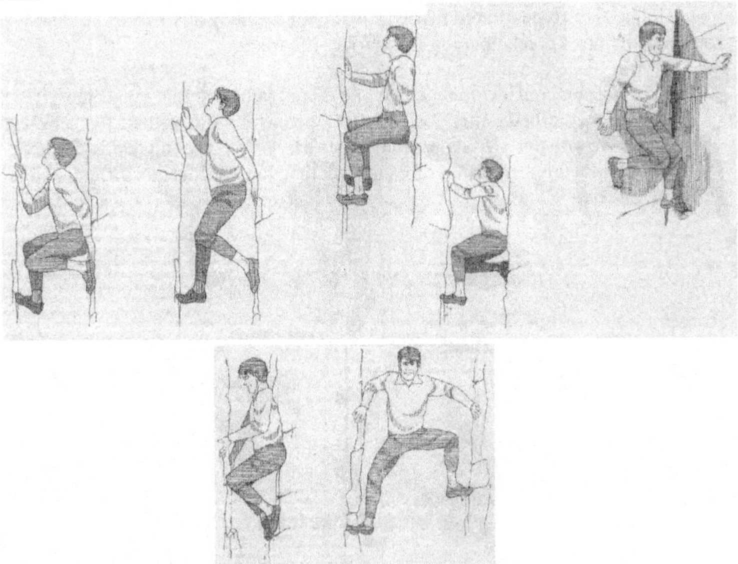

CHIMNEY CLIMBING TECHNIQUES

b. Lie-back. This is a combination of both pull holds with your hands and friction holds with your feet.

LIE BACK TECHNIQUE

c. Push-Pull. As the name implies, this is when you use a push hold and a pull hold together.

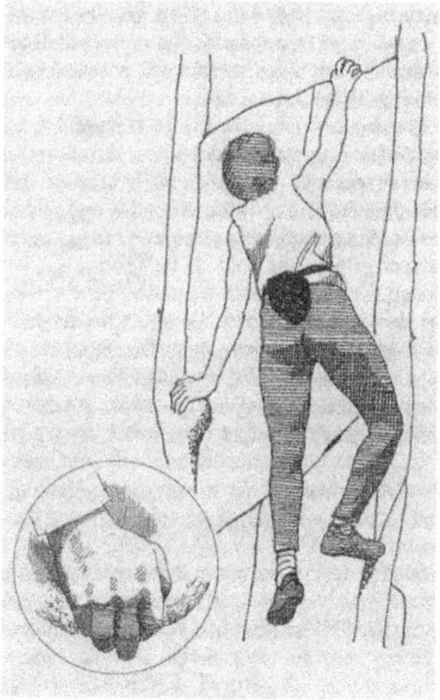

PUSH-PULL

d. Mantling. This is a technique where you continue to climb without moving your hands off a projection by pulling yourself up until your hands are at chest level and then invert your hands and push on the same projection.

MANTELING

MANTELING WITH FACE HOLD

e. Cross-Pressure in Cracks. This is a technique of putting both hands in the same crack and pulling your hands apart to hold/raise yourself.

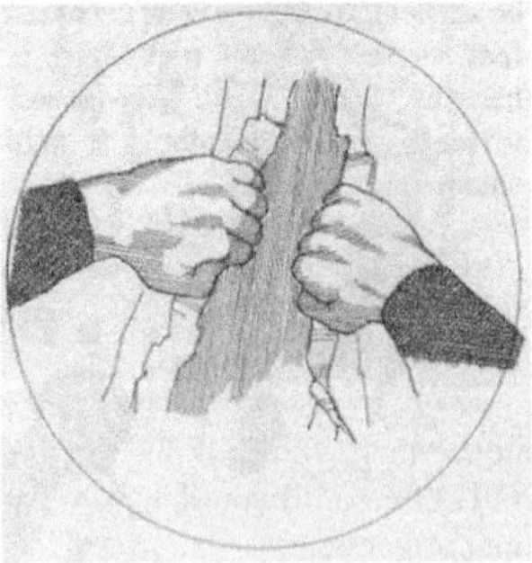

CROSS-PRESSURE METHOD

f. Inverted. Pull or push.

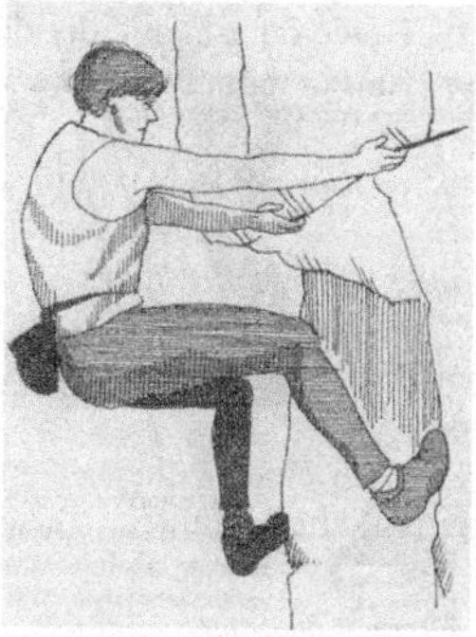

INVERTED TECHNIQUE USING COUNTERFORCE PRESSURE BETWEEN FEET AND HANDS

g. Pinch. As the name implies this is a grip used on tiny little nubbins.

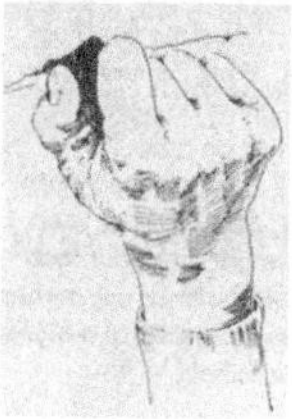

PINCH

h. Stemming. The spreading of arms or legs to maintain a proper body position. (i.e. usually used in a book or chimney.)

STEMMING

8. **GENERAL USE OF HOLDS**. How you use an individual hold is dependent on your experience level, or sometimes, your imagination. Here are some general guidelines.

a. Most handholds can be used as foot holds as you move up the rock.

b. Use all holds possible in order to conserve energy.

c. Even small projections may be used as holds.

d. Do not make use of your knees or elbows due to the reason that it is skin on bone and a slip could occur if pressure is exerted on them. Knees and elbows can be used with the extension of a limb jam.

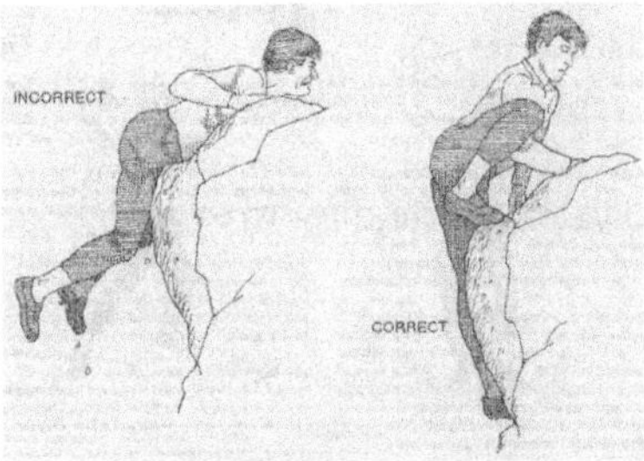

PROPER AND IMPROPER

9. **MOVEMENT ON SLAB**. Movement on slab is based on friction holds.

a. Use any and all irregularities in the slope to gain additional friction.

b. Traversing requires both hands and feet.

TRAVERSING

c. Descending steep slab may require turning inboard to face the slab and backing down.

d. The biggest mistake in slab climbing is leaning into the rock. Maintain maximum friction by keeping weight centered.

SLAB CLIMBING

10. BODY POSITION

a. The climber should climb with his body in balance by keeping his weight centered over and between his feet. Don't hug the rock. Don't over extend and become "spread-eagled". While climbing, keep in mind the acronym "CASHWORTH" for proper body position and movement. (WMC.C3F)

C - Conserve energy.

A - Always test holds.

S - Stand upright, on flexed joints.

H - Hands kept low; handholds should be waist to shoulder high.

W - Watch your feet.

O - On three points of contact.

R - Rhythmic movement.

T - Think ahead.

H - Heels kept low, lower than the toes.

UNITED STATES MARINE CORPS
Mountain Warfare Training Center
Bridgeport, California 93517-5001

WMC.C4
04/11/02

STUDENT OUTLINE

NATURAL AND ARTIFICIAL ANCHORS

TERMINAL LEARNING OBJECTIVE In a summer mountainous environment, establish an anchor system in accordance with the references. (WMC.C4)

ENABLING LEARNING OBJECTIVES.

(1) Without the aid of references and from a given list, choose the two types of anchors, in accordance with the references. (WMC.C4a)

(2) In a summer mountainous environment, establish a natural anchor point, in accordance with the references. (WMC.C4b)

OUTLINE

1. **TYPES OF ANCHORS.** (WMC.C4a)There are two types of anchors that we use. These two types are:

1) NATURAL
2) ARTIFICIAL

C4

2. **<u>GENERAL CONSIDERATIONS FOR ANCHORS</u>**. Anytime you employ natural or artificial anchors; there are special considerations that you must apply. These considerations apply both to the anchor itself and to the material being used to build the anchor. Some examples of these considerations are:

a. Whether using natural or artificial anchors, the installing unit must insure that the anchor is suitable for the load.

b. The anchor position must be relative to the direction of pull on the anchor.

c. The angle between the anchor points should not exceed 90 degrees. This is to ensure that no added stress is put upon the anchors, as well as the equipment being used to construct the anchor.

(1) To decrease the angle between anchor points, materials (i.e. sling ropes, web runners, Prusik cord, etc..) could be used to extend the anchor which will decrease the angle between anchor points.

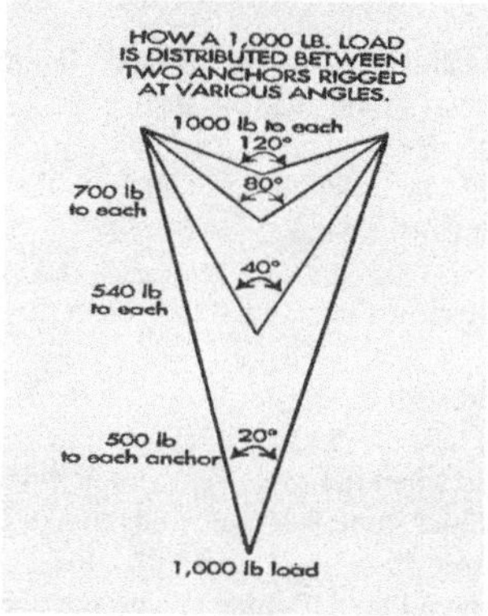

3. **<u>NATURAL ANCHORS.</u>** A natural feature is the preferred type of anchor point. Some examples and considerations are as followed:

a. <u>Types of natural anchors.</u>

(1) Trees.

(a) Select a tree that has not been chopped, burned or is rotten.

(b) The tree should be at least 6" in diameter and strong enough to support the intended load.

(c) Trees growing on rocky terrain should be treated with suspicion, since the roots normally are shallow and spread out along a relatively flat surface.

(2) Shrubs and Bushes.

(a) Select a shrub or bush that is alive and is not brittle, charred or loose.

(b) To avoid leverage, locate the central root and construct the anchor as near to the base as possible.

(3) Rocks and Boulders.

(a) Stability is of prime importance when considering a rock or boulder for an anchor. It must be strong enough and secure enough for the intended load.

(b) All surfaces of the rock or boulder should be inspected for any rough or sharp points. These areas must be padded to protect the rope from being abraded or cut.

(4) Spikes and Flakes.

(a) To check stability of a spike or flake, thump it with the heel of your hand. Anything that sounds hallow is suspicious.

(b) They should be checked for cracks or other signs of weathering that may impair their firmness.

(c) Sharp edges must be padded to protect the rope against cuts and abrasion.

(5) Threads and Chockstones.

(a) A thread is when the rock weathers or cracks to form a hole in the main wall. A chockstone is a rock wedged in a crack.

(b) Check a thread by thumping it with the heel of your hand. Make a visual inspection for cracks and weather. Common sense will prevail for choosing this type of anchor.

(c) When choosing a chockstone, ensure that it has substantial contact with the crack and that the stone's symmetry corresponds with the intended direction of pull.

b. <u>Types of Natural High Tension Anchors</u>. The following are types of Natural High Tension Anchor Systems used to construct installations. These can be tied on any suitable natural anchor point:

(1) Tree Wrap Anchor. The tree wrap is an anchor used to relieve tension on the actual knot itself. This system requires more rope when tied around a large anchor point.

(a) Tie a Figure 8 Loop on the standing end of the rope and wrap the 4-6 times around an anchor point.

(b) Attach a locking carabiner through the knot's loop and clip it onto the running end of the rope.

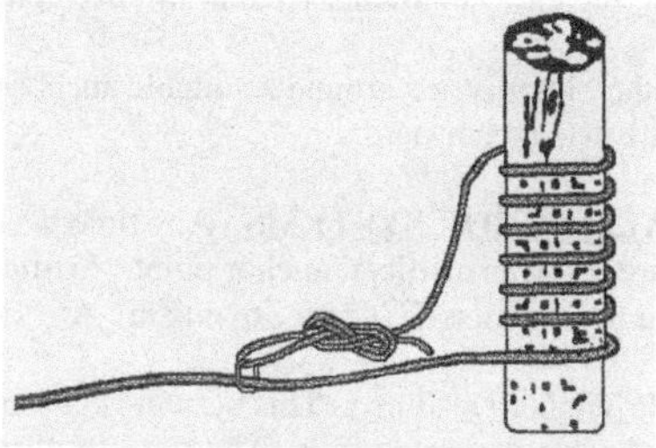

NOTE: If the anchor point to be wrapped is larger then 18' around, then three wraps will suffice.

(2) Swami Wrap Anchor. When tying the Swami Wrap, ensure that the joining knot is kept on the side of the anchor point. The Swami Wrap can be loosened under load if need be. This system can be substituted for a tree wrap when the installation's rope length is an issue.

(a) Select a suitable anchor point and tie a Swami Wrap around it.

(b) Clip a locking carabiner into as many wraps as possible. This will serve as the attachment point for the anchor system.

(3) Figure of 8 Anchor. The Figure of 8 Anchor is a quick and efficient system. This system cannot be loosened under load. It is tied in the following manner:

(a) Tie a Figure of 8 Loop on the standing end of the rope and wrap the knot around a suitable anchor point. Attach a locking carabiner through the knot's loop and clip it onto the running end of the rope.

(b) Before tensioning the anchor system, adjust the rope so the running end of the rope runs smoothly through the carabiner towards the direction of pull. This will prevent any lateral tension.

(b) If two ropes are used, the upper rope's carabiner has its gate upwards, and the lower rope's carabiner has it's gate downward.

c. Primary and Secondary Natural Anchor Point System. The two anchor points should be in line with the direction of pull. The primary anchor is the point nearest the running end while the secondary anchor point is directly behind the primary. It is constructed in the following manner:

(1) The primary anchor knot should be an around the object clove hitch. This is chosen for ease of untying the system after tension has been placed on the rope.

(2) The secondary knot is tied around a suitable anchor point ensuring that the rope is taut between the two anchor points.

4. **ARTIFICIAL ANCHOR SYSTEMS**. Any time we use anything other then a natural feature, we are using an artificial anchor point. Artificial anchors can be constructed in the ground, or on the rock itself. The following are artificial anchor systems:

a. Single Timber Deadman. This system is constructed in the ground and it requires considerable time and effort. The steps of it's construction are as follows:

(1) Dig a trench 6 feet long and 3 feet deep and wide enough to work in at a 90 degree angle to the direction of the pull.

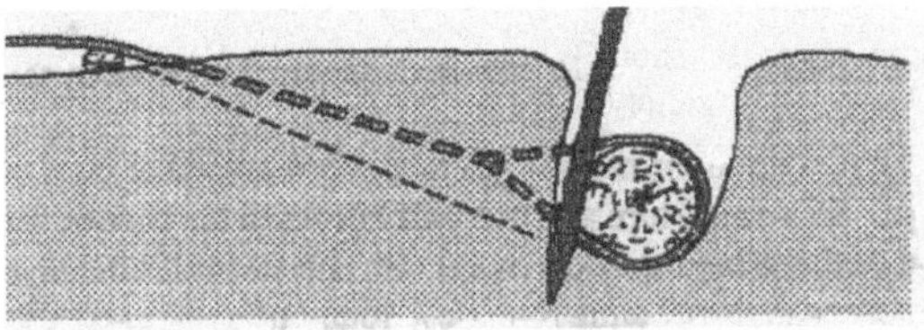

(2) Dig another trench about 12 inches wide. This trench is dug so that it intersects the main trench at a right angle in the middle. The bottom of this trench should be parallel to the direction of pull and should join the bottom of the main trench.

(3) Take an anchoring device (i.e. log, engineer stakes, bundled up branches etc.) that is strong enough to support the intended load. The anchor is placed into the main trench and covered with dirt with the exception of that part of the anchor that joins the second trench. Stakes approximately 3 feet long should be driven approximately 1-1/2 feet into the ground between the dead-man and the slanted side of the trench to assist in holding the dead-man in place if the soil is soft.

b. Picket Hold Fast. The picket hold fast is an easier anchor to construct than the Deadman, and can be used almost anywhere. The strength of the picket system depends on the pickets and the soil or snow conditions. The picket system can be used for both high and non-tension systems. Construction is as follows:

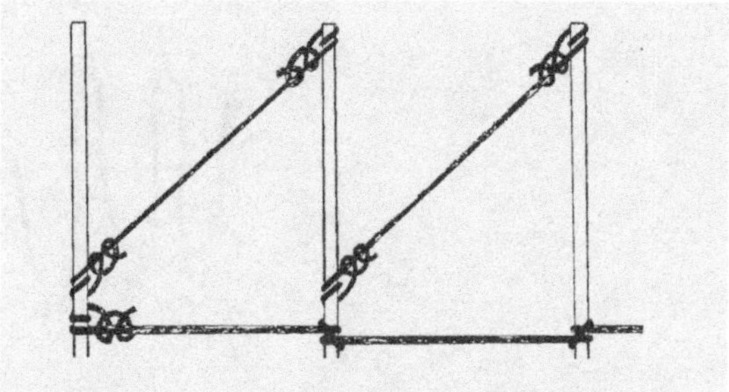

(1) Three stakes (i.e. logs, engineer stakes, etc.) are driven into the ground at a 30 degree angle against the direction of pull.

(2) The line of pickets should be driven in line with the direction of pull. The distance between the pickets can be anywhere from 3-12 feet apart depending on the terrain and soil conditions.

(3) Before tying any rope to the picket anchors, the base of the pickets must be buffed or padded.

(4) To tie a rope to the pickets, go to the furthest picket away from the cliff edge. Tie a round turn and two half hitches at the base of the picket. Take the running end of the rope and tie an over the object clove hitch to the base of the middle picket. Then with the running end of the rope tie an over the object clove hitch to the picket closest to the edge of the cliff face. Ensure that there is tension on the rope in between each picket.

(5) To tie off the pickets to themselves, go to the furthest picket from the cliff edge. And with a sling rope, tie a round turn and two half hitches at the base of that picket. With the running end of that sling rope tie a round turn and two half hitches at the top of the middle picket. Using a second sling rope tie a round turn and two half hitches to the base of the middle picket. Then with the running end of that sling rope tie a round turn and two half hitches to the top of the picket closest to the cliff edge.

c. Equalized Anchor. This system is built with a minimum of three pieces of protection. It can be tied with the standing end of a rope or by utilizing a practice coil.

(1) Construction with the Standing End of the Rope:

(a) Tie a Figure of 8 Loop in the standing end of the rope.

(b) Place a carabiner in each artificial anchor point and attach the knot in either of the outside carabiners.

(c) Clip the rope into the remaining carabiners.

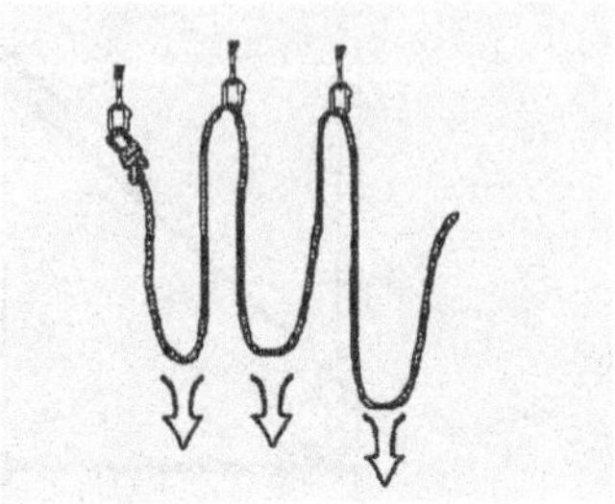

(d) A bight of rope is then pulled down after each carabiner into the anticipated direction of pull. With all three bights, tie an overhand knot around itself to include the running end of the rope.

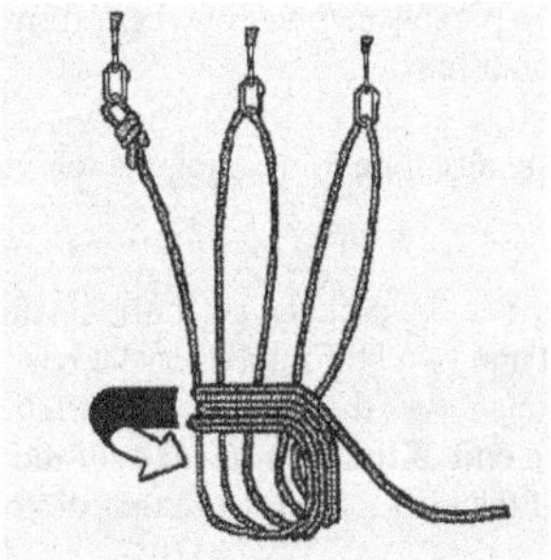

(e) With the running end of the rope coming from the overhand knot, tie a Figure of 8 Loop and attach it to the overhand knot's loop with a locking carabiner.

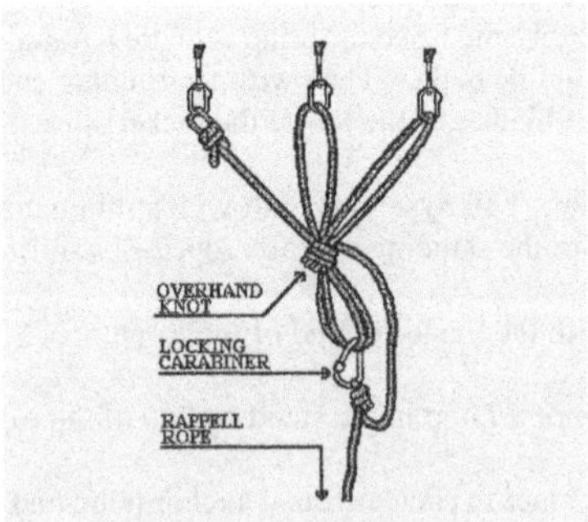

(f) Ensure that there is sufficient slack in the dead rope to prevent disfiguring the knot thus ruining the structural integrity of the knot.

(2) Construction with the Cordalette Method:

(a) Create an endless loop with a practice coil and clip it into each of the artificial anchor points.

(b) A bight of rope is then pulled down between each carabiner into the intended direction of pull.

(c) With all three bights, tie an overhand knot on itself.

(d) Place a locking carabiner into the overhand knot's loop. This will serve as the attachment point for the anchor system.

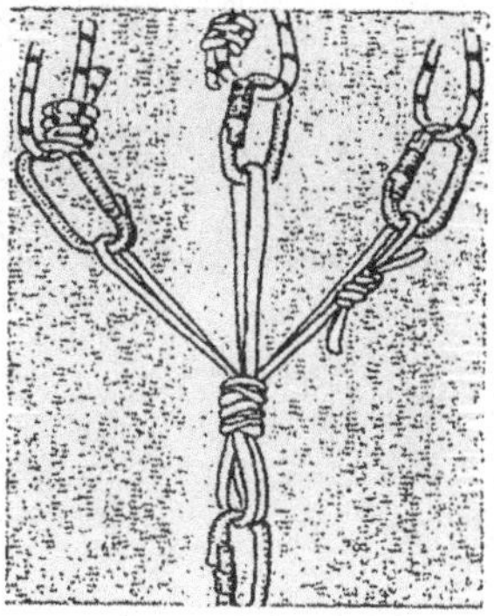

NOTE: If constructing a two-rope high-tension installation, a minimum of three artificial anchors per rope will be used.

UNITED STATES MARINE CORPS
Mountain Warfare Training Center
Bridgeport, California 93517-5001

SML
ACC
02/11/02

LESSON PLAN

SIT HARNESS

LEARNING OBJECTIVES

TERMINAL LEARNING OBJECTIVE: Without the aid of references, properly wear and maintain the MWTC sit harness, in accordance with the references.

ENABLING LEARNING OBJECTIVES

(1) Without the aid of references and given a diagram, name in writing the parts of a sit harness, in accordance with the references.

(2) Without the aid of references, wear a sit harness, in accordance with the references.

BODY

1. (5 Min) **NOMENCLATURE**

   a. Waist belt.

   b. Leg loops (adjustable).

   c. Buttocks straps (adjustable).

   d. Fastex buckle - 2.

   e. Doughnut.

   f. Equipment loops.

   g. D-ring (older models).

   h. Waist belt tie-in point.

   i. Crotch strap.

WEARING OF THE SIT

C5

HARNESS

a. First, disconnect the fastex buckle at the rear of the harness.

b. Hold the harness in front of you, put your feet through the leg loops ensuring that the buckles on the leg loops are outboard, on your thighs.

c. Fasten the waist belt into the buckle, ensuring that it is a tight but comfortable fit. **You must ensure that the waist belt is threaded back through the buckle, this action locks the waist belt to the buckle. Failure to do this will cause the waist belt to slip through the buckle when it is under load.**

d. Adjust the leg loops so that they are high on your thigh, once you have them adjusted, get your buddy to clip the fastex buckle male to the most comfortable female fastex. Then adjust the buttock straps so as the leg loops are held up.

e. If all the above are done correctly, the harness should now be a comfortable but snug fit, after you have fitted it for the first time there is no need to go through the same procedure each time you put it on. Simply hold the harness in front of you, step into it and attach the waist belt to the buckle in the approved manner.

2. **TYING IN THE ROPE (END OF ROPE)**

a. Pass the end of the climbing rope up through the crotch strap, then through the doughnut and through the waist belt tie-in point.

b. Tie the rope using a retraceable figure eight, adjusting the knot to get it as close as possible to the body.

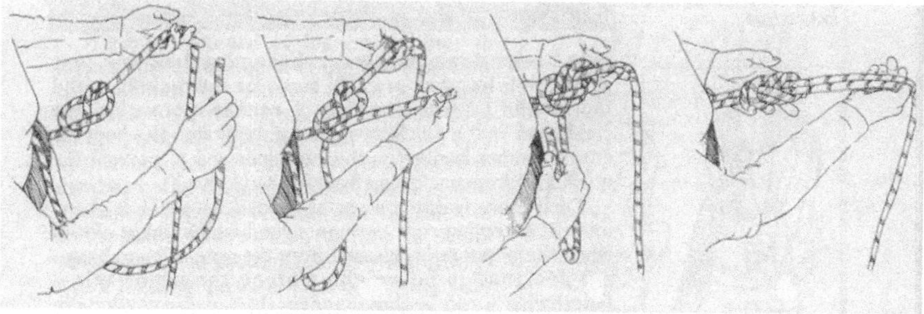

RETRACE FIGURE 8 IN SIT HARNESS

3. **TYING IN THE ROPE (MIDDLE OF THE ROPE)**

a. Take up a bight of rope and tie a figure eight loop, then take a steel locking carabiner and attach the carabiner to the harness by securing the crotch strap, doughnut, and the waist belt tie-in loop. Now attach the figure eight loop to the carabiner.

4. (5 Min) **CARE AND MAINTENANCE**

a. Avoid contact with chemicals, as this will damage the nylon.

b. Regularly inspect for signs of abrasions and normal wear. Pay particular attention to wear points such as the tie-in loops, buckles, and sewn joints.

c. Keep away from heat such as open flames, cigarettes, etc.

d. If soiled by grit and seawater, wash in lukewarm water with pure soap and allow to dry in a warm room away from direct heat.

e. Two to three years of life can be expected during normal climbing use.

f. It is recommended that a harness that has experienced a serious fall should be discarded.

g. Under no circumstances will you ever tie into the equipment rack of any sit harness as a belay, anchor point, etc.

UNITED STATES MARINE CORPS
Mountain Warfare Training Center
Bridgeport, California 93517-5001

FMST.07.33
05/02/01

STUDENT OUTLINE

ESTABLISHMENT OF RAPPEL POINTS AND RAPPELLING

TERMINAL LEARNING OBJECTIVE Given a mountainous environment, mountaineering equipment, helmet, sit harness, rappel gloves, 165 ft static rope, rescue 8, prusic cord, and one or more assistants, conduct rappelling operations to support mission requirements per the reference. (FMST.07.33)

ENABLING LEARNING OBJECTIVES.

(1) Without the aid of references, select from a given list the three criteria involved in site selection for a rappel site, in accordance with the references. (FMST.07.33a)

(2) In a summer mountainous environment, execute the commands used between a rappeller and his belay man when conducting a rappel, in accordance with the references. (FMST.07.33b)

(3) Without the aid of references, select from a given list three of the nine duties of the rappel point NCOIC, in accordance with the references. (FMST.07.33c)

(4) Without the aid of references, select from a given list three of the six duties of the first man down a rappel rope, in accordance with the references. (FMST.07.33d)

(5) Without the aid of references, select from a given list the three types of rappels, in accordance with the references. (FMST.07.33e)

(6) In a summer mountainous environment, rappel down a cliff face using a seat-hip rappel, in accordance with the references. (FMST.07.33g)

C6

OUTLINE

1. **<u>INHERENT DANGER OF RAPPELLING</u>**. Rappelling is inherently dangerous because rappellers rely totally on the equipment.

   a. To ensure a safe training evolution, two ropes will be utilized at MWTC.

   b. All the rappels taught at MWTC can be completed with a single rope if the situation arises.

   c. If utilizing a one-rope system with the carabiner wrap, the rope will be attached to the carabiner with two wraps vice one.

2. **<u>SITE SELECTION</u>**. (FMST.07.33a) when selecting a rappel site consider these three factors.

   a. <u>There must be a good anchor</u>. As previously taught in *NATURAL* and *ARTIFICIAL ANCHORS*, natural anchors are preferred.

   b. The rappel route down should be as free of obstacles (i.e., vegetation, debris) as possible.

   c. There must be suitable loading and unloading platforms.

NOTE*:* The evaluation of a site for the above factors should be made by the rappel point NCOIC, who should be the most experienced rappeller in the unit.

3. **<u>DUTIES OF THE RAPPEL POINT NCOIC</u>**. (FMST.07.33c)Once a rappelling site has been selected, one person will be appointed to each rappel lane as a rappel point NCOIC. These individuals should have experience as rappellers. The rappel point NCOIC has ten duties and responsibilities, which are:

   a. Ensures that the anchor points are sound and that the knots are properly tied.

   b. Ensures that loose rock and debris is cleared from the loading platform.

   c. Allows only one man on the loading platform at a time and ensures that the rappel point is run in an orderly manner.

   d. Ensures that each man is properly prepared for the particular rappel; i.e. gloves on, sleeves down, helmet secured, rappel seat tied correctly and secured properly.

   e. Attaches the rappeller to the rope and ensures the rappeller knows the proper braking position for that particular rappel.

   f. Ensures that the proper commands or signals are used.

   g. Dispatches each man down the rope.

h. The rappel point NCOIC will be the last man down the rope.

i. The rappel point NCOIC will ensure that the ropes are inspected after every 50 rappels.

j. The rappel point NCOIC will maintain a rope log.

4. **TYPES/USE**. (FMST.07.33e) The three types of rappels and when they are preferred to be used are:

a. Hasty Rappel: It is used when carrying loads down moderate slopes.

b. Seat-Shoulder Rappel: It is used for heavily laden troops over vertical to near vertical cliff faces.

c. Seat-Hip Rappel: It is used when carrying loads over vertical to near vertical faces.

5. **HASTY RAPPEL**. The hasty rappel is the easiest type of rappel to prepare for. It requires no equipment other than a rope and gloves.

a. Conduct. A hasty rappel is conducted in the following manner:

(1) Sleeves will be rolled down and gloves will be put on.

(2) Face slightly sideways.

(3) Place the rappel rope across your back, grasping it with both hands, palms forward, and arms extended.

(4) The hand nearest the anchor is the guide hand. The hand farthest from the anchor is the brake hand.

(5) Lean out at a moderate angle to the slope.

(6) Descend down the hill facing half-sideways, taking small steps and continually looking downhill while leading with the brake hand.

(7) Feet should not cross and the downhill foot should lead at all times.

b. Braking. The steps for braking during a hasty rappel are as follows:

(1) Bring the lower (brake) hand across the front of the chest to brake.

(2) At the same time, turn to face up toward the anchor point.

HASTY RAPPEL

6. **SEAT-SHOULDER RAPPEL**. The seat-shoulder rappel relies on friction as the main effort of controlling the descent. It is very efficient for men with heavy packs because it provides support for heavy loads on the back.

a. Conduct. A seat-shoulder rappel is conducted in the following manner:

(1) Put on your rappel-seat, roll down your sleeves and put on your gloves.

NOTE: To avoid causing a possible injury, it is advised that the rappel seat be constructed at the rear of the rappel site loading platform. If the rappel seat is worn for too long, the rappel seat could loosen up enough to cause you to slip out of your seat while rappelling.

(2) The steel locking carabiner is placed on the rappel-seat so that the gate opens down and away, to prevent the gate from opening once the wraps are placed into the carabiner.

(3) Step-up to the rope with you're left shoulder facing the anchor.

(4) The rappel rope is attached to the rappeller's hard point carabiner as follows:

(a) Snap the rope into the locking carabiner.

(b) Taking slack from the standing (anchor) end of the rope, make one wrap with the rope around the body of the carabiner and back through the gate.

(c) Ensure that the locking nut of the carabiner is fastened to lock the carabiner closed.

NOTE: If you are using only one rope to rappel with due to the tactical situation or equipment availability, the procedures are the same, EXCEPT the individual will make two wraps around the body of the carabiner instead of one.

(5) Take the rope across your chest, over your left shoulder, diagonally across your pack and down to the right (brake) hand.

(6) Descend by walking down the cliff using the braking procedure to control the rate of descent. Look under your brake arm for possible obstacles to avoid.

b. Braking. The steps for braking during a seat-shoulder rappel are as follows:

(1) Lean back.

(2) Face directly uphill while bringing the brake hand across the chest.

7. **SEAT-HIP RAPPEL**. The seat-hip rappel is the most commonly used rappel.

a. Conduct. A seat-hip rappel is conducted in the following manner:

(1) Construct the rappel seat; roll down sleeves and put gloves on.

(2) The steel locking carabiner is placed on the rappel-seat so that the gate opens up and away.

(3) Step up to the rope with your left shoulder facing the anchor.

(4) The rappel rope is snapped into the carabiner as follows:

(a) Snap the rope into the locking carabiner.

(b) Taking slack from the standing (anchor) end of the rope, make one wrap with the rope around the body of the carabiner and through the gate again.

(c) Ensure that the locking nut of the carabiner is fastened to lock the carabiner closed.

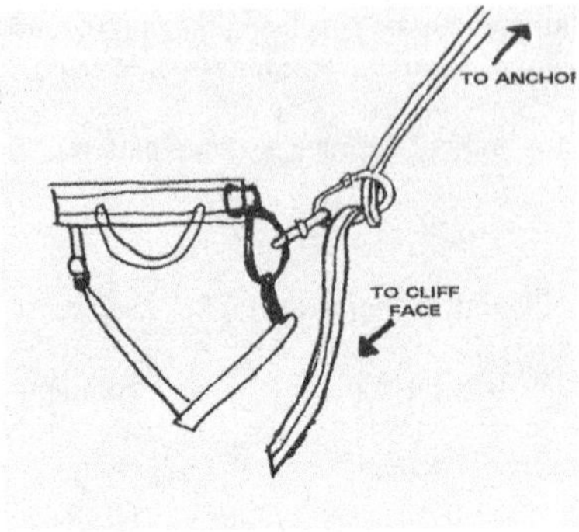

SEAT-HIP RAPPEL

NOTE: If you are using only one rope to rappel with due to the tactical situation or equipment availability, the procedures are the same, EXCEPT the individual will make two wraps around the body of the carabiner instead of one.

(5) The rappeller will grasp the running end of the rope with the brake (right) hand, palm down and turned slightly inboard, near the hip.

b. Braking. The steps in braking for a seat-hip rappel are as follows:

(1) Grasp the rope tightly with the brake hand.

(2) Take the brake hand and place it in the small of the back. This will create enough friction to stop all momentum.

NOTE: At no time will you bound or jump while you are descending. You "walk down" the cliff face using the proper braking procedure to control your rate of descent.

8. **SAFETY OF THE FIRST MAN DOWN**. To ensure the safety of the first man down, the rappeller will:

a. Before deploying the ropes, tie the two ends of the ropes together with an overhand knot.

b. Tie a friction knot on the standing end of the rappel rope with a Prusik cord.

c. Attach the rappel ropes to the harness's hard point with a locking carabiner.

d. Attach the Prusik cord to the harness's hard point with a locking carabiner. This cord is also referred to as a safety Prusik.

NOTE: If the Safety Prusik is too short, it can be extended with a web runner.

9. **DUTIES OF THE FIRST MAN DOWN**. (FMST.07.33d) The first man down the rope has specific duties. They are as follows:

a. Selects a smooth route down for the ropes.

b. Clears the route of loose rocks and debris.

c. The first man down will untie the overhand knot and straighten the ropes out once he reaches the bottom.

d. The first man down belays the next man down the rope.

(1) To belay correctly, the belay man will stand facing the cliff face, arms up with palms facing upward and over lapping. The rope will pass through the opening of the over lapping hand's thumb and index fingers.

(2) To stop a fallen rappeller, the belay man will grab the rope with closed fists and pull straight down bringing the forearms parallel to the deck.

e. Take charge of personnel as they arrive at the bottom to include appointing a belay man.

10. **RAPPELLING COMMANDS.** (FMST.07.33b) In order to conduct rappelling operations safely, it is essential that everyone understands the sequence of events. The following voice commands or rope tugs will be utilized:

VOICE COMMANDS	GIVEN BY	MEANING
"Lane # on Rappel"	Rappeller	I am ready to begin rappelling
"Lane # on Belay"	Belayer	I am on belay and you may begin to rappel
"Lane # off Rappel"	Rappeller	I have completed the rappel and am off the rope
"Lane # off Belay"	Belayer	I have completed the belaying of the rappeller

ROPE TUGS	GIVEN BY	MEANING
Three Tugs	Rappel Point NCOIC	The rappeller is ready to begin rappelling
Three Tugs	Belayer	I am on belay and the rappeller may begin to rappel
Three Tugs	Belayer	The rappeller is off the rappel rope

11. **TYING-OFF**. Occasionally, it may be necessary to stop during a rappel before reaching the bottom of a cliff. The following sequence is used:

a. The commands used in tying-off are as listed below:

COMMAND	GIVEN BY	MEANING
"Lane # Tying-off"	Rappeller	I am ready to tie-off, give me some slack.
"Lane # Tying-off"	Belayer	I have given you enough slack and you may tie-off.
"Lane # On rappel"	Rappeller	I have completed tying-off and I am ready to resume rappelling.
"Lane # On rappel"	Belayer	I am on belay, you may rappel.

b. The procedure used when tying-off is as follows:

(1) The rappeller gives the command, "Lane #, tying-off."

(2) The Belayer gives the rappeller slack and gives the command "Lane # Tying-off". He continues to hold the rope in the belay position, remaining alert and watching the rappeller.

(3) The rappeller quickly brings his brake hand to the twelve o'clock position, so that the running end and standing end are parallel.

(4) The rappeller grasps all ropes with his guide hand as close to the carabiner as possible.

(5) The rappeller steps over the rope so that the running end is going between his legs.

(6) The rappeller releases the rope with his brake hand then reaches under the running end, over the standing end. He then takes up a bight from the running end about two feet long and pulls it over the standing end and under the running end, forming a half hitch.

(7) He pulls the half hitch tightly against the guide hand.

(8) He works the half hitch down snugly against the carabiner while maintaining contact with the guide hand as long as possible.

NOTE*:* A loose half hitch could bind into the carabiner causing difficulties in clearing the knot out. Make sure that the first half hitch is dressed down tightly before moving it against the carabiner.

(9) Place another half hitch above the one already tied.

(10) To untie, reverse the steps. Remove the safety half hitch, then shrink the first half hitch down to a small loop by grabbing the running end of both ropes and pulling them straight out to the left. Place your right hand in the middle of the two bights until they are snug on all four fingers. Then remove one finger and make the bights snug on three fingers. Repeat this process until it's down to one finger and both the bights are equal.

(11) Now, grasp the running end with both hands and smartly jerk the running end of the ropes upward to pop the small loop out from the first half hitch. From this position, keep the guide hand around both of the ropes next to the carabiner; step back over the rope so that the running end is to your right side. Grab just the running end with your brake hand and quickly set the brake behind the small of your back then, readjust you guide hand onto the standing end. You now should be in the seat-hip rappel position.

NOTE: The wraps may bind up some after untying, where no further movement down is possible. To alleviate this, keep the brake on and force your body weight down to pop the wraps in the carabiner back to their intended position and then continue on with the rappel.

(12) The rappeller gives the command "Lane #, on rappel."

(13) The Belayer takes up the slack and gives the command "Lane #, on belay."

(14) The rappeller resumes to rappel down.

12. **RESCUE TECHNIQUES**. When conducting rappelling operations, the possibility of a rappeller getting caught on the rope due to either his clothing or equipment can occur. There are two types of rescues that can be performed to free the rappeller from this situation.

a. The Self-Rescue Technique.

(1) After realizing that you are caught up on the rope, check with your belay man to ensure that he has a solid brake set.

(2) Using a safety Prusik, tie a friction knot onto the rappel rope approximately an arms length above the area that is fouled. Anchor the other end of your safety Prusik to your rappel seats hard point. This is the same technique used as discussed earlier with the first man down the rope.

(3) Work the friction knot up until there is no tension on the rappel device. This will give you the necessary slack to free the malfunction.

(4) Once you have corrected the problem, continue to rappel down using the first man down method.

b. The Buddy Rescue Technique. Himself uses this method when the rappeller is unable to correct the problem.

(1) A rescuer will rappel down on another rope to the disabled rappeller and tie-off.

(2) The rescuer will then establish a safety Prusik onto the victim as taught in the self-rescue technique.

(3) After clearing the malfunction, continue with rappelling operations.

c. Rappelling Casualty Rescue. There are three different methods in which to get an injured rappeller to the bottom.

(1) The first method is to allow the belay man to lower the casualty by slowly releasing the tension on the rappel rope.

(2) The second method is used for critical injuries or when the belay man cannot properly control the casualty's descent.

(a) The rescuer will rappel down using another rope. Once the rescuer gets to the casualty, the rescuer will have his belay man brake him off, enabling the rescuer to use both hands.

(b) If necessary, the rescuer will perform the basic first aid needed. Once the casualty is ready to be lowered, the rescuer will call down to both of the belay men to simultaneously lower both men. The rescuer will hold onto the casualty the entire way down so that he doesn't bounce off the rock face.

(3) The third method is called a tandem rappel. Its used to rescue a casualty who has sustained an injury serious enough that he cannot operate the rappel device himself and requires the assistance of a second rappeller. This type of rescue can be conducted from the top of the cliff face or while on rappel, depending on where the injury occurred.

(a) If the injury occurs while on top of the cliff face, the following steps should be taken:

1. Take either a long sling rope or Prusik cord and tie a figure-eight knot offset so that one length of the cord is longer then the other.

2. Tie a figure-eight knot at each end.

3. Take a rappel device and attach it to the offset figure-eight loop with a carabiner and attach it to the rappel rope.

4. Next, attach the casualty's hard point to the short end of the sling rope/Prusik cord and the rescuer to the other end.

5. The rescuer will then utilize a safety Prusik in the same method as the first man down technique, except that the friction knot will be tied below the rappelling device and controlled with the brake hand.

6. The rescuer will then maneuver himself under the casualty to provide assistance and support.

7. At this point you are ready to rappel both rescuer and casualty at the same time.

NOTE*:* Your rappel device should be far enough in front of you so that the casualty will not be able to reach it.

(b) If the injury occurs during the rappel, the following steps are taken:

1. The rescuer will preset the same system on a separate rappel rope and rappel down to the casualty.

2. The rescuer will then take the short end of the sling rope and attach it to the casualty's hard point.

3. With a knife, the rescuer will cut the casualty's rappel rope away.

4. Rappel down to the bottom of the cliff face.

NOTE*:* THIS IS THE ONLY TIME THAT A KNIFE WILL BE USED DURING A RAPPEL. EXTREME CAUTION SHOULD BE USED. IT IS NOT TO BE DONE IN TRAINING, BUT ONLY DURING AN ACTUAL RESCUE AND AS A LAST RESORT.

13. **RETRIEVABLE RAPPELS**. Once a unit has rappelled down a vertical obstacle, it may be necessary to retrieve the rope(s). Depending on the height of the obstacle either one or two ropes will used to construct the rappel lane.

a. <u>One-Rope Retrievable Rappel</u>:

(1) Find the middle of the rope and place it directly behind a suitable anchor point.

(2) Join the pigtails of the rope with an overhand knot and deploy the rope down the obstacle.

(3) On one side of the rope in front of the anchor point, tie an over-the-object clove hitch onto a locking carabiner.

(4) On the other side of the rope, tie another over-the-object clove hitch and attach it to the same carabiner.

(5) The first man down will utilize a safety Prusik and untie the overhand knot after he reaches the bottom.

(6) All others conduct proper rappel procedures.

(7) The last man down will disconnect the carabiner from the rope, and ensuring that the middle of the rope is directly behind the anchor point, rappel down the rope.

(8) Pulling on either end of the rope then retrieves the rope.

b. Two-Rope Retrievable Rappel. This type of system is identical to the one rope retrievable rappel except for a few considerations.

(1) The reason for using two ropes vice one rope is that the height of the rappel is greater.

(2) When two ropes are used they should be joined together using a square knot. This knot will be placed directly behind the anchor point.

(3) All else remains the same as the one rope retrievable rappel except:

(a) The last man down will disconnect the carabiner from the rope and will move the joining knot as close to the vertical obstacle's edge as possible. This will prevent the knot from possibly getting caught up while retrieving.

(b) The last man down will then place a carabiner on the rope below the knot. This will enable him to know which line to pull for retrieval.

14. **EQUIPMENT**. Equipment should be worn in accordance with unit SOP. The unit has the responsibility to determine which methods it feels are most beneficial to the mission. Weapons should be worn across shoulder; muzzle down and away from the brake hand with a tight sling securely attached to the weapon. Weapons should also be dummy corded to the individuals.

UNITED STATES MARINE CORPS
Mountain Warfare Training Center
Bridgeport California, 93517-5001

WMC
04/11/02

STUDENT OUTLINE

TOP ROPING

LEARNING OBJECTIVES.

TERMINAL LEARNING OBJECTIVE. In a summer mountainous environment, and given a rock face and a belay man, conduct top roping with cartridge belt and rifle, in accordance with the references. (WMC)

ENABLING LEARNING OBJECTIVE

(1) In a summer mountainous environment, establish a belay point from the top of a cliff for use in a top rope climb, in accordance with the references. (WMC.a)

(2) In a summer mountainous environment, demonstrate the use of climbing commands while conducting a top rope, in accordance with the references. (WMC.b)

OUTLINE

1. **ESTABLISHING A BELAY STANCE FROM THE TOP**.

a. The Belayer will establish a sitting belay stance on the cliff head by:

(1) Constructing a suitable anchor with the standing end of the rope.

(2) With the direction of pull away from the anchor, tie a directional figure 8 loop near enough to the cliff's edge so that the climber can be observed, if possible.

(3) Tying a swami Wrap around oneself, clip a locking carabiner through all the rear wraps and clip a large locking carabiner through all the front wraps.

(4) Secure the rear locking carabiner into the directional figure of 8 loop.

C7

(5) Secure the running end of the rope to the large locking carabiner utilizing a suitable friction belay (i.e., Munter hitch, stitch plate, etc.).

NOTE: Gloves **will not** be worn while belaying a climber during a top rope.

2. **SECURING THE CLIMBER TO THE ROPE**.

a. The climber will tie into the end of the top rope by:

(1) Constructing a Swami wrap around oneself.

(2) Tying a retrace figure 8 loop through all of the Swami's wraps or;

(3) A figure of 8 loop clipped into a locking carabiner secured through all the Swami's wraps.

b. Alternative methods.

(1) A bowline on a coil as discussed in ROPE MANAGEMENT or;

(2) Tie into the hard point of the harness as discussed in SIT HARNESS.

3. **ESTABLISHING A BELAY STANCE FROM THE BOTTOM**.

a. Belaying from the bottom is commonly referred to as a "yo-yo" or "Sling Shot" belay. This system can be constructed for either a direct or indirect belay.

(1) Construct a suitable top anchor utilizing either a pulley or one steel locking or two non-locking carabiners as the attachment point for the anchor.

(2) The rope will travel up from the bottom and pass through the attachment point then return to the bottom.

(3) The climber will be secured to one end of the rope and the Belayer from the other end.

4. **CLIMBER'S RESPONSIBILITIES**.

a. The climber ensures that the Belayer is anchored and on belay, by use of commands or prearranged signals prior to beginning the climb.

b. The climber will not out climb the belayer; this will cause slack in the rope between the belayer and the climber.

c. Avoid placing excess pressure/weight on belay man.

d. Weapons will be worn across shoulder, muzzle down and to the left, with a tight sling securely attached to the weapon.

5. **CLIMBING COMMANDS AND SIGNALS.** In order to conduct top rope operations safely, it is essential that everyone understands the sequence of events. The following voice commands or rope tugs will be utilized:

VOICE COMMANDS	GIVEN BY	MEANING
"Lane #, Up Rope"	Climber	Belayer needs to take in the slack.
"That's Me"	Climber	Excess slack has been taken up between us.
"Lane #, On Belay"	Belayer	I am on belay.
"Lane #, On Climb"	Climber	I am ready to climb.
"Climb Climber"	Belayer	You may begin to climb.

ROPE TUGS	GIVEN BY	MEANING
Three Tugs	Climber	Belayer needs to take in the slack.
Three Tugs	Belayer	I have taken up the slack and you may begin to climb.
Three Tugs	Climber	Climbing.

UNITED STATES MARINE CORPS
Mountain Warfare Training Center
Bridgeport, California 93517-5001

FMST.07.40
03/02/02

STUDENT OUTLINE

MOUNTAIN CASUALTY EVACUATIONS

TERMINAL LEARNING OBJECTIVES. In a summer mountainous environment and given a simulated/actual casualty, evacuate casualties, in accordance with the references. (FMST.07.40)

ENABLING LEARNING OBJECTIVES

(1) With the aid of references, state orally the general considerations for medevac procedures, in accordance with the references. (FMST.07.40a)

(2) In a summer mountainous environment, and with the aid of references, construct an expedient litter, in accordance with the references. (FMST.07.40b)

(3) In a summer mountainous environment, and with the aid of references, secure a simulated/actual casualty in a SKEDCO litter, in accordance with the references. (FMST.07.40c)

(4) In a summer mountainous environment, and with the aid of references, attach a belay line to a SKEDCO litter for movement up/down moderate slopes, in accordance with the references. (FMST.07.40d)

(5) In a summer mountainous environment, with the aid of references, and given a simulated/actual casualty, conduct a medevac using the barrow-boy method, in accordance with the references. (FMST.07.40e)

C8

OUTLINE

1. **GENERAL CONSIDERATIONS** (FMST.07.40a) the general considerations are a set of guidelines that can be used no matter how serious the casualty is. They are remembered by a simple acronym, **APASSNGG.**

a. Apply Essential First Aid. (i.e. splints, pressure bandage, etc.)

b. Protect the Patient from the Elements. Provide the casualty with proper insulation from the ground. Ensure that he is warm and dry. If there are any natural hazards (i.e. rock fall, lighting, etc.) either move the casualty as quickly as possible or ensure that he is well protected.

c. Avoid Unnecessary Handling of Patient.

d. Select Easiest Route. Send scouts ahead if possible, to break trails.

e. Set Up Relay Points and Warming Stations. If the route is long and arduous, set up relay points and warming stations with minimum amount of medical personnel at warming stations to:

(1) Permit emergency treatment. Treat for shock, hemorrhage, or other conditions that may arise.

(2) Reevaluate the patient constantly. If patient develops increased signs of shock or other symptoms during the evacuation, he may be retained at an emergency station until stable.

f. Normal Litter Teams Must Be Augmented in Arduous Terrain.

g. Give Litter Teams Specific Goals. This litter teams job is extremely tiring, both physically and mentally. The litter teams must be given realistic goals to work towards.

h. Gear. Ensure all of the patient's gear is kept with him throughout the evacuation.

2. **ONE -MAN CARRIES**

a. Sling Rope Carry. The sling rope carry requires two men and a 15-foot sling rope. One as the bearer, and an assistant to help in securing the casualty to the bearer. Conscious or unconscious casualties my be transported in this manner:

(1) Bearer kneels on all fours and the assistant places casualty face down on bearer's back ensuring the casualty's armpits are even with the bearer's shoulders.

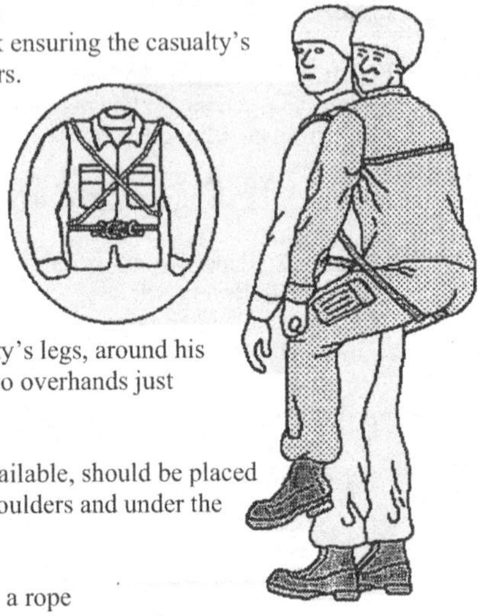

(2) He then finds the middle of the sling rope and places it between the casualty's shoulders and the ends of the sling rope are run under the casualties armpits, crossed and over the bearer's shoulders and under his arms.

(3) Then the ropes are run between the casualty's legs, around his thighs, and tied with a square knot with two overhands just above the bearer's belt buckle.

(4) Ensure the rope is tight. Padding, when available, should be placed where the rope passes over the bearer's shoulders and under the casualties thighs.

b. <u>The rope coil carry.</u> This requires a bearer and a rope coil. It can be used to carry a conscious or unconscious casualty.

(1) Position the casualty on his back.

(2) Separate the loops of the mountain coil into two approximately equal groups.

(3) Slip ½ of the coil over the casualty's left leg and ½ over his right leg so that the wraps holding the coil are in the casualty's crotch, the loops extending upward the armpits.

(4) The bearer lies on his back between the casualty's leg and slips his arms the loops. He then moves forward until the coil is extended. When using the rope coil the bearer ties the coil to himself vice slipping his arms through the loops.

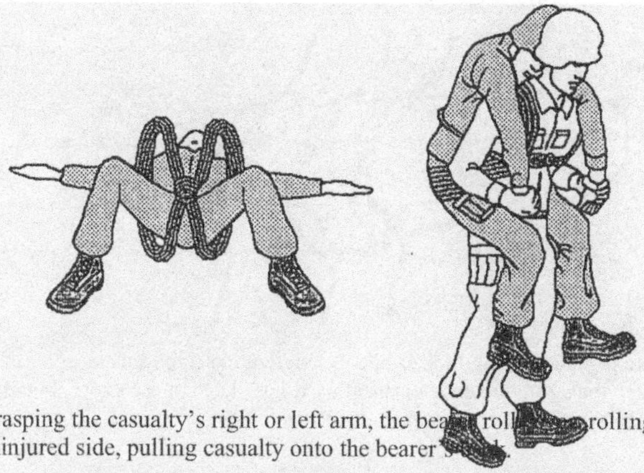

(5) Grasping the casualty's right or left arm, the bea[illegible] rol[illegible] rolling to the casualty's uninjured side, pulling casualty onto the bearer [illegible].

(6) Holding the casualty's wrists, the bearer carefully stands, using his legs to lift up and keeping his back as straight as possible.

NOTE: The length of the coils on the rope coil and the height of the bearer are to be considered. If the coils are too long and the bearer happens to be a shorter person, it will require the coils to be uncoiled and shortened. If this is not done, then the casualty will hang too low on the bearer's back and make it a very cumbersome evacuation. A sling rope harness can be used around the victim's back and the bearer's chest, which will free the bearer's hands.

3. **TWO-MAN CARRY**

a. Pole Carry. The pole carry method is a field expedient method and should be considered as a last resort only, when narrow ledges must be traversed; vegetation limits the bearers to a narrow trail. This method is difficult for the bearers and uncomfortable for the casualty. Two bearers, four sling ropes and a 12 foot pole - 3 inches in diameter, are required for this carry.

(1) The casualty is placed on his back in a sleeping bag or wrapped in a poncho or blanket, then placed on an insulated pad.

(2) One sling rope is placed under the casualty below the armpits and tied with a square knot across the casualty's chest.

(3) The second sling rope is tied in the same manner at the casualty's waist.

(4) The third sling rope is placed at the casualty's legs below the knee.

(5) The fourth sling rope is tied around the ankles.

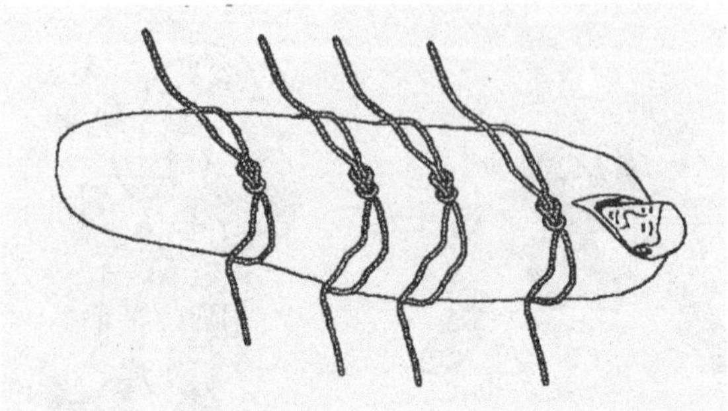

(6) The pole is placed along the casualty's length and secured using square knots with two overhands with the ends of the sling ropes. The square knots should be so tight that the overhands are tied onto themselves

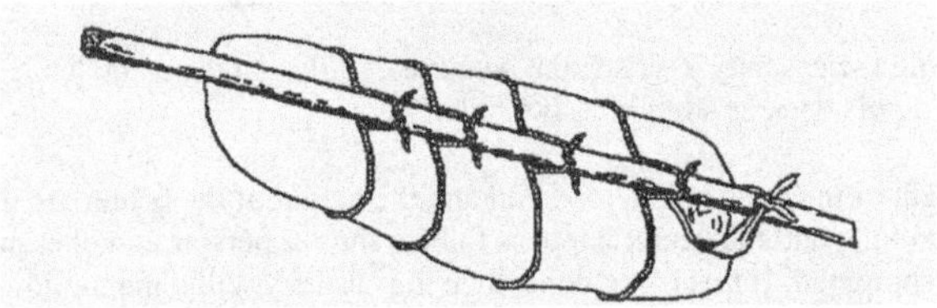

(7) The casualty should hang below the pole, as close to the pole as possible, to prevent swinging during movement.

(8) The casualty's head may be supported using a triangular bandage or a cartridge belt passed around the pole.

(9) For additional support and of movement, two additional bearers may be required, as well as a mountain coil.

(a) Mountain coil is split into two equal coils.

(b) Place knot of mountain coil under casualty's lower back.

(c) Additional bearers slip into each half of the hasty coil one on each side of casualty, aiding in support and movement of the casualty.

4. **EXPEDIENT LITTERS.** (FMST.07.40b) The two types of expedient litters that we will talk about are the alpine basket and the poncho litter.

a. <u>Alpine Basket.</u> To belay the alpine basket, the pre-rigs are attached to the bights formed coming through the loops.

(1) If barrow boy is to be used, the procedure previously discussed will be adhered to.

(2) If barrow boy is not used, then a tag line from the bottom must be implemented to keep the casualty away from the cliff face on the decent.

(3) Construction of the Alpine Basket:

(a) Start by making the same amount of bights as in the rope litter, but start from one end and tie a figure-of-eight loop to run the first bight through.

(b) Place padding i.e., isopor mat, on top of the bights and then lay the casualty on the padding and bights.

(c) Start at the casualty's feet and pull the first bight up around the casualty's ankles and through the figure-of-eight loop tied into the starting end of the rope.

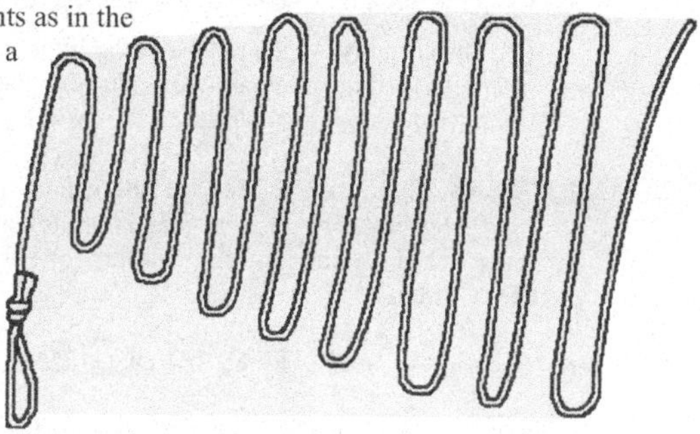

(d) Go to the opposite side of the casualty and pull up the second bight and pull it through the loop formed by the bight that was pulled through the figure-of-eight.

(e) Continue until you get to the casualty's armpits, bring the second to the last bight up over the casualty's shoulder and into the bight and then bring the last bight up over the casualty's other shoulder and into the last bight formed.

(f) Secure the end of the bight with an appropriate anchor knot.

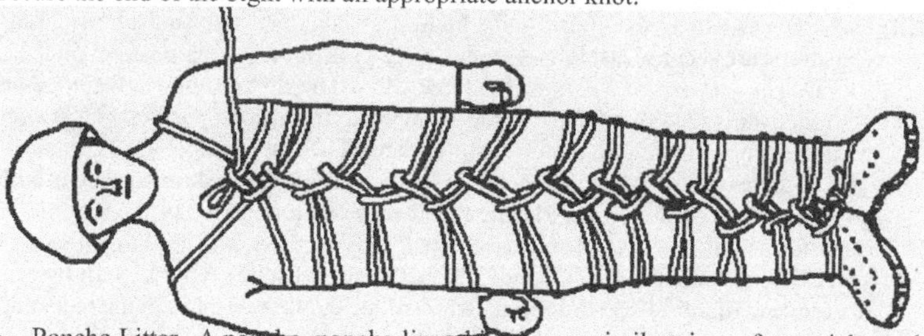

b. <u>Poncho Litter</u>. A poncho, poncho liner, bivy bag or similar piece of material may be used. In addition you will need six individuals with sling ropes.

(1) Lay poncho litter flat on the ground.

(2) Select six rocks about the size of a golf ball. Place one rock in each of the corners and one in the middle on each side in the middle of the litter. The rocks are placed on the under side of the poncho or like material. If a bivy bag is used the casualty should be zipped inside the bivy bag. The rocks should then be arranged in the same manner only on the inside below the zipper.

(3) Tie the sling rope together with an overhand knot. Take the middle of the rope and secure it around the rock with a clove hitch.

(4) An isomat may be laid on the poncho to help make the litter firmer.

(5) The casualty is then placed in the litter. The sling ropes are adjusted by feeding the pigtails of the over hand knot through itself to adjust for length. The loop is then put over the inboard shoulder of the carriers. Insure that the casualty is carried level.

5. **LITTERS.** There are two kinds of litters that are used for casualty evacuations in moderate to vertical terrain. The SKEDCO and the Stokes litter. Each has a variation of procedures for securing a casualty and rigging the litters for either raising or lowering. Let's first look at the SKEDCO litter.

(1) Securing a Casualty to the SKEDCO (FMST.07.40c)

(a) First unroll the litter. The litter must be re-rolled the opposite way to allow the litter to lay flat. Then lay the litter next to the casualty.

(b) If the casualty has any possibility of a spinal injury the Oregon Spine Splint must be used. Secure the splint to the casualty by use of the color-coded buckles. Experienced medical personal are recommended if spinal immobilization is necessary.

(c) Once the casualty is on the SKEDCO use the four body straps to secure the casualty to the litter. Unless injuries prevent, the casualty's arms should be at his sides to prevent further injuries to himself or the rescuers.

(d) Once the casualty is secured with the body straps the feet straps must be secured. The feet straps are secured last to ensure the casualty is in the proper position on the SKED. The feet can be positioned in three ways. The first position is feet together with the straps running on the outside of the feet. The second position is feet apart with the straps running on the inside of the feet. The last position is the feet stacked. This is the most uncomfortable position and not recommended for casualties with possible spinal cord injuries. Placing the heel of one foot on top of the toes of the other forms this position. This position will only be used for a casualty in confined spaces. Once the feet are positioned the feet straps must be secured. Start by bending the feet end of the SKED to form a plat form for the feet. Then loop the feet straps through the second grommets on each side.

(e) The last thing to do is to form the head end to protect the casualty's head. If possible the casualty should wear a helmet. Form the head end tying the pull strap up and secure it to the first body straps.

(2) Medevacing a Casualty. There are many ways to move a casualty once in the SKEDCO. However the medevac team must keep the general considerations in mind. The two methods that we will talk about next are the simplest and require the least amount of additional rigging

(a) The first way is to drag the casualty by the drag strap located at the head end of the SKED. You can also use the SKEDCO's carrying bag as a harness in conjunction with the pull strap and towing harness. If additional people are required, cordage can be added to the pull strap or the front carrying handles.

(b) The second way is to carry the casualty using the carrying handles. By using the set of four removable webbing the litter team can be augmented. To do this each pieces of webbing is tied to make an endless loop. Then pass a bight of the loop through one of the grommets to create additional handles.

(3) Rigging the SKED for Vertical Terrain and Helicopter Lift. In this type of terrain special requirements must be taken to raise or lower a casualty in the SKEDCO.

(a) Rigging the SKED for a Vertical Employment. A vertical raise or lower is used when moving a casualty on steep earth to avoid any further injury to the casualty. On vertical terrain the vertical raise or lower can be used if the terrain is not uniformed or there is a chance of rock fall. Ensure that the casualty's head is always above his feet.

(1) Identify the 30-foot piece of cordage that comes with the SKEDCO. Then tie a figure eight in the middle of the rope. If the rope is worn or missing, the same process can be done with two sling ropes.

(2) Next pass each end of the rope through the grommets at the head of the SKEDCO. Leaving approximately 1-2 feet of rope between the stretcher and the knot.

(3) Continue to feed each end through the grommets and the carrying handles towards the foot end of the SKEDCO. Pass the ends of the rope through the last grommets at the foot end and secure the two ends with a square knot without over hands.

(4) Bring the pigtails up and over the casualty's feet and pass the ends through the
a carrying handles towards the middle of the casualty. Then tie a square knot
b with two over hands.

c

(4) Rigging the SKEDCO for a Horizontal Employment. A horizontal raise or lower is preferred on uniformed vertical terrain. The horizontal employment allows the rescuer to assist the casualty on either a raise or lower. It also allows the rescuer the ability to monitor the casualty's condition and can easily treat the casualty if the need arises.

(a) Identify the two 4 inch nylon straps. They should be two lengths; one four inches shorter than the other. The shorter strap should be marked HEAD STRAPS.

(b) Insert one end of the head strap into a slot near the head of the litter. Then wrap the rest of the straps under the SKEDCO and pass the other end through the opposing slot. Do the same at the foot end of the SKEDCO with the other strap. Ensure that the strap runs smoothly under the SKEDCO.

(c) Connect the strap ends with the large locking caribiner that comes with the SKEDCO. If the caribiner is worn or missing, opposing Stubai 85 locking carabiners will suffice.

6. **STOKES LITTER**. The stokes litter is a litter that is constructed of metal tubing with a plastic covering. The litter is formed in a rectangular basket shape with mesh attached to the frame. Using the stokes for an evacuation (as with any evacuation) it should be padded for the casualty.

a. Securing the casualty to the stokes litter. In the event that the "seat belts" are missing from the stokes litter, sling ropes can be used to the lash the casualty. The steps involved are:

(1) Tie two sling ropes together using square knots and two overhands.

(2) Tie a stirrup hitch around ankles and feet, feed the two pigtails through the right angles of the stokes. Do not cross the ropes at the ankles.

(3) Lace the sling rope towards the casualty's head by passing the rope through the right angles (not over the top of the rails of the stokes).

(4) Secure the ends of the sling ropes by tying a clove hitch with two half hitches on the thick vertical bar located by the victim's shoulder.

7. **YOSEMITE PRE-RIG.** The title of this method, pre-rig implies the meaning ready to use, but in this case the rig must be constructed. Its purpose is for attaching a belay line to a litter and to make the litter easily adjusted. This is the one method that is used to secure a litter to a belay line.

a. Construction: This method normally requires four sling ropes. The steps are: (1) using one sling rope, tie a figure-of-eight loop with one tail.

(1) Take the remaining tail and run it through the window of the stokes litter or in the stirrups of the collapsible litter.

(2) Tie a Kragur knot onto the same sling rope.

(3) Repeat steps (1), (2), and (3) with the three remaining sling ropes.

(4) Suspend the litter to ensure that the comers are balanced.

8. **ASCENTS OR DESCENT OVER STEEP TO MODERATE SLOPES.** When the litter team is ascending or descending a slope they must consider the potential for further injury to the casualty or to themselves. If the risk of injury is high a belay line may be used to prevent injury to the casualty and the rescuers.

a. Preparing Casualty for Ascents or Decent over Steep to Moderate Terrain. This procedure will be depending on several things. Initially, site selection should contain the following features.

(1) Suitable anchor points.
(2) Clearance for casualty along the route
(3) Loading and unloading points.

b. Additional considerations.

(1) The casualty will always be rigged for vertical employment when on steep to moderate terrain.
(2) The smoothest possible route must be selected.
(3) Ensure that the casualty's head is above his feet.

c. Rescuers positions. There are two methods that can be used for the rescuers for moving a casualty in steep to moderate terrain.

(1) Two to four men will position themselves on each side of the litter. They can then carry the litter by the carrying handles. In steep terrain a second belay line may be used to assist the rescuers. We will discuses the belay line later in the chapter.

(2) The Caterpillar method will require as many personnel as possible. The personnel will split in half and position themselves on each side of the litter forming a tunnel. As the litter is raised or lowered each member will hand the litter to the next member in the tunnel. As the litter passes each person in the tunnel he will peel off and assume the lead either at the top or bottom of the tunnel. This will continue until the litter reaches its desired destination.

d. Belay Line. (FMST.07.40d) For belaying of a casualty, one rope will be used and from the top using one of two methods depending on the application.

(1) Body Belay. This method should only be used over moderate terrain. The belay man will establish a sitting position behind a suitable anchor (i.e., rock, tree, etc.) and pass the standing end of the rope behind his back. The running end of the rope will feed out from the belay man's right side. A figure of eight loop is tied to the end of the running end of the rope. It is then attached to the litter's figure eight loop with a locking carabiner. The belay man will then remove all of the slack between himself and the litter. The standing end of the rope should be stacked on the belay man's left side and run through his left side. As the casualty is lowered, the belay man will feed the rope from behind his back allowing it to run through his right hand. If the belay man needs to stop the casualty, he will clench the rope in his left hand, and

bring the rope to the center of his chest.

(2) Direct Belay. This method is the safest for either raising or lowering a casualty in either moderate to steep terrain.

(a) To Lower a Casualty. First a swami wrap will be tied around a suitable anchor point. Two locking carabiners will be clipped into all of the wraps of the swami wraps, gates up. A figure of eight loop is tied into the end of the static rope and attached to the litter with a locking carabiner. After all the slack has been taken up between the litter and the anchor, the rope must be tied through an appropriate belay device. The belay device is attached to the anchor through one of the two locking carabiners on the anchor. A safety (French) prussic will be tied to the running end of the rope and clipped into the second locking carabiner on the anchor. While the casualty is being lowered, one person will control the rope running through the belay device. The safety prussic will be controlled by a second person. Should the primary belay man lose control, the person operating the safety prussic simply lets go and the prussic will bind onto the rope, stopping the casualty.

(b) To Raise a Casualty. The anchors will be established in the same manner as discussed in lowering the casualty with one minor change. The one change is that instead of running the rope through a belay device, the rope will only run through a locking carabiner. The load will be raised by the use of a mule team. The mule must consist of as many people as possible. The mule team will raise the load in as straight a line from the anchor as possible. If the space does not permit a ninety-degree angle away from the anchor is also an option. The mule team will walk backward until the last man reaches his limit of advance. Once he reaches that limit he will peel off the end and return to the front of the mule team. This process is continued until the casualty reaches the top. If the load becomes unmanageable, the safety prussic will be allowed to bind on the rope while the mule team repositions themselves. If the person operating the safety prussic can not see the casualty a Point NCO will be in charge of communicating with the mule team.

(c) To Belay the Rescuers. If the route is too steep or the footing is poor the rescuers may need some assistance either on the raising or lowering of a casualty. If this is the case a separate belay line will be established for the rescuers. The anchor and the belay line are established in the same manner. The same anchor can be used if it is suitable for the load. The rescuers will then tie either around the chest bowlines or swami wraps. A figure eight loop will be tied into the end of the static rope and connected to the bottom rescuer with a locking carabiner. The other rescuers will connect themselves to the same rope with middle of the line prussic. They will be connected in this manner so that they can adjust their position to the casualty.

9. **BARROW BOY.** (FMST.07.40e) A barrow boy is no more than an assistant to the litter on vertical to near vertical cliff faces. The barrow boy can be used for either the Stokes or the

SKED litters. For this situation the Stokes litter should only be used in the horizontal position. However the SKED can be employed in either the horizontal or the vertical positions.

NOTE: Static load, lowered slowly, little friction, so don't need 2 ropes.

a. Rigging the Barrow Boy.

(1) First the rescuer must ensure that a suitable anchor has been established, a proper belay has been constructed, and that a safety prussic has been constructed.

(2) Then the rescuer must ensure that if an A-frame is used that it has been constructed and anchored properly.

(3) Next the rescuer will tie a rappel seat on. (a sit harness can also be used) Then he will ensure that an around the body bowline is tied onto the casualty. A figure eight will be tied on to the end to act as the casualty's safety.

(4) Then after running the running end of the rope though the carabiners or pulley of the A-frame he will tie a figure eight loop at the end of the static rope. He will then attach the figure eight loop to his hard point with a locking carabiner. Then he will tie a middle of the line prussic above the figure eight and attach it to the same locking carabiner in his hard point. This is called the adjustment prussic. It is used to adjust the position of the Barrow Boy in relation to the litter. Next he will take six to eight feet of slack from the end of the line figure eight and tie a directional figure of eight with direction of pull down. (Note: the prussic should be between the end of the line figure of eight and the directional figure of eight.) The directional figure of eight is the attaching point for the litter and the casualty's safety.

(5) Once the rescuer and the litter are secured, the belay man must take all the slack out of the system. The rescuers will the maneuver the litter through the apex of the A-frame with the help of the point NCO.

(6) Once onto the cliff face the rescuer will then position him self with his adjustment prussic so that he can be of most assistance to the litter on the raise or lower. The rescuer will pull the litter out away from the cliff face so that the casualty rides smoothly up or down the cliff face.

(7) The Point NCO will be in charge of the belay men or the mule team. He will also communicate with the rescuer about the rate of speed, if the rescuer needs to be stopped along the route, and when he reaches the top or bottom of the cliff.

10. **TANDEM LOWERING.** The tandem lowering system can be used for the walking wounded, POW's, or more serious casualties when situation would not permit using the barrow boy.

a. The assistant to the casualty should first tie a rappel seat on himself and then assist the casualty with his.

b. The assistant will take only one belay line and tie a end of the line figure 8 loop, and clip this into his rappel seat.

c. A directional figure 8 will be tied approximately 12 inches up the rope from the figure 8 loop (with the loop pointed down) and this will be clipped to the casualty's rappel seat.

d. If needed, safety prussic cords should be tied above the casualty's directional figure 8. These safety cords should be attached in the same manner as the first man down a rappel.

e. The casualty and the assistant will lower as one, with the assistant helping on the way down the cliff.

11. **OTHER CONSIDERATIONS.** All of the techniques we have discussed for the evacuation of a casualty from top to bottom can also be used on a suspension traverse or rope bridge, with a slight variation in the belay line. Two belay lines may be used for rope bridges and the suspension traverse, if they are available. No matter what type of litter is used, the individuals involved in the evacuation must ensure that the head is always uphill or not lower than the feet.

UNITED STATES MARINE CORPS
Mountain Warfare Training Center
Bridgeport, California 93517-5001

FMST.0
7.34
04/12/02

STUDENT HANDOUT

ONE-ROPE BRIDGE

LEARNING OBJECTIVES

TERMINAL LEARNING OBJECTIVES. In a summer mountainous environment, conduct bridging, in accordance with the references. (FMST.07.34)

ENABLING LEARNING OBJECTIVES

(1) Without the aid of references, list in writing the criteria for site selection for a one-rope bridge, in accordance with the reference. (FMST.07.34a)

(2) In a summer mountainous environment with an assault load , cross a one-rope bridge using the rappel seat method, in accordance with the reference. (FMST.07.34b)

(3) Without the aid of references, list in writing the rescue techniques, in accordance with the reference. (FMST.07.34c)

(4) In a summer mountainous environment transport a SKED stretcher across a one rope bridge, in accordance with the reference. (FMST.07.34d)

BODY

1. **SITE SELECTION**

a. The two criteria for site selection for a one-rope bridge are: (FMST.07.34a)

1) There must be suitable anchors on both sides of the stream.

2) The anchors must offer good loading and unloading platforms.

b. Other considerations involved are:

1) The site chosen for the initial crossing does not have to be at the location for the construction of the bridge, just as long as the rope can be taken to the selected site the

crossing.

2 .) The site chosen for the lead swimmer to cross should be as free as possible from obstacles in the water, such as large boulders, stumps or logs as discussed in *STREAM CROSSING.*

3 .) The anchors must be close enough for the 150 foot coil to reach both near side and far side anchors. Keep in mind that it will take approximately 1/3 of the 150 foot rope for tightening and anchoring of the bridge.

2. **DISTANCE ESTIMATION**. The follow methods can be used to determine the distance between anchor points:

a. Azimuth Method. Shoot an azimuth to a point on the far side of the intended obstacle to cross. Then move LEFT or RIGHT (perpendicular to the azimuth) until you get a 15 degree offset of your previous azimuth. Next, measure the distance that you paced in feet from your first azimuth to your second azimuth, and multiply that distance by three. This total will give you the approximate distance across the obstacle in feet.

b. Unit Average Method. Get three Marines to judge (best guess) the distance across the intended obstacle. Add up the total accumulated distance, then divide by three. This will give an estimation of the distance across the obstacle.

Example:	Marine 1	240 ft
	Marine 2	230 ft
	Marine 3	250 ft
		720 ft

Divided by 3 = 240 ft

3. **ORGANIZATION**. The organization of construction is broken down into four groups. They are as follow:

a. The Bridging Team - consisting of the bridge NCO and another Marine.

b. The Safety Line Team - consisting of the lead swimmer and his belay man.

c. The Mule Team - consisting 3 Marines for a 9:1, and 6 Marines for a 3:1.

d. The Security Team – consisting of the rest of the party.

4. **CONSTRUCTION**. This bridge is ideal for squad and platoon sized units, for its quick construction and minimal amount of equipment required.

a. Once the site for the bridge has been designated by the bridge NCO, the Safety Line Team will move up stream from this site to enter the water. Their first step is to flake out the rope. Once this is accomplished, the lead swimmer's will take a bight of the rope and tie a figure 8 loop at the end that will be going to the far side, ensuring the knot has a 18"-24" loop. The lead swimmer's upstream arm will go in it, with the belay man tending his rope the lead swimmer will cross the never using the flow to assist him. (Ferry angle as taught in STREAM CROSSING)

b. At the same time as the lead swimmer is crossing, the bridging team will be preparing the near side anchor. The bridging team will tie a swammi wrap around the anchor, using a sling rope or practice coil, and ensuring the square knot is behind or on the side of the anchor. Once the anchor is secure, flake out the bridging line in order to send it across.

c. Once the lead swimmer is across he will move to the site where the far side anchor will be established and secure his rope. The belay man will then move down to the near side anchor, and as close to the waters edge as possible. He will attach the bridging line to the safety by tying a middle of the rope in a figure 8 knot to the safety line and an end of the rope figure 8 knot to the bridging line connecting the two ropes with an 85 carabiner. The lead swimmer will then pull the rope across. The belay man with the help of another Marine will belay the bridging line across keeping tension on the rope to keep the rope out of the water.

d. After the bridging line is across the lead swimmer will detach the bridging line from the safety line and secure it to the far side anchor using a tree wrap, wrapping from right to left, ensuring there are 3-5 wraps.

e. Once the bridging line is secure. The lead swimmer and the belay man will secure the safety line on both banks, ensuring the rope is creating a ferry angle (minimum 45 degrees). This will be the safety line for the Marines crossing the bridge.

NOTE: If a fall from a rope bridge would result in death or serious injury, then utilize a safety line that will be anchored parallel with the bridge and hand tightened. The Marine crossing will clip into both ropes prior crossing.

f. On the near side bank:

1. The bridging team will need the following gear to build the mechanical advantage:

   (a) Four steel locking carabiners & 3 -Three foot prusiks

2. Once the bridging line is secured on the far side, the bridging team will take one 85 and clip it into the near side anchor (swammi wrap). Then they will take the bridging line and clip into that 85, then using one three foot long prusik (16" loop) the bridging line will be secured to the 85 by tying a French Prusik, which will act as a braking knot. The bridging team will now pull the rope taut, and begin construction of the mechanical advantage.

3. The first step is to take one three foot prusik (16" loop) and tie a French prusik on the bridging line as far away from the anchor as possible. Then take one 85 and clip it to the loops of the French prusik.

4. Taking a bight from the running end of the rope clip it into the 85 hanging from the tails of the French prusik.

5. Then bring a bight back from the clipped in rope to the anchor, and taking one 85 clip it into the 85 created in step (2), take the bight of rope and clip into this 85.

6. Take one three foot prusik and tie a French prusik onto the bottom end of the bight clipped into the 85 created in step (3), take one 85 and clip it into the tail loops of the French prusik.
7. From the anchor, take a bight of rope from the 85 created in step (6) and clip it into the 85 created in step (6).

8. You have made approximately a 9:1 mechanical advantage.

NOTE: If the manpower is available the squad/platoon has the option of building a 3:1 mechanical advantage vise a 9: 1, the mule team will just have to add another body having no more then 3 bodies on a 9:1 and 6 bodies on a 3:1

g. Tensioning of the Bridge

1. Now that the bridge is built, the bridge NCO can call up the mule team to tension the bridge.

2. This procedure is facilitated by the mechanical advantage system that was just put into place. The braking knot (French prusik) is used to hold tension on the bridging line while the bridge is being tensioned.

3. The mule team will begin to pull on the running end of the rope coming out of the mechanical advantage. They will pull the rope straight back, trying to keep it in line with the bridge as best as possible.

4. The mule team will tension the rope as much as possible, the bridge NCO will monitor the system. Once the mule team cannot pull anymore tension, the bridge NCO will then have them hold the rope in place, and will reset the braking knot. Once the braking knot is set the bridge NCO will ask for slack from the mule team and cycle the system out. This process will continue until the bridge is tight.

5. On the last cycle the bridge NCO will set the brake knot and with the help of another Marine they will make a bight out of the running end and bring it around the tree while keeping tension.

6. Take a bight and make a complete round turn on the body of the 85 created in step (2). Last, the bridge NCO will tie two half hitches encompassing all the ropes just behind the anchor.

7. At this stage the bridge is tight and secured, the bridge NCO will now call up the remainder of the squad/platoon to cross. The bridge NCO will monitor the crossing of all Marines, and will be the last one to cross.

TRANSITION: Now that we have discussed the construction of the bridge, are there any questions over the material? Next we will discuss the crossing of the bridge.

5. **CROSSING.** (FMST.07.34b) The method used to cross is known as a horizontal traverse. This traverse can be accomplished in the following ways.

a. Rappel Seat Method

1) The Marine ties himself into a rappel seat and inserts a large steel locking carabiner with the gate facing down and away.

2) The Marine faces the bridge with his right hand towards the near anchor to snap into the bridge, once the 85 is locked onto the bridge rope, a helper flips the 85 over so that the locking nut screws down. A carrier rope with a girth hitch may be added to assist the Marine in mounting the rope.

3) The Marine hangs below the bridge from his rappel seat with his head pointing in the direction of the far anchor and allows his legs to hang free.

4) Pulling with his arms makes progress.

5) This method is the safest and therefore the preferred method.

6) If the Marine must take a pack across, he may wear it with the waistband secured: However, the preferred method is to have another carabiner (aluminum non-locking is sufficient) attached to the pack frame at the top and attach this to the bridge behind the Marine putting his legs through the shoulder straps and pulling the pack across with him.

7) One man at a time will cross, although one can load and another can unload concurrently.

8) Weapons will be worn across the shoulder; muzzle down with a tight sling securely attached to the weapon.

b. <u>Pulley Method.</u> This method is used when the one rope bridge is long, uphill, or speed is vital and the Marines crossing it have a lot of heavy personal equipment; M24OG's, radios, etc.

1) Equipment required. One pulley, four steel locking carabiners, and a hauling line twice the length of the obstacle.

2) Setting up the system. Construction of the first suspension point is done by attaching a pulley to the one rope bridge, one 85 carabiner is attached to the pulley, gate down, and a second 85 is attached to the first 85 with the gate facing the near side bank. Into this second 85, attach the hauling rope with a figure-8 loop. The figure-8 is placed halfway along the hauling rope and attached to the 85, which is then locked. Four feet down the line another figure-8 on a bight is placed into the second suspension point.

3) Connection. The Marine clips his 85 into the lower 85, his equipment is clipped into the second 85. The mule team starts to pull.

c. Helmets will be worn for all methods of crossing the bridge. Gloves are optional.

NOTE: There are additional methods of crossing a one rope bridge such as the Commando crawl, Monkey crawl and hand over hand techniques, however, they are not taught at MWTC due to safety reasons.

6. **<u>RESCUE TECHNIQUES</u>** If an individual is unable to complete the crossing on the one rope bridge using the rappel seat method of crossing (i.e. from a fall or exhaustion), a rescue will have to be made in the following manner: (FMST.07.34c)

a. Reach. First try to reach the victim by using an object such as a pole, your hand, if the victim is close enough, etc ...

b. Throw. If reaching the victim is impossible, try throwing a rope to the victim and have him attach the rope to himself, preferably his seat, and pull him back to the desired side of the installation.

c. Tow. If the victim is unable to catch or reach a rope being thrown to him or the victim is unconscious, tie a figure-of-eight loop into the middle of the safety line and then connect the rope to the rope bridge with a steel locking carabiner. The mule team will pull the carabiner up against the victim's seat and begin towing the victim to the desired side of the installation. The rope used to tow the victim should be twice the length of the span of the rope bridge (If necessary, ropes can be tied together to accommodate the span). This will allow towing from either side without having to throw or have a Marine carry the safety line back across the bridge each time a towing rescue is performed.

d. Go. If all else fails, the last option will be to go after the victim. The rescuer will move out onto the rope bridge with a safety line attached to himself at approximately eight feet from the end of the safety line. In the end of the safety line tie a figure-of-eight loop and insert a steel locking carabiner in the end of it. Once the rescuer has made contact with the victim. He will attach the steel locking carabiner to either the victim's seat (preferably the victim's carabiner that is attached to the rope bridge), or to the bridge itself. To ensuring that the carabiner is placed so that it pulls against the victim's carabiner. The mule team now starts to pull both the victim and the rescuer to the desired side of the installation.

e. Cut. If while crossing the rope over water the individual goes underwater and no other rescue technique can be employed, the rope will be cut.

7. RETRIEVING THE ONE ROPE BRIDGE

a. Before the bridge NCO sends the last Marine to cross, they must make the bridge retrievable.

b. The first step is to break down the mechanical advantage, ensuring the braking knot is set before doing so. The system will be broken down until the bridging line is attached only to the brake knot.

c. They will now take the rope around to the back of the anchor, and tie a slip figure 8, with the loop of the slip 8, take it and attach it to the bridging line just in back of the braking knot with a 85.

d. Now pull the running end coming out of the slip 8 until it tightens, and bites on the rope. Once the rope bites, tie a thumb knot behind the slip 8 as close to the slip figure 8 as possible.

e. The bridge NCO, with the help of another Marine, will now untie the swammi wrap, until all of the tension is on the bridging line.

f. The safety line will now be brought up, and attached to the end of the bridging line with an overhand knot.

g. The bridge NCO and the other. Marine will now cross the bridge.

NOTE: It is possible that the bridge may lose tension after breaking the bridge down. The packs of the last two Marines across can be sent over with other Marines, prior to the slip figure being tied.

h. After both Marines have crossed, the safely line will be brought up from down stream.

i. The tree wrap will now be taken off. Once the tree wrap is off, all knots and carabiners will be taken out.

j. The safety line will now be pulled, the bridge is retrieved.

UNITED STATES MARINE CORPS
Mountain Warfare Training Center
Bridgeport, California 93517-5001

WMC
04/12/02

STUDENT OUTLINE

SUMMER MOUNTAIN WARFIGHTING LOAD REQUIREMENTS

LESSON PURPOSE. The purpose of this period of instruction is to introduce the student to the different levels of gear requirements in accordance with MWTC's uniform codes. This lesson relates to mountain walking and tactical evolutions

OUTLINE

1. **BASIC UNIFORM REQUIREMENTS.**

a. Marines operating in mountainous terrain will generally wear the standard issued utility uniform. However, due to the wide range of temperatures, and sudden changes in weather in the mountains, Marines will find the need to be continually adding and removing layers. The amount of clothing worn may vary depending upon the severity of the weather, the activity level of the Marine, and the individual metabolism of the Marine. Every person in that unit should still maintain the same outer camouflage layer.

b. As part of the basic uniform, each man will be required to have in his possession, at all times, seven required pocket items. These seven items should be carried in the pockets of your utility uniform:

(1) Pocketknife

(2) Whistle

(3) Pressure Bandage

(4) Chapstick and sunscreen

(5) Sunglasses

C10

(6) Survival Kit and rations (fire starting material, signaling material, food gathering material, water procuring material, 1st aid, shelter material, and some high energy, lightweight snacks.)

(7) Notebook with pen/pencil

c. Some additional items that should be carried in your pockets at all times:

(1) Contact gloves

(2) Paracord (10 meters)

(3) Flashlight with tactical lens and spare batteries

(4) Chemlights

2. **ASSAULT LOAD.**

a. The Assault Load is equipment in addition to the basic uniform requirements. It is carried in the load-bearing vest (LBV), butt pack and the pack system. This is the equipment carried for short duration missions such as security patrols or during the assault. It is also carried at all times when away from the bivouac site.

(1) An extra insulating layer (Field jacket liner, wooly pully, etc.)

(2) Protective layer (ECWCS parka and trousers)

(3) LBV with 2 quarts of water and first aid kit

(4) Helmet

(5) Rations for the time away from your bivouac site

(6) Cold weather hat or balaclava

(7) Individual mountaineering gear

a. Sling rope

b. Carabiners

c. Rappelling Gloves

(8) Specialized mountaineering equipment

(9) Mission essential gear

3. **COMBAT LOAD**. The Combat Load is the equipment carried for longer duration missions.

The following items are in addition to the items already being carried in the Assault Load:

a. Sleeping bag with bivy bag

b. Isopor mat

c. Stove

d. Fuel bottle

e. Poncho (for expedient shelters or medevac purposes)

4. **EXISTENCE LOAD**. The Existence Load is to be packed for longer duration combat missions, where it is necessary to replace worn out items. These items could be uniforms, boots, personal hygiene gear, etc. The Existence Load should be brought to the forward elements once the situation allows.

5. **PACKING CONSIDERATIONS.** Because most Marines are familiar with how to pack a pack, these are general guidelines only.

a. When not wearing your protective layer or insulating layers, keep it handy at the top of the pack. When taking a break during a movement, you will be able to quickly don a layer to prevent getting chilled.

b. Keep your stove and fuel bottle in the outside pockets of your pack. They may leak and soak your equipment with fuel if stored inside your pack.

c. Keep the climbing rope in the rope bag. This will reduce the rope's exposure to petroleum products, ultra-violet light, cuts, and abrasions.

d. Ice ax is placed through the ice ax loops with spike up, adze outboard, and secured in place.

e. Crampons are attached to the outside of the pack with the tips covered and facing outboard.

TABLE OF CONTENTS

PART I. SCOUTING

Chapter 7. Scouting Fire Team

PART 2. INFANTRY PATROLLING

Chapter 8. Fundamentals of Infantry Patrolling

Chapter 9. Patrol Organization

Chapter 10. Patrol Preparation

Chapter 11. Movement to and Return from the Objective Area

Chapter 12. Reconnaissance Patrols

PART I. SCOUTING

CHAPTER 1. FUNDAMENTALS OF SCOUTING

Scouting involves observing terrain and/or the enemy, and accurately reporting those observations. Scouting requires proficiency in the use of weapons, cover and concealment, route selection, and skill in unobserved day or night movement.

1001. PURPOSE

When an infantry unit is not actively fighting the enemy, it should be actively searching for the enemy, which is the primary purpose of scouting. The unit attempts to keep the enemy off balance while making preparations for further attacks. Physically locating and keeping the enemy off balance are normally accomplished by small units ranging from a two-person scouting party to a squad-size patrol.

Infantrymen are sent out as scouts or as members of a patrol because the commander needs information about the enemy, terrain, and the location of friendly troops. The lives of the entire unit may depend upon the success or failure of a scout or patrol and the accuracy and timeliness of the report. The success of the scout or patrol will depend upon their training, preparation by the commander, and understanding their mission and the commander's requirements.

To wage combat successfully, a commander must have accurate, detailed, and timely information about the enemy, the terrain, and adjacent friendly units. Well-trained scouts and capably led patrols are among the most effective means the commander has for acquiring the information necessary to plan tactical actions and make decisions in execution.

1002. REQUIRED SCOUTING SKILLS

To be effective, a scout must be able to—

- Recognize terrain features.
- Read a map and determine direction.
- Practice and implement the principles of cover and concealment.
- Fully utilize movement and route selection.
- Know the enemy (estimate enemy unit composition and strength).
- Observe and report information accurately.
- Select routes and move through numerous types of terrain.

CHAPTER 2. TERRAIN, MAPS, AND DIRECTION

A scout must understand map symbols, identify elevations from contour interval lines, scale distance on a map, relate natural and man-made features shown on the map to the actual features on the ground, plot a course from one point to another, and locate his current position. To relate a map to the actual terrain and its features, a scout must be able to orient it to the ground using a compass, two points, a watch, and the sun or the stars.

2001. TERRAIN FEATURES

Since the infantry works and fights on the ground, terrain ashore information that scouts gather and report on is of great importance. Hills, valleys, woods, and streams are the forms and growths commonly referred to as natural land features. Artificial or man-made features include houses, bridges, and railroads. Figure 2-1 shows some important terrain features.

2002. THE LENSATIC COMPASS

The best method of finding direction, during both day and night, is with a compass. The lensatic and M2 are the two types of compasses issued to infantry units. Both work on the same principle.

The standard compass for general use in the Marine Corps is the pivot-mounted lensatic compass, so called because azimuths are read through a magnifying lens in the eyepiece. Figure 2-2, on page 2-2, shows the lensatic compass and its nomenclature. The meter graphic scale on the side of the compass is 1:50,000, which is the most used scale in military mapping. The graphic scale is useful in the field as a straightedge, an aid in orienting the map, and a means of reading map grid coordinates. The plastic dial is graduated in both degrees and mils. Numbers on the dial are printed in black. There are luminous markings on the bezel, floating dial, and on both ends of the sighting wire, plus a 3-degree bezel serration and clicking device that permit reading azimuths at night. The compass is carried in a nylon case that may be attached to the cartridge belt.

Magnetic compasses are affected by the presence of iron or magnetic fields. Consequently, the scout should not be within the influence of local magnetic

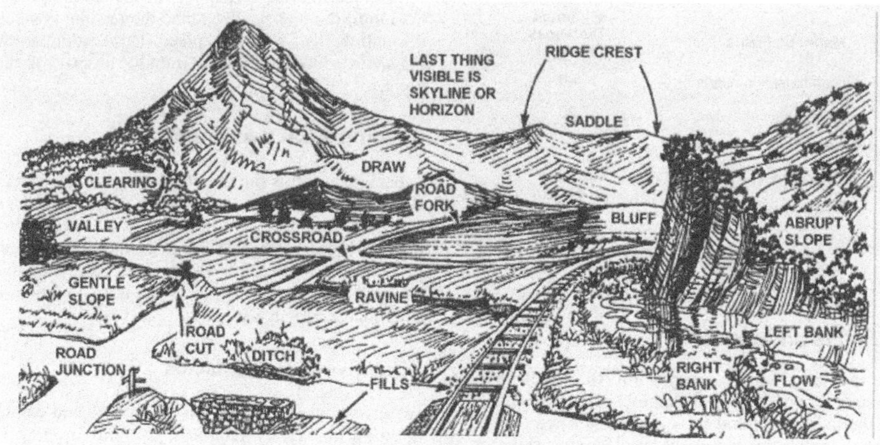

Figure 2-1. Natural and Artificial Terrain Features.

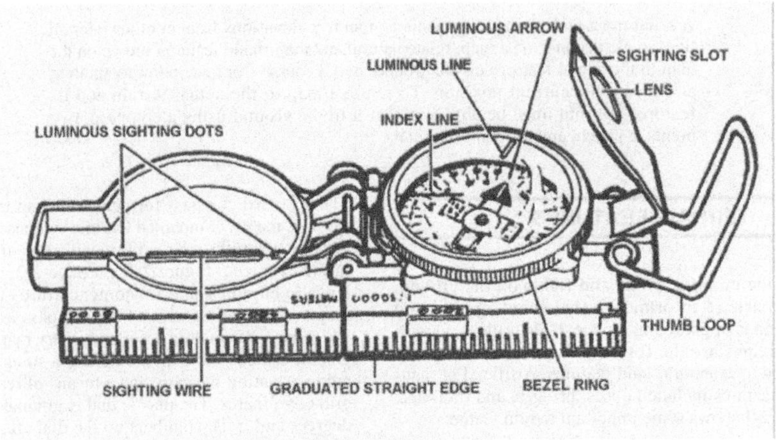

Figure 2-2. Lensatic Compass.

attraction while using a compass to determine direction. The rifle, pistol, and other metal objects must be laid aside when reading the compass. The minimum distances the scout should be from visible masses of iron and electrical fields of magnetism for the compass to provide accurate readings are provided below.

Magnetic Fields	Minimum Distances (meters)
High tension powerlines	60
Field guns	20
Vehicles (wheeled or tracked)	20
Telephone and telegraph wires	10
Barbed wire	10
Machine gun	3
Rifle, pistol	1

Reading an Azimuth

To read an azimuth to any point, the cover of the compass is raised to an angle of 90 degrees in relation to the index face, and the eyepiece is lifted to a 45-degree angle in relation to the bezel, or so the numbers on the dial can be seen. The thumb of either hand is placed in the thumb loop, the index finger extended along the side of the compass case, and the remainder of the hand closed. The closed hand and wrist are grasped with the other hand. The elbows are drawn in close to the body, forming a firm foundation for the compass. The eye is placed to the lens on the eyepiece. The compass is pointed at the object or point to which the azimuth is to be read. A sighting is taken through the sighting slot in the eyepiece, and the point is lined up with the sighting wire in the cover. The compass is held until the dial steadies; then the reading is taken through the lens of the eyepiece. This reading is the magnetic azimuth of the line from the observer to the point.

Reading a Back Azimuth

A back azimuth is the direction opposite the line of sight. If the azimuth is less than 180 degrees, the back azimuth is obtained by adding 180 degrees. If the azimuth is greater than 180 degrees, the back azimuth is obtained by subtracting 180 degrees. Back azimuths are used to determine a return route or to resection to determine a current position.

Circumventing Obstacles

When a scout is traveling on an azimuth and comes upon an obstacle—such as a contaminated area, minefield or swamp—the following steps (sometimes referred to as the *90-degree offset* method) is employed to go around or circumvent the obstacle and resume movement along the original azimuth (see fig. 2-3). The steps are as follows:

1. Move up to the obstacle and make a full 90-degree turn to the right (or left).

2. Walk beyond the obstacle, keeping track of the distance in paces or meters.

3. Stand at the end of the obstacle, face in the original direction of march, and follow that azimuth until the obstacle has been passed.

4. Make a 90-degree turn to the left (or right) and move the distance previously measured to return to the original line of march.

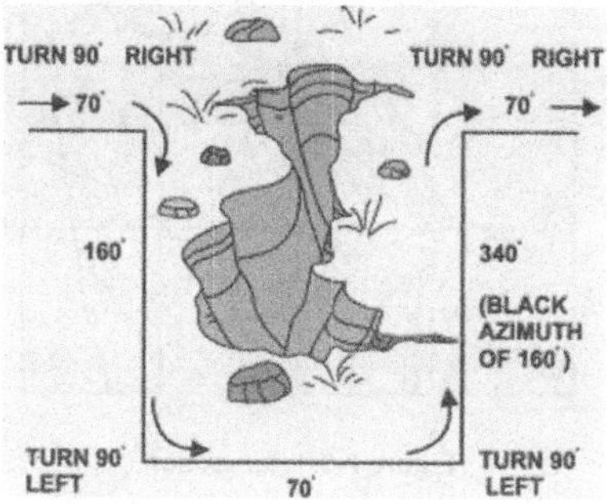

Figure 2-3. Circumventing an Obstacle.

Following an Azimuth During the Day

The eye is placed to the lens in the eyepiece and the compass moved until the desired azimuth reading is visible beneath the fixed index. Without moving the compass, the vision is shifted from the lens through the sighting slot in the eyepiece, and a sighting is taken out beyond the sighting wire in the cover. A prominent terrain feature on this line of sight is selected, the compass closed, and the landmark approached. When the scout reaches the landmark, the procedure is repeated.

Following an Azimuth at Night

It is necessary to prepare and set the compass before departing on a night movement because at night only the luminous parts of the compass can be seen. To prepare the compass for night use, the luminous parts must be fully charged by sunlight or artificial light, such as a flashlight. To set a compass—

1. Move the compass so that the desired azimuth on the dial is directly under the index line on the lower glass.

2. Rotate the upper movable glass so that the luminous line is directly above the north arrow of the dial.

3. Set the compass for marching at night on the specified azimuth.

—OR—

1. Face the general direction of movement.

2. Line up the north arrow and the luminous line on the bezel with the luminous sighting dots.

3. Hold the compass still with one hand and grip the knurled bezel ring with the other hand.

4. Turn the bezel ring the prescribed number of clicks in the proper direction, remembering that each click equals 3 degrees. For example, to set an azimuth of 21 degrees, the bezel ring would be turned seven clicks to the left.

5. Turn the whole compass until the north needle lines up with the luminous line. The compass is then set on the desired azimuth. The azimuth is the line formed by the two luminous sighting dots on the inside of the cover.

To march on a preset azimuth during night movement, open the compass and move it so the north arrow is directly below the luminous line (see fig. 2-4). Move

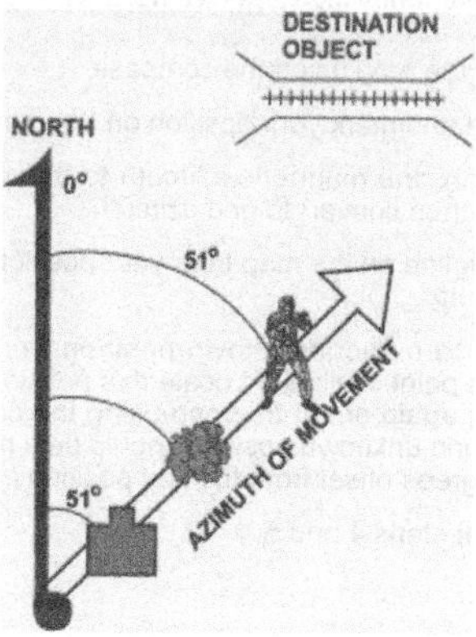

Figure 2-4. Following a Night Azimuth.

in the direction of the line formed by the two luminous sighting dots. It is necessary to refer to the compass more frequently at night than during the day. If stars are visible, find a prominent star along the azimuth of movement to use as a reference point. When the view of the sky is restricted by overcast conditions or vegetation, send a scout forward along the azimuth of movement to the limit of visibility. This scout is guided along the azimuth of movement by a stationary navigator. When the scout reaches the limit of visibility, the navigator moves to the scout's location. This process is repeated until the destination is reached.

A more rapid method for reaching the scout's destination is to equip the navigator with a compass. The navigator can set the compass as explained earlier and the scout proceeds providing security 180 degrees to the front on the specified azimuth, receiving right and left corrections from the navigator while both are on the move. The point scout must stay within visual range of the navigator. If available, a strip of white or luminous tape on the back of the point scout's helmet will assist.

Intersection

Intersection is the location of an unknown point by successively occupying at least two, preferably three known positions and sightings on the unknown point. It is used to locate features not depicted on the map or not readily identifiable. To determine an intersection, perform the following steps (see fig. 2-5):

1. Orient the map using the compass.

2. Locate and mark your position on the map.

3. Measure the magnetic azimuth to the unknown position; then convert to grid azimuth.

4. Draw a line on the map from your position on this grid azimuth.

5. Move to a second known position from which unknown point is visible. Locate this position on the map and again orient the map using the compass. The second unknown position should be a minimum of 30 degrees offset from the first position.

6. Repeat steps 4 and 5.

Figure 2-5. Intersection.

To check accuracy, move to a third position and repeat steps 1 thorough 4. Where the lines cross is the location of the unknown position. Using three lines, a triangle is sometimes formed—called the *triangle of error*—instead of an intersection. If the triangle is large, recheck your work to find the error. Do not assume that the position is at the center of the triangle.

Resection

Resection is the location of the user's unknown position by sighting on two or three known features that are identifiable on the map. To determine a resection, perform the following steps (see fig. 2-6):

1. Orient the map using the compass.

2. Locate two or three known positions on the ground and mark them on the map.

3. Measure the magnetic azimuth to a known position then convert to grid azimuth.

4. Change the grid azimuth to a back azimuth and draw a line on the map from the known position back toward the unknown position.

5. Repeat step 3 and step 4 to determine a second known position.

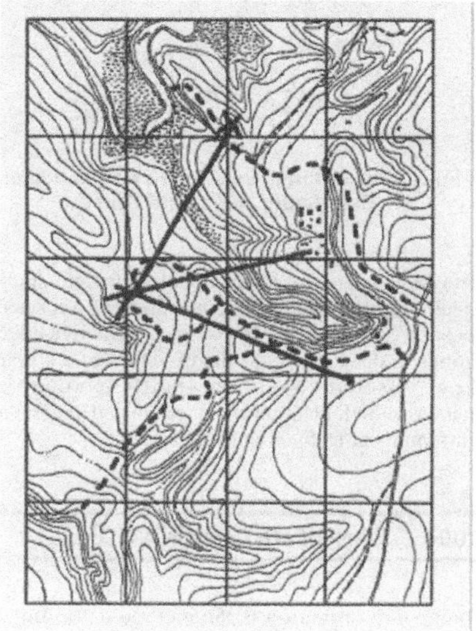

Figure 2-6. Resection.

To check accuracy, repeat the steps above for a third known position. The intersection of the lines is your location. Using three lines, a triangle of error may be formed. If the triangle is large, recheck.

2003. STARS, SUN, AND OTHER FEATURES

In rare cases when a scout is without a compass, the following examples are alternate means to determine direction. When using constellations to determine direction, identify your location's Temperate Zone. A Temperate Zone is the area between the tropics and the polar circles.

At night, the stars provide an excellent means of maintaining a line of march. In the North Temperate Zone (north of the equator), the Big Dipper constellation is one key to determining direction of true north. It is made up of seven fairly bright stars in the shape of a dipper with a long curved handle (see fig. 2-7). The two stars that form the side of the cup farthest from the handle, used as pointers, are situated in the direction of a bright star that is about five times the distance between the two stars of the dipper cup. This bright star is the North Star and is directly over the North Pole. The pointers always designate the North Star, which is the direction of true north.

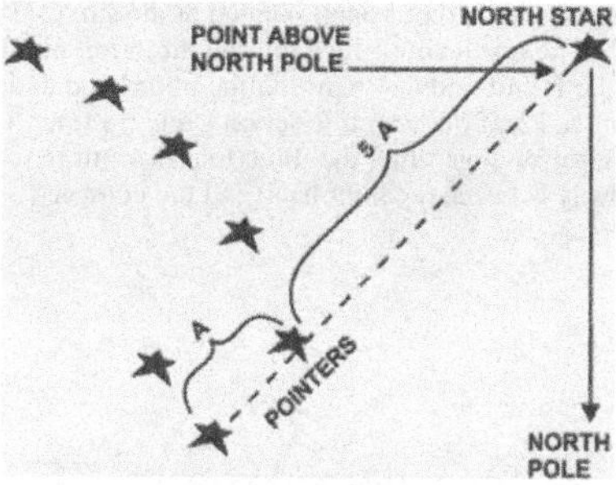

Figure 2-7. Locating the North Pole.

In the Southern Hemisphere, true south is determined in relation to the Southern Cross, a constellation composed of five stars. Two bright pointer stars in the vicinity of the Southern Cross serve as locators to help locate true south (see fig. 2-8). The outer four stars are

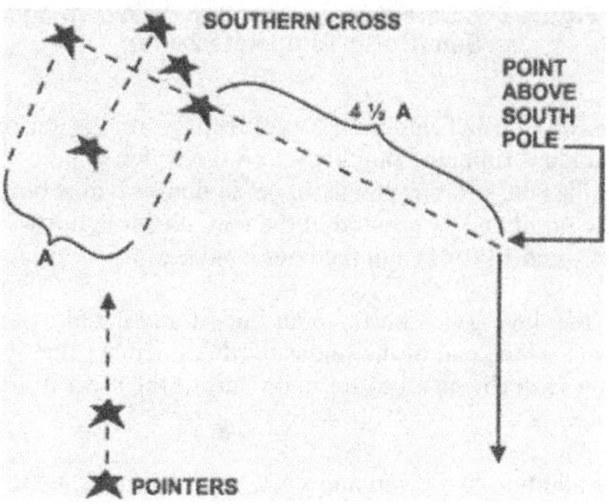

Figure 2-8. Locating the South Pole.

fairly bright and form a cross. This cross is imagined as the frame of a kite. A straight tail, four and one half times as long as the length of the kite itself, is put on the kite using finger widths for a measuring stick. The end of this tail will be close to a position directly over the South Pole. Usually, it will not be possible to see a star in the immediate vicinity, because there is no bright star visible directly above the South Pole.

During daylight hours, a watch and the sun can be used to determine direction within 8 degrees. In the North Temperate Zone, the watch is held horizontally, face up, and the hour hand pointed at the sun (see fig. 2-9). The north-south line and the direction of south can be found midway between the hour hand and the number 12, if the watch is set on standard time. If in daylight savings time, the direction of south is found midway between the hour hand and the number 1.

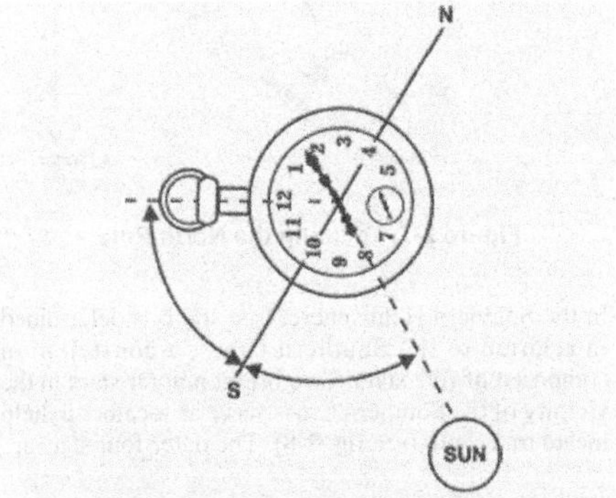

Figure 2-9. Determining Direction by Watch and Sun (North Temperate Zone).

In the South Temperate Zone, if the watch is set on standard time, the number 12 on the watch is pointed at the sun; if the watch is set on daylight savings time, the number 1 is pointed at the sun. North is midway between 12 (or 1) and the hour hand (see fig. 2-10).

When laying in a north-south line, if any doubt exists as to which end of the line is north, remember that the sun is in the east before noon and in the west in the afternoon.

In addition to the sun and stars, other methods a scout without a compass can use to determine direction include determining prevailing wind direction and

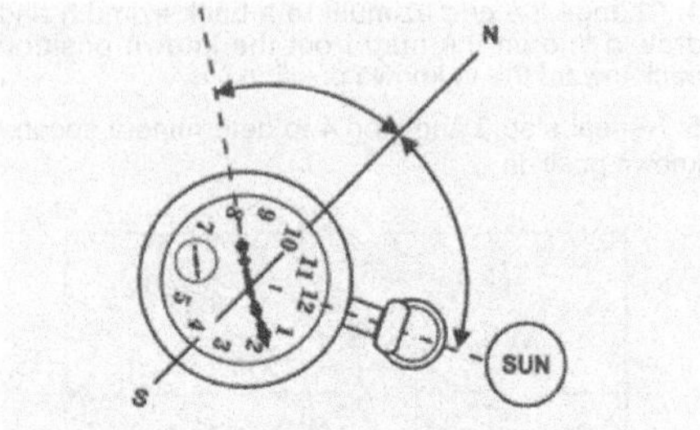

Figure 2-10. Determining Direction by Watch and Sun (South Temperate Zone).

using a mountain for orientation. By previous study of maps and photographs, a scout can keep informed of location and direction by using a distinctive edge of woods, a deep ravine or the direction of a stream's flow. A scout should constantly evaluate and memorize both the immediate terrain and general area for prominent features and landmarks.

2004. RANGE DETERMINATION

Range determination is the method of finding the distance between an observer and an enemy target or an object. By accurate range determination, the members of a given unit can set their sights correctly and place effective fire on enemy targets. The degree of accuracy is dependent on several factors, such as terrain relief, time available, and experience of the observer.

Mental Estimation

A mental distance estimate is made using a known unit of measure. Distance is estimated to the nearest 100 meters by determining the number of known units of measure between the observer's position and a target. For example, a football field, which is 100 yards, can be used as a known unit of measure for determining the distance between an observer's position and a target. For longer distances, progressive estimation may be necessary. To do this, the observer determines the number of units of measure to an intermediate point and doubles the value. The observer should consider the effects in table 2-1 in estimating distances.

Table 2-1. Effects to Consider in Mental Estimation of Distances.

Objects Appear Nearer	Objects Appear More Distant
In bright light.	In poor light or in fog.
In clear air at high altitude.	Only a small part of the object can be seen.
The background is in contrast with the color of the object.	The background is similar in color to that of the object.
The observer is looking down from a height.	The observer is looking over a depression, most of which is visible.
The observer is looking over a depression, most of which is hidden.	The observer is kneeling or sitting, especially on a hot day, when the ground is moist.
The observer is looking down a straight feature such as a road.	
The observer is looking over water, snow, or a uniform surface such as a cultivated field or desert.	

Estimating in Good Visibility

When visibility is good, distances can be estimated by using the appearance of tree trunks, branches, and foliage (as seen by the naked eye) in comparison with map data. Table 2-2 is a guide for wooded terrain. Table 2-3 is a guide for urban environments.

Estimating From a Terrain Study

The Marine should always use terrain/map analysis to assist in estimating distances. When the Marine is looking in a specific direction, the estimation of distance can be enhanced by studying the terrain and comparing it with the map. Particular emphasis should be given to color contrasts of terrain features seen along the observer-target line (OTL). For example, the distance across successive ridge lines or depressions in the distance may be identifiable by only slight changes of color to the eye. Different colors of grass might reveal a hidden terrain feature such as a stream.

Table 2-2. Estimating Distance in Wooded Terrain.

Distance in Meters	Tree Description
1,000	Trunk and main branches are visible. Foliage appears in cluster-like shape. Daylight may be seen through the foliage.
2,000	Trunk visible, main branches distinguishable, foliage appears as smooth surface. Outline of foliage of separate trees distinguishable.
3,000	Lower half of trunk visible. Branches blend with foliage. Foliage blends with adjoining trees.
4,000	Trunk and branches blend with foliage and appears as a continuous cluster, smooth in appearance. Movement of foliage due to wind cannot be deleted.
5,000 and beyond	Whole area covered by trees and appears smooth and dark.

Table 2-3. Estimating Distance in Urban Terrain.

Distance in Meters	Object Identified by the Unaided Eye
1,000	Lone tree trunk
1,500	Individuals and horsemen
3,000	Chimneys on rooftops
4,000	Windows in houses
4,000–5,000	Individual houses in populated area
8,000–9,000	Villages and individual houses
15,000–18,000	Large houses, towers, and steeples

Chapter 3. Enemy Activity

A commander often acts on information furnished by scouts. Therefore, scouts must aim at absolute accuracy in reporting enemy activity. This chapter discusses estimating enemy strengths, interpreting signs and tracks, and knowing the enemy.

3001. ESTIMATING ENEMY STRENGTHS

If troops cannot be counted, their strength may be estimated by: noting the length of time it takes various types of moving columns to pass given point, the area required of a unit in camp or bivouac, or the front on which they are deployed. When the ground is dry, infantry on the march raise a low, thick cloud of dust, and motor vehicles or mechanized units raise a thick, rapidly moving cloud. Additionally, through practice, a scout may gain information as to the strength and composition of enemy forces by listening to noises and observing lights, fires, and smoke. A scout gains valuable experience in estimating enemy strengths by observing friendly forces in camp, on the march, and deployed. The knowledge scouts gain during field exercises of the appearance and tactical dispositions of squads, platoons, companies, and larger units will be of great assistance in estimating the strength and composition of enemy units observed under various conditions.

3002. INTERPRETING SIGNS AND TRACKS

In addition to estimates made through direct observation, a scout may often be able to estimate size, composition, direction, rate of movement, condition, discipline, state of training, and morale of enemy forces through signs and tracks left behind.

Signs

The examination of vacated enemy positions provides valuable information. The size of a bivouac or defense area ordinarily indicates the number of enemy occupants. Clothing, ration containers, dumps, etc., further indicate the quantity of the departed enemy force. The condition of the bivouac area and amount of material abandoned give an indication of the enemy morale, training, and discipline. A well-policed area indicates good discipline. Rubbish, ration and smoking residue, and nonessential personal items of equipment adrift indicate a lower state of morale, training, and discipline. Stores and material left behind in good condition may indicate a hasty movement or withdrawal. Burned or destroyed materials indicate a deliberate, orderly withdrawal or movement. Letters, insignia, and other articles may reveal the identity of the enemy unit.

In the case of a moving enemy, the distance between periodic halts indicates the rate of march if enemy habits relative to marches and halts are known. Condition of the halt areas indicates the state of morale, training, and discipline.

The physical condition of enemy dead and wounded and their personal equipment and weapons are reported. The general condition and state of maintenance of destroyed or abandoned vehicles should also be reported.

Tracks

A track is a mark left on the ground by the passage of a person or object. Examination of tracks reveals information about the enemy.

Troops

A few tracks overlapping each other on both sides of a road or trail may indicate a patrol in staggered formation. A large number of tracks indicates troops in column formation. A large column will wear a dry road smooth and flat. In damp terrain, a freshly made track will have sharp edges; ordinarily, signs of moisture will disappear in about 15 minutes. A runner's toes are dug into the ground; a walker's footprint is fairly even.

Vehicles

The type of track indicates whether the vehicle is wheeled or tracked. A scout acquires the necessary experience to make the proper determination by observing vehicle tracks during training.

The direction of travel can be determined by the way tracks pass across ruts, by impressions on the edges of holes in the ground, how water is splashed from puddles, or by the way grass, twigs, and branches are bent; for example—

- A vehicle (wheeled or tracked) entering a rut pushes dirt into the rut and leaves an indentation on the exit side of the rut.
- A wheel going over holes in the ground leaves a deeper impression on the edge toward the direction of travel.
- The side of a puddle with the greater splash indicates the direction of travel of the vehicle.
- When traveling cross-country, the direction in which grass is bent and/or twigs, branches, and bushes are broken indicates direction of travel.

A general rate of speed can be estimated by the amount of water or mud splattered. A fast-moving vehicle will throw larger amounts of water or mud a greater distance to the front and sides than a slow-moving vehicle, and it leaves a deeper impression on the exit edges of holes. The faster the travel, the deeper the impression.

3003. KNOWING THE ENEMY

A scout should learn as much as possible about enemy psychology, habits, organization, and tactics. The more knowledge gained about the enemy, particularly the enemy's normal security measures, the better the scout's chances are to observe and obtain accurate information with minimum risk to the mission's success. Scouts gain much of this knowledge through experience, but they also gain a great deal of their preliminary information, particularly that pertaining to enemy organization and tactics, during training and may be updated by unit commanders and intelligence officers.

CHAPTER 4. DAYLIGHT SCOUTING

A scout must be able to operate in all types of terrain and under all conditions of visibility. He must be thoroughly familiar with the principles for using cover and concealment, camouflage, individual movement, and route selection, both to and from the objective.

4001. COVER AND CONCEALMENT

Cover is protection from the fire of hostile weapons. Concealment is protection from observation or surveillance from hostile air and ground observation, but not from hostile fire. Both cover and concealment are divided into two main categories: natural and artificial. Natural cover includes small hills, ditches, rocks or vegetation. Fighting holes, bunkers, and brick walls are examples of artificial cover. Some features, such as buildings, provide both cover and concealment. In deciding whether to seek cover or concealment, a scout must make the best choice to complete the mission (see fig. 4-1).

Figure 4-1. Correct Use of Cover.

Concealment Principles

Concealment principles are as follows:

- Remain motionless while observing. Anything in motion attracts the eye.
- Use all available concealment.
- Observe from the prone position (it offers a low silhouette and makes detection by the enemy difficult.
- Expose nothing that reflects light.
- Blend with the background because contrasting colors are noticeable.
- Remain in the shade because moving shadows attract attention.
- Distort or change the regular outline of objects. Most military objects have distinctive shapes that make obvious shadows and silhouettes.
- Avoid the skyline. Figures on the skyline can be seen from great distances and are easily identified by their outlines.

Concealment Techniques

Concealment techniques are as follows:

- When observing, the scout looks around an object's side (unless it is transparent) and prepares to fire, if necessary, around the side of or, if possible, through an object.
- Looking or firing over an object can make the scout an easily visible target for the enemy. If the scout must fire over the top of concealment or cover, the outline of the head or helmet should be broken or distorted.
- Upon the approach of an airplane, the scout takes a prone position, turns face-down, and remains motionless. If surprised by an airplane, the scout remains in place and does not look up.

- The scout covers exposed body parts such as the face, back of the neck, and hands with grease paint, mud or other materials to reduce sun reflection.
- Camouflage for equipment can be improvised from garnishing or sandbags to prevent sun reflection.
- In snowy terrain, white overgarments are worn.
- The helmet cover outline should be distorted.

4002. CAMOUFLAGE

Camouflage is the use of concealment and disguise to minimize the possibility of detection and/or identification of troops, material, equipment, and installations. The purpose of camouflage is to provide concealment of military objects from enemy observation. Camouflage is also used to conceal an object by making it look like something else. A scout's mission usually requires individual and equipment camouflage. If natural camouflage is not adequate, the position is camouflaged. In using camouflage, remember that objects are identified by their form (outline), shadow, texture, and color. The principal purpose of camouflage in the field is to prevent direct observation and recognition.

Individual Camouflage

Successful individual camouflage involves the ability to recognize and take advantage of all forms of natural and artificial concealment available (vegetation, soil, debris, etc.) and knowledge of the proper use of artificial camouflage materials.

Aids to Individual Camouflage

A scout must recognize the terrain's dominant color and pattern and must change the appearance of clothing and equipment accordingly in order to blend and not contrast with the terrain (see fig. 4-2).

The helmet is camouflaged by breaking up its shape, smooth surface, and shadow. Use of a helmet cover works best. In the absence of a helmet cover, mud can be irregularly blotched on the helmet to disguise its form and dull the surface. A helmet cover may be improvised from irregularly colored cloth or burlap to blend with the background. Foliage can be draped to prevent the visor of the helmet from casting a dark shadow across the face. Foliage should not stick up like plumes because any head movement will give away the position.

A small, thin bush in the shadow of a large bush makes a good observation point. Lone trees, rocks, fence corners, and outstanding landmarks are easily picked up by the enemy as obvious observation posts.

If camouflage clothing is not available, other available clothing can be attached in irregular splotches of appropriate colors.

Exposed skin reflects light and attracts the enemy's attention. Even very dark skin will reflect light because of its natural oil. The buddy system is recom-mended when applying camouflage. Standard Marine Corps issue camouflage face paint sticks are two toned:

- Loam and light green for light-skinned troops, in all but snow regions.
- Sand and light green for dark-skinned troops.
- Loam and white for troops in snow-covered terrain.

Shiny areas (forehead, cheekbones, nose, and chin) are painted with a dark color. Shadow areas (around the eyes, under the nose, and under the chin) are painted

Figure 4-2. Avoid Contrasting Backgrounds.

with a light color. Skin that is exposed on the back of the neck and hands is painted with a two-color combination in an irregular pattern (see fig. 4-3).

When standard issue face paint sticks are not available, burnt cork, charcoal or lamp black can be used to tone down exposed areas of skin.

Mud is used only in an emergency because it changes color as it dries and may peel off, leaving the skin exposed. Since mud may contain harmful bacteria, mud should be washed off as soon as possible.

Any equipment that reflects light should be covered with a nonreflective material that aids in the concealment of the weapon (for example, black electrical tape or mud). The straight line of the rifle or other infantry weapons may be very conspicuous to an enemy observer. The barrel and hand guard should be wrapped with strips of contrasting colored cloth or tape to break the regular outline. Mud or dirt dulls the reflecting surface of the stock, barrel, and bayonet where coloring has been worn. Lamp black may also be used on metal parts. The function of the weapon must not impaired.

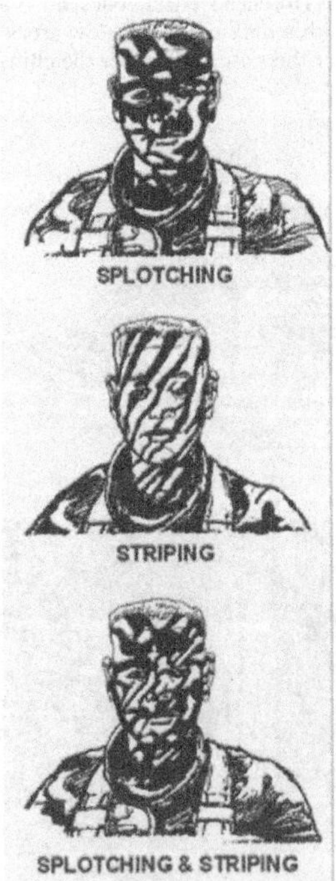

Figure 4-3. Face Camouflage.

If time, material, and surroundings permit, a ghillie suit should be constructed. (Refer to MCWP 3-15.3, *Scout Sniping*.)

Aids to Camouflage a Position

To successfully camouflage a position, the scout must remember to—

- Camouflage the position as soon as it is occupied.
- Avoid using too much material for camouflage. Even though natural materials are used, too much may make the object and its shadow stand out from its surroundings, thus attracting the attention of a hostile observer.
- Inspect completed camouflage work from the enemy's point of view to check effectiveness.

Continuous Camouflage

Camouflage around and on the scout's position must be maintained in a fresh condition as wilted and dead foliage can give the position away. If the mission dictates that the position should be occupied for longer periods, wilted foliage should be replaced during periods of reduced visibility.

4003. INDIVIDUAL MOVEMENT

Principles

The principles of individual movement are as follows:

- Scouts move from one concealed position to another. When not changing positions, they remain motionless.
- The scout's head is lifted slowly but steadily, without abrupt movements, to search for a new position.
- Scouts select the next stopping place before moving and ensures it is not contained by the enemy.
- Scouts change position on the run: springs up, runs with the body bent low, zigzags, quickly drops to the ground slightly to the right or left of the objective, then rolls or crawls to the desired position.

(Remember the phrase, "I'm up—he sees me—I'm down.")

Rushing

When starting from the prone position—

- Raise the head slowly and steadily and select a new position.
- Lower the head slowly, draw arms inward, cock right leg forward, and prepare to rush.
- Use one movement to raise the body by straightening both arms.
- Spring to your feet, stepping off with the left foot.
- Bend forward as low as possible when running. Never advance directly to the next position; always zigzag.

When hitting the deck—

- Stop.
- Plant both feet in place.
- Drop quickly to the knees and slide the hand to the heel of the rifle.
- Fall forward, breaking your fall with the butt of the rifle. (To confuse the enemy, roll over after hitting the deck and roll into firing position with feet, knees, and stomach flat on the ground.)
- Keep head down if you do not intend to fire.

When rolling over—

- Hit the deck and assume the prone position.
- Bring the rifle in close to the body, placing the rifle butt in the crotch.
- Roll over swiftly to confuse any enemy observers as to final intended location. Never reappear at the same place you went down.

Low Crawl

The low crawl is used when—

- Cover and concealment are scarce.
- The enemy has good observation over the area in which the scout is moving.
- Speed is not essential.

To perform the low crawl, keep the body as flat as possible against the ground. Grasp the rifle sling at the upper sling swivel. Let the balance of the rifle rest on the forearm and let the butt of the rifle drag on the ground. Keep the muzzle off the ground.

To start forward, push arms forward and pull right leg forward. To move forward, pull with arms and push with right leg. Change the pushing leg frequently to avoid fatigue.

High Crawl

The high crawl is used when—

- Cover and/or concealment are available.
- Poor visibility reduces enemy observation
- Greater speed of movement is required.

To perform the high crawl, keep body off the ground. Rest weight on forearms and lower legs. Cradle rifle in arms, keeping the muzzle off the ground. Keep knees well behind the buttocks to stay low.

Move forward, alternately advancing right forearm and left knee; then left forearm and right knee.

Movement Aids

Aids to movement include—

- Carrying only necessities. Additional weight causes premature fatigue and impedes free movement.
- Not disturbing birds or animals whose flight would betray your presence. If birds or animals are alerted, remain motionless under cover for a few minutes, as the enemy's attention may also be attracted.
- Moving during an incident that diverts attention, such as an airplane flight, a distant disturbance or sudden bursts of fire.
- Fog, smoke, or even light haze offer concealment for movement; however, the enemy may have thermoimagery and night vision devices. Therefore, darkness and smoke cannot be used as easily.
- Following a stream or road by staying as far away from them as possible while still keeping them in sight. Keep close to the dune line when moving along a beach.
- When moving through tall grass or similar growth, move when the wind blows, changing direction frequently. A straight route will be more readily noticed.
- Whenever possible, avoid areas of soft ground so as not to leave tracks.

- When crossing a road or water obstacle, choose crossing sites where the enemy's observation is restricted (an area in shadows or near a bend) and cross rapidly.

4004. ROUTE SELECTION

Prior to Movement

A scout and the immediate commander conduct a map reconnaissance before starting on a mission. This assists them in selecting the route according to available cover and concealment and any indicated enemy activity.

Prior to and during the course of the mission, move to an observation point to visually reconnoiter the terrain for movement and select the tentative route. It may be necessary to make wide detours around open spaces or those containing enemy patrols or other enemy activity.

Carefully study the country to be traversed and pay close attention to the general features, streams, ridges to be crossed, and their relation to the general direction to be taken (see fig. 4-4).

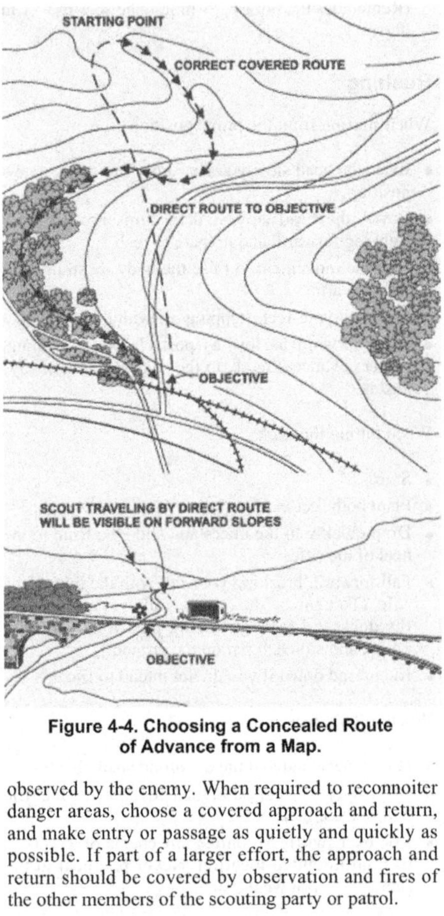

Figure 4-4. Choosing a Concealed Route of Advance from a Map.

Make notes of terrain features and landmarks along the proposed route and rely on notes for guidance (see fig. 4-5 on page 4-6). Additionally, determine the compass direction and readings for each change of direction at the start. Finally, learn the location of unit boundaries and observation/listening posts as well as general location of other friendly or scouting parties. Be sure to avoid man-made and natural obstacles as they will slow progress and overall success of the mission. If possible, use the local populous as a source of intelligence. When returning to friendly lines, avoid using the same route.

En Route

En route, the actual advance will be a series of movements from one observation point to the next. The distance and route will depend on cover and terrain. Assess the cover, terrain, and any enemy or civilian activity to determine whether or not to modify the approach or return routes. Unless the mission requires it, avoid danger areas (for example, houses, villages, potential assembly or bivouac areas, roads, and streams) that may give away your position by being observed by the enemy. When required to reconnoiter danger areas, choose a covered approach and return, and make entry or passage as quietly and quickly as possible. If part of a larger effort, the approach and return should be covered by observation and fires of the other members of the scouting party or patrol.

Stream Crossings

When the crossing does not appear to be held by the enemy, advance upon it rapidly. If there are two or more scouts, one crosses while the other(s) provide protection. Note the length, width, depth, and approaches to a crossing. Observe the condition of the road or trail that crosses the stream, and report on the suitability of the crossing for use by tracked and wheeled vehicles. If the crossing is under observation by enemy, seek another crossing site or dash across to avoid detection.

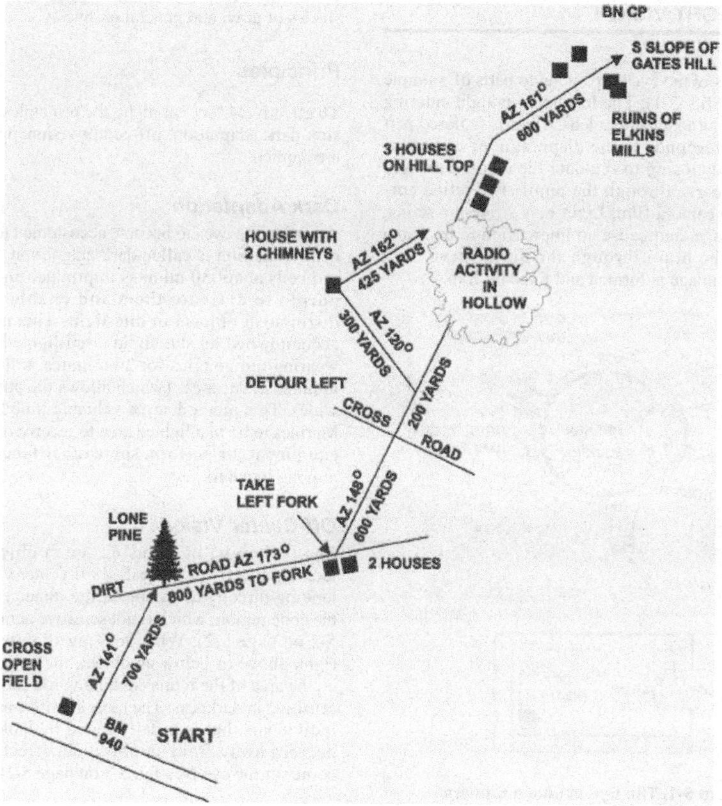

Figure 4-5. Proposed Route Sketch.

CHAPTER 5. NIGHT SCOUTING

Night scouting presents many of the same problems encountered in day operations-such as cover, concealment, movement, and camouflage-as well as additional considerations. Knowledge of human eye construction and operation will enable maximum advantage under night conditions or poor visibility.

5001. NIGHT VISION

Certain parts of the eye correspond to parts of a simple camera (see fig. 5-1). The lens focuses light entering the eye similar to a camera lens. The iris (colored part of eye) corresponds to the diaphragm of a camera, opening and closing to regulate the amount of light entering the eye through the pupil. The retina corresponds to camera film. Light rays strike the retina, form an image, and cause an impression to be transmitted to the brain through the optic nerve. In a camera, the image is formed and fixed on film.

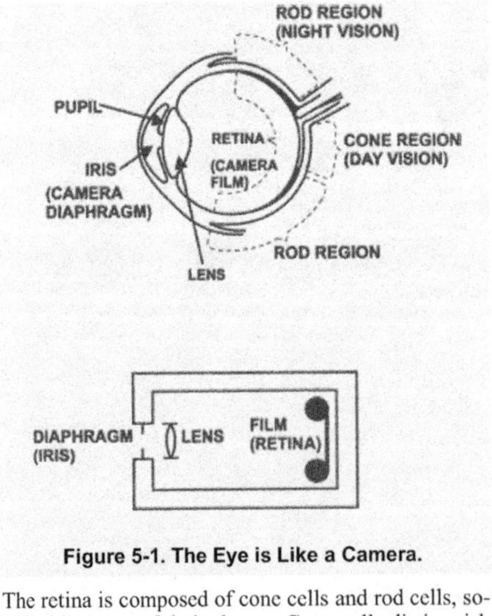

Figure 5-1. The Eye is Like a Camera.

The retina is composed of cone cells and rod cells, so-called because of their shapes. Cone cells distinguish color, shape, and sharp contrast. Because they are activated by light conditions, they are blind during periods of low illumination. Rod cells produce a chemical substance called visual purple that makes them active in darkness, low illumination or night conditions. Rod vision distinguishes black, white, shades of gray, and general outlines.

Principles

To effectively "see" at night, the principles of night vision dark adaptation, off-center vision, and scanning are applied.

Dark Adaptation

Allowing the eyes to become accustomed to low levels of illumination is called dark adaptation. It takes the rod cells about 30 minutes to produce enough visual purple to activate them and enable the eye to distinguish objects in dim light. This may also be accomplished by staying in a red-lighted area, or by wearing red goggles for 20 minutes, followed by 10 minutes in darkness (which allows the pupils to open wide). This method saves valuable time by allowing Marines to be in a lighted area to receive orders, check equipment, or perform some other function before moving into darkness.

Off-Center Vision

The technique of focusing on an object without looking directly at it is called off-center vision. When looking directly at an object, the image is formed on the cone region, which is not sensitive at night (see fig. 5-2 on page 5-2). When looking slightly to the left, right, above or below an object, the image is formed on the area of the retina containing rod cells, which are sensitive in darkness. The most sensitive area varies in individuals, but is usually found by looking 6 to 10 degrees away from an object; in effect, out of the corner of the eye (see fig. 5-3 on page 5-2).

Scanning

Off-center vision used to observe an area or an object is called scanning. When using rod vision, the visual purple in the rod cells bleaches or blacks out in 4 to 10 seconds and the object observed disappears. As the visual purple in the rod cells in one area bleaches out, the eyes must slightly shift to use fresh rod cells. Eyes

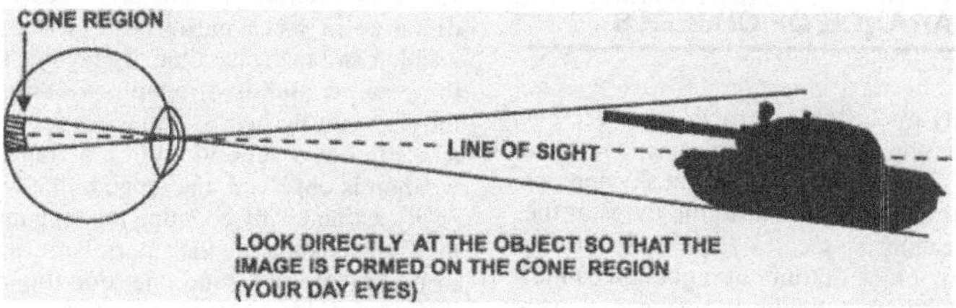

Figure 5-2. Day Vision.

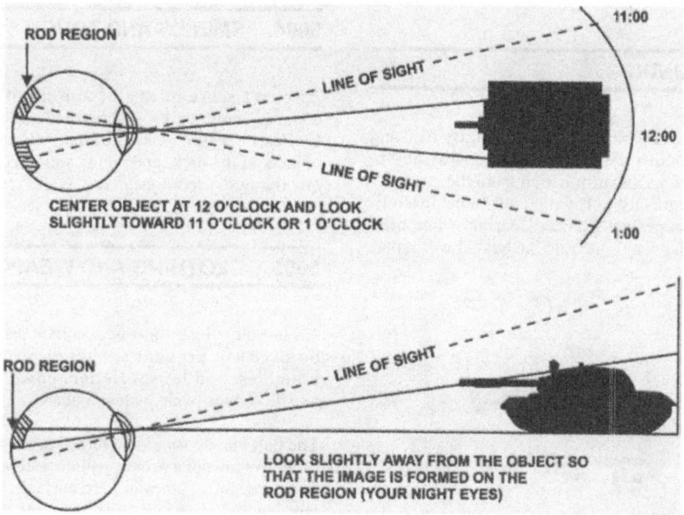

Figure 5-3. Night Vision.

should be moved in short, abrupt, irregular movements over and around the target (see fig. 5-4).

Preserving Night Vision

Night vision is quickly destroyed if bright light is allowed to enter the eye. When entering a lighted area or when observing in a temporarily lighted area (illumination, flares), one eye should be closed and covered to preserve its night vision. When the light goes off, fades or the lighted area is exited, the night vision retained by the protected eye enables it to see until the other eye adapts to the darkness. Red light helps preserve night vision, but like white light, it can be observed at long distances.

Factors that decrease night visual acuity include fatigue, lack of oxygen, long exposure to sunlight, alcohol, nicotine (within the past 48 hours), and age. When night vision has been attained, straining will not improve effectiveness; however, practice identifying objects at night will improve perception.

5002. APPEARANCE OF OBJECTS

Darkness not only makes objects difficult to see but also changes their appearance, distorts size, and blots out details. A tree visible against the night sky appears smaller than in the daytime because the twigs at the end of branches cannot be seen. A scout must train to identify objects by block outlines at night and cannot rely on details visible in daylight. Binoculars enlarge objects or parts of objects otherwise too small to be seen and help identify objects already spotted. Night observation devices increase night visibility and should be used whenever possible.

5003. SOUNDS

At night, sounds become very important. By listening, a scout gains information about the enemy and by exercising care, keeps information from the enemy. A scout stops frequently to listen. Scouts must listen for long periods in perfect silence. Hearing is amplified with the mouth open; removing the helmet will reduce sound distortion. Sounds are transmitted a greater distance in wet weather and at night than in dry weather and in the daytime. By holding the ear close to the ground sounds of people walking and vehicles moving can be heard. Sound travels approximately 370 meters a second. When a flash from a fired weapon is observed, the range to the weapons can be easily estimated by counting the time interval between the flash and hearing the report. For example, counting to three (one thousand one, one thousand two, one thousand three), indicates the distance is 1,110 meters. The cadence is determined by actual practice at known ranges.

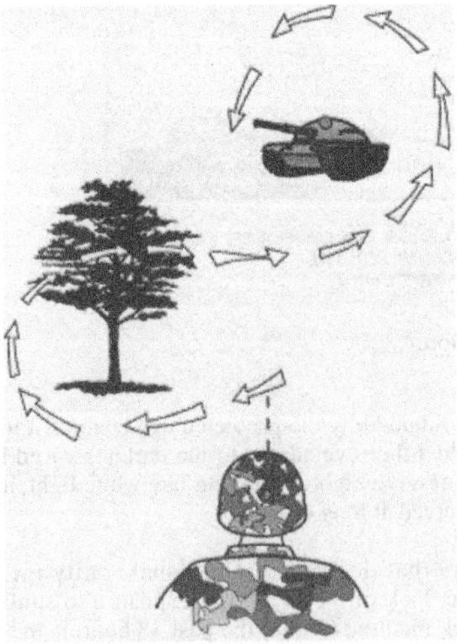

Figure 5-4. Scanning

5004. SMELLS AND TOUCH

A scout's sense of smell can warn of enemy fires, cooking, motor parks, gasoline and diesel engines, and bodies of water. A scout must feel and recognize objects in the dark, and adjust and operate equipment quietly by sense of touch.

5005. CLOTHING AND WEAPONS

All loose clothing must be secured (string or tape can be used) to prevent snagging on barbed wire, brambles, and brush. Helmet covers are worn to muffle sounds made by low branches.

The belt buckle should be turned around to the side in order to move in a prone position without scraping the buckle against stones or hard surfaces. Identification tags can be taped together to prevent rattling. Hands, face, and neck can be blackened so skin does not reflect light or appear as white spots in the darkness. (Refer to para. 4002.)

Scouts are normally armed with rifles. Rifle slings should be taped to prevent rattling. All weapons parts should be checked for glare elimination measures.

5006. CONCEALMENT

Although total darkness provides concealment, scouts must observe the same principles of concealment during moonlight conditions as in the daytime. Scouts should assume enemy employment of night observation devices and observe the principles of night movement such that presence will not be disclosed by noise when close to the enemy.

5007. AIDS TO NIGHT SCOUTING

Aids to night scouting include the following:

- Carry out scouting missions close to or within hostile positions on dark or rainy nights.
- Stifle a threatening sneeze by pressing fingers upward against the nostrils.
- Stifle a threatening cough by applying slight pressure with the finger on the Adam's apple.
- Stop a ringing sound that interferes with hearing by yawning.
- Speak softly rather than whispering if voice communication is necessary.
- Move boldly and rapidly when taking advantage of any sound-such as shelling, rustling wind or distant firing-to push forward if firing is taking place.
- Avoid shell craters and depressions in damp weather conditions if the enemy has employed chemical munitions.
- Move the eyes constantly; concentrating on one object too long will strain them.
- Take notice of the enemy's use of flares. When the enemy employs flares, few enemy patrols are apt to be out; when flares are not employed, the enemy's patrols are likely to be numerous.
- Drop to a prone position upon hearing a flare being fired and before it illuminates. Remain motionless while it is burning. If open or moving when a flare bursts in the air, freeze or drop quickly in the split second after the flare illuminates while the enemy is blinded. You are an easy target for the enemy if the flare bursts in the air or on the ground behind you. Never look at a flare. If you activate a trip flare, drop to the ground and crawl away from the illuminated area.
- Consider all patrols or individuals encountered as hostile until proven friendly. If encountering someone, crouch low, silhouetting the approaching person against the sky. At the same time, make yourself an indistinct target in case the person encountered is an enemy.
- Return fire only to avoid capture if fired on when close to enemy positions.

5008. AIDS TO NIGHT MOVEMENT

- Aids to night movement include the following:
- Move silently.
- Advance in stealthy legs. Each leg should follow some terrain feature that serves as a guide. When there are no terrain features to serve as guides, move in a straight or nearly straight line from one defined point to another, or maintain direction by using a compass.
- Avoid running, except in an emergency.
- Take advantage of sounds that may distract the enemy.
- Fall silently without making an outcry.

Walking

When walking at night—

- Place the heel down first. Balance the weight of the body on the rear foot until a secure spot is found.
- Lift the forward foot high to clear any stiff grass, brush, or other obstruction.
- Continue to balance body weight on the rear foot, lower the forward foot gently, toe first, to explore the ground for objects that might make noise. Step over fallen logs and branches, not on them.
- Lower the heel of the forward foot slowly; gradually transferring body weight to that foot.

Creeping

The low crawl and high crawl are not suitable at night when very near the enemy because an easily heard shuffling noise results. Creeping is the recommended method of movement:

- Creep at night on the hands and knees.
- Use your hands to feel for twigs, leaves or other substances that might make a noise. Clear a spot to place your knee. Keeping your hand at that spot,

bring your knee forward until it meets your hand. Then place your knee on the ground and repeat the action with the other hand and knee.

- Lay the rifle on the ground at your side and clear an area for it. Lift the rifle up and move it forward. Movement is slow and tedious, since it must be done silently.

Hitting the Deck at Night (Right-Handed Shooter)

To safetly hit the deck at night from the standing position—

- Advance your left leg, place the butt of the rifle in your right armpit with the hand remaining on the pistol grip, and grasp it with the right hand at the balance.
- Quietly drop down on the right knee and left hand.

- Move the left leg carefully to the rear, and then move the right leg to the rear.

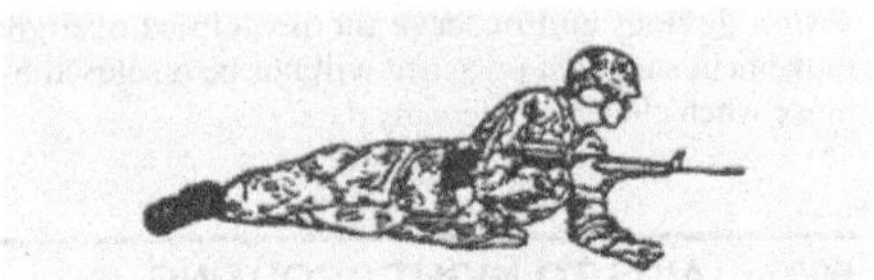

- Lie flat on the ground, or take up a firing position if necessary.

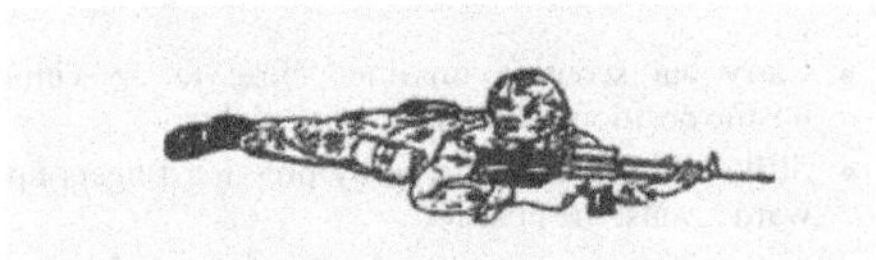

Wire Obstacles

A mission often requires a scout to pass through and work behind enemy positions. To accomplish this, the scout must be able to quietly pass through enemy wire obstacles and cross trenches. Cutting a gap in wire is time-consuming. If possible, walk over the low bands of enemy wire and crawl under the high bands (see fig. 5-5). Avoid movement along wire barriers, as enemy

Figure 5-5. Crossing Wire Silently at Night.

covering fires are generally planned parallel to them to take advantage of canalization and enfilade fire.

To step over low wire at night, crouch low to view the strands against the sky. Grasp the top strand with one hand; with the other hand, reach forward and feel for a clear spot for foot placement without stepping on other strands or any object apt to make a noise. Raise the body up, still grasping the top strand of wire. To avoid catching the foot in another strand, lift the foot up and over, passing it close to the hand grasping the wire.

If a high wire obstacle is encountered at night and wire cutters are not available, pass under the wire with your back on the ground. Grasp the lowest strands in your hands and hold them clear of the body while you slide under them.

When cutting wire and working solo, cut a wire near a post (see fig. 5-6), then dispose of all but one loose end. Grasp the wire close to a post and cut between your hand and the post, muffling the sound and keeping the loose wire in your grasp. When cutting wire in tandem, one firmly holds the wire with the hands positioned close to the cutters, in order to muffle the sound and prevent the loose ends from flying back, while the other one cuts. In both instances, the loose ends of the wire are bent back to form a passage.

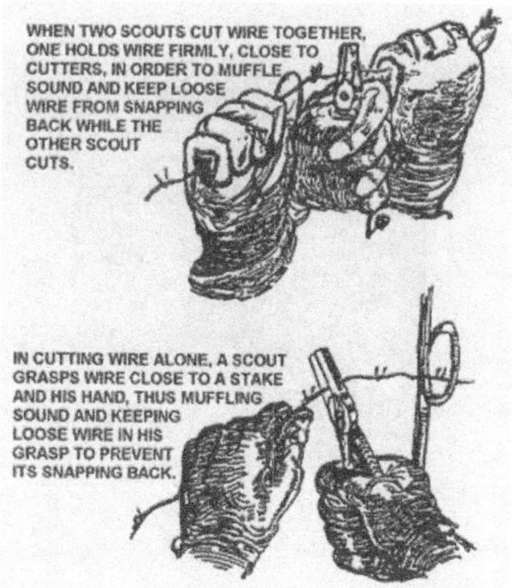

Figure 5-6. Cutting Wire Silently at Night.

Wrap a sandbag around the wire cutters and wire to deaden the sound.

Do not cut a complete gap in the wire; cut only the bottom wire(s). Leave the top wire(s) intact to lessen the chance of discovery by the enemy.

Crossing Trenches

Before approaching a trench, wait outside the trench for awhile and listen. Do not enter or cross a trench near its junction with a communication trench. Crawl silently up to the edge of the trench and look into it. Remove all loose dirt and rocks from the edge. If it is a narrow trench, spring up and jump across, sinking quietly to the ground on the other side and remaining there a moment to listen before proceeding. If the trench is wide, climb silently and slowly down into it and out the other side, using the revetment for support (see fig. 5-7). Do not enter enemy trenches unless it is absolutely necessary in order to accomplish the mission. Ordinarily, work is better accomplished from outside the trench. Sentries usually pay more attention to sounds in front of them; therefore, if it is necessary to enter a trench, cross it first at the place where enemy observation is restricted, then approach from the rear.

5009. LOCATING AND PLOTTING THE ENEMY AT NIGHT

For night work, a scout must understand the use of a lensatic compass. (Refer to para. 2002.) Using the lensatic compass, the scout can guide the platoon into position, locate adjoining elements of the command, keep direction when on patrol, determine the location of gaps in the enemy wire and the position of enemy out guards.

Locating Gaps in Enemy Wire

When searching for gaps in enemy wire, at least two lensatic compasses are needed: one to register the gap in the wire and the other for navigation. When a gap in enemy wire is located, lie outside the gap, keeping a distance of 10 meters from the barbed wire. Sight with the lensatic compass on a prominent point on the skyline in line with the gap. In selecting the prominent point in the skyline, pick one that appears on the map (i.e., hill mass, house, road junction). If the only prominent point available is one not identifiable on a map (i.e., a tree, destroyed vehicle, enemy position),

Figure 5-7. Crossing Trenches Silently at Night.

use it. The exact location of this point can be fixed the next day by visual reconnaissance of the area from an observation point. When the needle rests—

- Clamp it in place by lowering the eyepiece to the closed position.
- Rotate the movable bezel ring until the luminous line is directly over the north end of the needle. The azimuth of the gap from the prominent point is now registered.

The compass is carried back without further adjustment. The azimuth setting can be recorded later on a map (see fig. 5-8).

Locating Enemy Out Guards

At night, locate enemy guards by their sounds and failure to maintain light discipline. As sounds of the enemy are heard, and/or observations of the enemy made, shoot an azimuth with one compass. When the needle comes to rest, clamp the compass by lowering the eyepiece to the closed position.

Figure 5-8. Locating a Gap in Enemy Wire at Night.

Note the time and nature of each sound, the estimated distance, and which compass was used to fix the location. Plot this data on a map then wait until the debrief to turn in the notes and compasses. See figure 5-9.

Figure 5-9. Locating Enemy Outguards at Night.

50010. ROUTES OF MOVEMENT

Prior to beginning a night missions, a scout studies the ground in detail from an observation point, air photos, and a map during daylight. The route of advance should be below the skyline. Avoid becoming a silhouette (see fig. 5-10).

Unless the moon is bright, avoid, if possible, passing through woods, ditches, ravines, and brush, because noises of movement may lead to discovery. If the enemy is known to have night observation device capability, avoiding these kinds of terrain may not be possible. To avoid enemy ambushes, return by a different route from the advance and change routes on successive nights.

Figure 5-10. Difference Between Correct Routes by Day or Night.

CHAPTER 6. OBSERVING AND REPORTING

When sent out on a mission, a scout's duties are to observe and report. the first five chapters covered scout protection measures and movement in enemy territory. This chapter discusses performing the assigned mission and reporting the mission after its completion.

6001. OBSERVATION POSTS

Positions

The following guidelines apply to observation posts. See figure 6-1.

- When selecting observation posts, scouts should chose the least prominent position.
- Scouts may occupy one or more observation posts.
- An observation post should not be manned for more than 24 hours.
- A selected observation post should be observed for 10 to 15 minutes to ensure it is not occupied.
- Scouts move to the chosen observation post by a concealed route.
- If the post is located on a hill, crawl to a position where the skyline is broken.
- If a tree is used, the position should have a background so as not to be silhouetted against the sky while climbing or observing.
- When leaving the observation post, a different route from that of the approach should be used.
- If a radio is used, its antenna should be located to provide clear communication to the controlling commander but masked from enemy observation and direction-finding equipment. Upon departure, scouts should remove the antennae from the observation post so as not to give away the position.

Observing

Using all senses available, be particularly alert for movement, objects, sounds, and smells inappropriate to the surroundings. While observing, avoid all unnecessary movement. If observing from a building, keep back from doors and windows.

In daylight, look first at the ground nearest you. Begin observing close to your post and search a narrow strip 50 meters or less deep, going from right to left parallel to your front; then search from left to right a second and similar strip farther away but overlapping the first. Continue to observe until the entire field of view has been searched (see fig. 6-2 on page 6-2).

At night, use a night observation device. If one is not available, search the horizon with short, jerky movements, and short pauses. Look a little to one side of an object and then to the other. Lower the head close to the ground to view the object more clearly. Use low-powered field glasses to increase sight range.

Figure 6-1. Method of Approaching an Observation Post.

Figure 6-2. Method of Searching Ground.

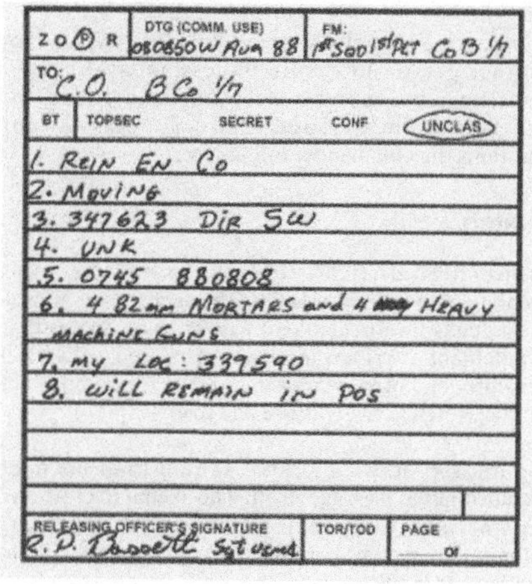

ZO (P) R | DTG (COMM. USE) 080850W Aug 88 | FM: 1st SQD 1st PLT Co B 1/7

TO: C.O. B Co 1/7

BT | TOPSEC | SECRET | CONF | (UNCLAS)

1. REIN EN Co
2. MOVING
3. 347623 DIR SW
4. UNK
5. 0745 880808
6. 4 82mm MORTARS and 4 HEAVY MACHINE GUNS
7. MY LOC: 339590
8. WILL REMAIN IN POS

RELEASING OFFICER'S SIGNATURE R.P. Bassett Sgt USMC | TOR/TOD | PAGE ___ OF ___

Figure 6-3. Sample Field Message.

6002. REPORTING

It is imperative that the scout accurately and completely report who, where, when, and what was observed to the proper authority upon mission completion. In most cases, a scout will not be equipped with a radio.

Verbal Reports

Verbal reports should be made when writing is impractical, when the information is not complicated or when the enemy is likely to intercept a messenger carrying a written message.

Written Messages

Written messages, preferred to verbal reports, are recorded in message book blanks issued for that purpose (see fig. 6-3) and delivered to a higher authority as soon as possible. The NATO spot report (SPOTREP) should be reviewed as the written message is a NATO format.

The message body is brief, accurate, and clear; facts and opinions are distinguished. If secondhand information is reported, its source is included. Reports include all information of value, first about the enemy, and then about the originator. Information about the enemy should cover—

- Size and/or strength.
- Actions or activity.
- Location and direction of movement.
- Unit identification. (The designation of the enemy unit may be derived from unit markings, uniforms worn or through prisoner interrogation.)
- Time of observation.
- Equipment and weapons.

Messages are printed in block letters. Individual items of information are numbered and separated into paragraphs. If doubt exists as to message receipt by the commander, a summary of its contents is included in the next message. Information about the originator or writer should cover—

- Location at the time of enemy observation (reference to an important terrain feature, by map coordinates, by the back azimuth from each of two definitely located points, or the back azimuth and distance from one known point).
- Intentions. (Remain in position? Continue on the mission? Take other action?)

The message is carefully reread and if possible read by another person to ensure understandability. If a messenger is used, the messenger must read and understand the message in order to answer any questions the commander might have.

Sketch

Information difficult to describe may be given accurately on a simple sketch. The sketch may give all the necessary information or it may be used to supplement a written message. A military sketch is generally one of two types: simple or panoramic. The simple sketch is easily made and read.

Figure 6-4 shows a simple sketch that has been included in the message itself. The sketch may be on a separate sheet of paper, but all of the necessary information must be contained in either the sketch, the message or both.

The panoramic sketch is a picture of the terrain's elevation in perspective, as seen from one point of observation. Although a panoramic sketch is not difficult to create, skill and training are necessary to enhance usefulness. Figure 6-5 on page 6-4 illustrates how to make a panoramic sketch.

Overlay

The same information sent back to higher headquarters on the sketch may be sent on an overlay, if the sender and the person to whom the message is to be sent have copies of the same map. Figure 6-6 on page 6-6 illustrates a simple overlay. The overlay is drawn on transparent paper as follows:

- Orient the map and place it on a hard, flat surface.
- Place the transparent paper over the part of the map of the object or information to be transmitted and hold the paper in this position.
- Orient the overlay to the map by tracing in the intersecting grid lines at two opposite corners of the overlay. Write the correct number designation on the overlay. The cross made by the intersection is called a *tick mark* and enables the receiver to locate the exact area on the map covered by the overlay.
- Sketch the object seen or the information to be transmitted on the tracing paper (the sheet on top of the map) in the exact location it would appear on the map (the sheet underneath the tracing paper).

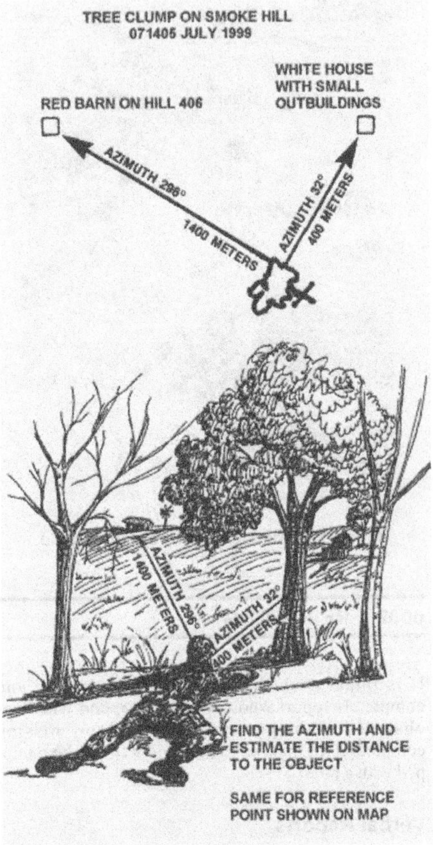

Figure 6-4. Making a Simple Sketch.

Explanatory notes are annotated in the overlay's margin, arrows point to the objectives.

- Indicate with an "X" and an appropriate explanation the position from which the observer saw the object or obtained information.
- Include title and scale of the map from which the overlay was made, date and hour the information was obtained, and signature of the observer on the overlay in the lower right-hand corner.

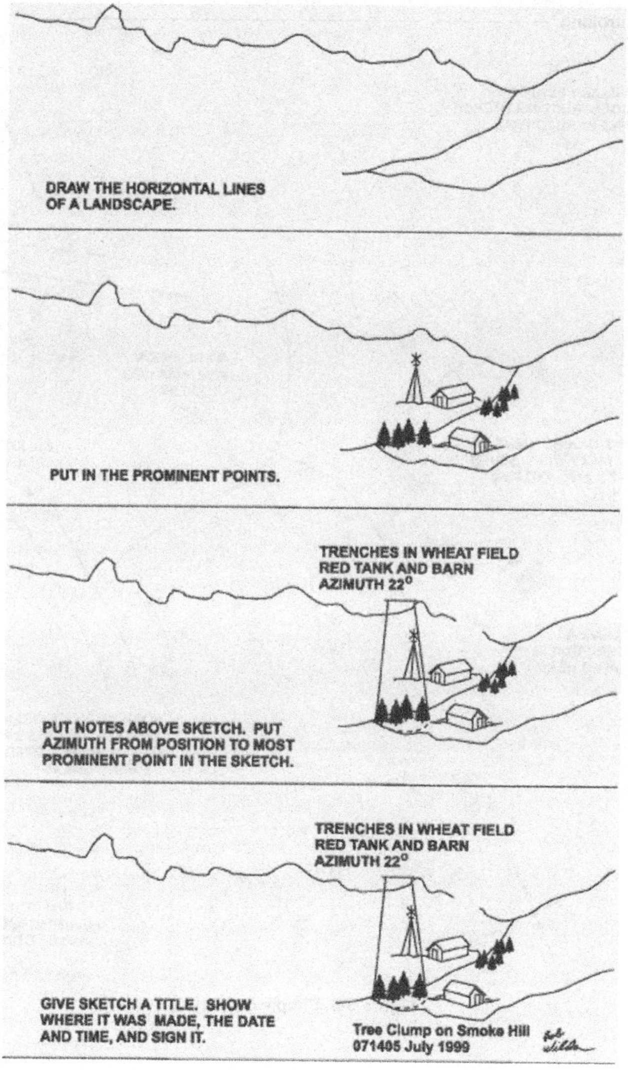

Figure 6-5. Panoramic Sketch.

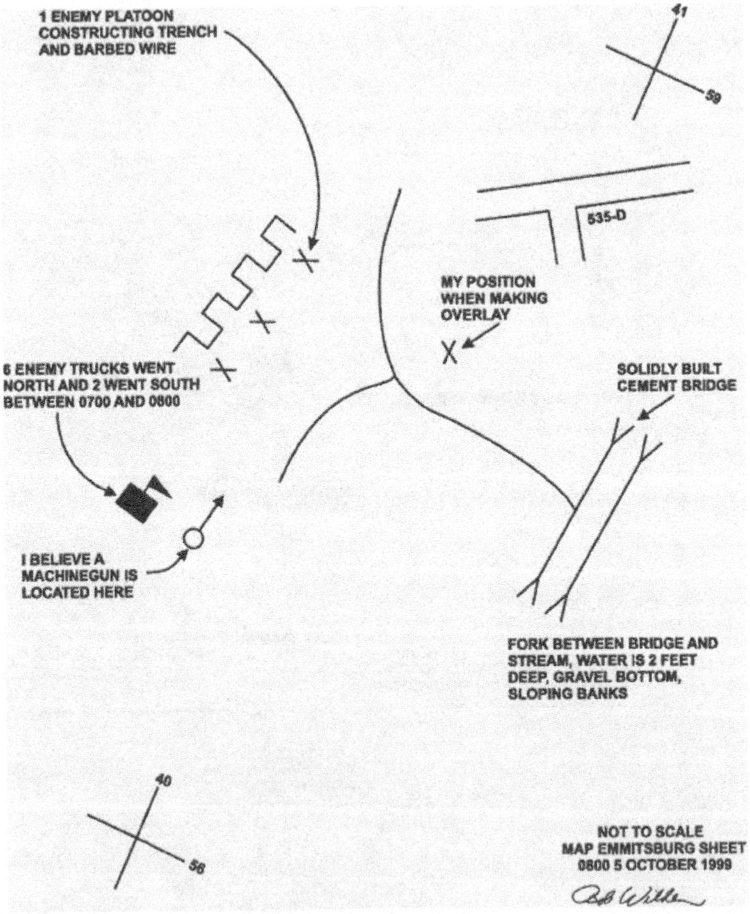

Figure 6-6. Simple Overlay.

CHAPTER 7. SCOUTING FIRE TEAMS

Upon completion of individual training, a Marine adapts this training to the functioning of the unit. The basic tactical unit for scouting is the fire team. Within the fire team, scouts normally work in pairs to furnish security and gather information. Information gathered by the individual scouts is passed to the fire team leader.

7001. POSITIONING

When a rifle platoon in the approach march is not preceded by adjacent units (left, right, front, and rear), it employs its own scouting elements. The scouting element for a platoon is usually one fire team; however, an entire squad may be used.

A fire team used as a scouting element is called a scouting fire team and is controlled by the platoon commander, assisted by the squad leader. A squad leader whose squad is providing the scouting fire teams normally marches near the platoon commander to assist in the control of the scouting fire teams.

A scouting fire team moves aggressively to cover the front of the advancing platoon and to locate the enemy's position(s). It generally moves in a wedge or skirmisher's formation. Normally, a scouting fire team is deployed on a frontage of 50 to 75 meters (10 to 17 meters between each individual scout). The entire squad may be employed to cover a wider frontage. The platoon commander coordinates the movement of the scouting fire team(s) so as to protect the main body of the platoon from enemy fire from points within 400 to 600 meters away, or in close terrain from points within the limits of enemy observation.

Scouting fire team(s) should have enough firepower to overcome resistance from small enemy advanced posts and patrols; the intent is to make enemy riflemen and machine gunners open fire and disclose their positions. Without scouts in advance, the platoon may move into areas where enemy fire may prevent further advance or maneuver and inflict heavy casualties (see fig. 7-1).

Scouting fire teams are covered by the platoon or, when the platoon is masked, the fire team leader ensures individual scouts are maneuvered and coordinated so that the fire team covers its own advance. The fire team leader constantly watches for signals from the platoon commander and remains in visual contact at all times.

The distance between the scouting fire team and the platoon is terrain-dependent. The scouting fire team should not be beyond visibility of the platoon. In open terrain, the platoon commander usually directs the scouting fire team to move by bounds along a succession of locations designated by the platoon commander as intermediate objectives.

Individual scouts should advance as stealthily as possible, while remaining consistent with their mission of reconnaissance to the front, taking advantage of cover without delaying the advance. An occasional glimpse of scouts constantly advancing over a wide front can make the enemy uneasy. It is this activity, and not the target the scouts offer, that may cause the enemy to open fire and disclose its location.

When fired upon, scouts must drop to cover and return fire only when necessary to complete the mission. If

Figure 7-1. Position of Scouts Preceding an Attacking Platoon.

necessary, one scout reports back to the fire team leader who informs the platoon commander.

7002. LOCATING ENEMY POSITIONS

The key terrain of defensive positions are those points that afford extended observation over the ground where the attack must advance. The enemy will place machine guns and infantry to defend critical points.

Members of a scouting fire team preceding an attacking platoon identify the probable enemy infantry and machine gun positions (see fig. 7-2). They use concealment and cover to conduct their advance in order to discover the exact location of enemy positions.

7003. ACTION WITH AN ATTACKING PLATOON

The scouting fire team reconnoiters to the front of the advancing platoon. As soon as the scouting fire team leader indicates the area is secure, the platoon advances and the scouting fire team moves forward. Squads within the platoon advance by bounds; at least one squad is positioned to support the other(s) by fire. Successive positions along the line of advance are selected and designated by the platoon commander as intermediate objectives, and reconnoitered by the scouting fire team before occupation. By conducting proper reconnaissance, surprise by the enemy or movement in the wrong direction may be prevented.

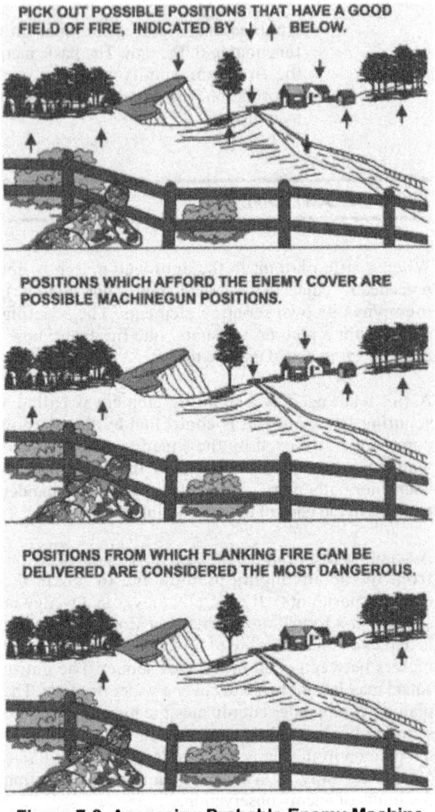

Figure 7-2. Assessing Probable Enemy Machine Gun Positions.

Movement

The distance between the scouting fire team and the front of the main body of the platoon is dependent on the mission, enemy, terrain and weather, troops and support available, and time available (METT-T). In close terrain, such as dense woods, the scouting fire team's movements closely resemble those used for night operations. In approaching houses, woods, and villages, one scout of each pair covers the other while the latter reconnoiters (see fig. 7-3).

A scouting fire team moves forward aggressively to cover the front of the advancing platoon, usually adopting either skirmisher's or wedge formation in order to be prepared to go into action immediately and to cover a wide frontage of 50 to 75 meters.

As a scouting fire team advances in open terrain, it is supported when possible by elements of the platoon; in close terrain, by mutual support within each fire team. Mutual support within the fire team is accomplished by the fire team leader and the automatic rifleman forming a team that supports by fire the advance of the rifleman and assistant automatic rifleman until they reach a location designated by the fire team leader. The rifleman and assistant automatic rifleman then support by fire the movement of the team leader and automatic rifleman. These successive points to which the sub elements of the fire team move are normally designated as fire team intermediate objectives by the fire team leader. The fire team leader sets as many fire team intermediate objectives as necessary to maintain mutual support within the team. This process is repeated until the team can be covered by other elements of the rifle platoon.

A scouting fire team takes advantage of available cover and concealment without delaying its advance. The orders of the platoon commander govern the distance at which it precedes the platoon. The terrain and the probable position of the enemy affect the scouting fire team's distance in front of the platoon. It may be as much as 400 to 600 meters in advance of the platoon. In open terrain, the platoon commander usually directs that the scouting fire team move by bounds to a succession of intermediate objectives. In close terrain or conditions of limited visibility, the scouting fire team is normally ordered to precede the platoon at the limit of visibility, maintaining visual contact with the platoon commander.

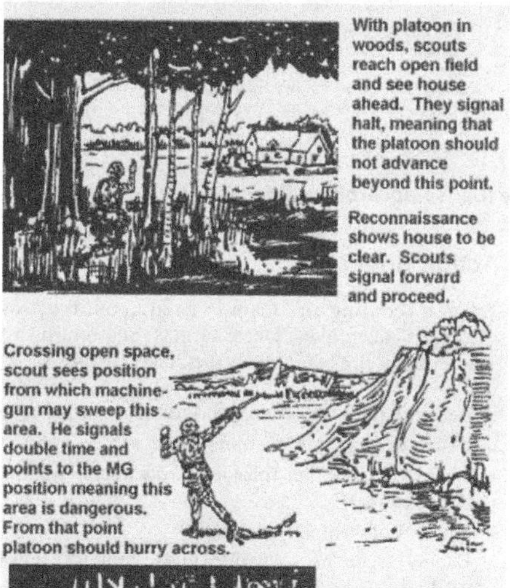

Figure 7-3. Conduct of Scouts Preceding a Platoon.

Action in Woods

When a scouting fire team is directed to advance over open ground to the edge of a woodline, two members of the team, preferably the rifleman and assistant automatic rifleman, reconnoiter inside the woodline while the remainder of the fire team covers them. It is not recommended they separate until finished with their reconnaissance of the far side of the danger area. Both members staying together can cover the same area using a zigzag reconnaissance and they are better equipped to overcome any opposition.

In heavy underbrush and/or poor visibility, the rifleman and assistant automatic rifleman proceed into the woodline together for 50 to 60 meters. The two then separate, searching out either flank to the first high ground or limits of observation, probably 50 to 100 meters (see fig. 7-4 on page 7-4). After the initial search and out posting the limit of advance, the remaining scout signals the fire team forward.

In light underbrush and/or good visibility, the assistant automatic rifleman remains at the edge of the woodline while the rifleman searches the woodline. (see fig. 7-5 on page 7-5). The rifleman searches the woodline in a zigzag pattern, reports back to the assistant automatic rifleman, then moves to an outpost position at the limit of advance. In turn, the fire team leader signals the platoon commander that it is safe for the platoon to move forward.

The fire team leader then moves the remainder of the fire team into the woods, joining up with the forward scout manning the outpost. The scouting fire team occupies and holds a line 50 to 75 meters within the woods and observes toward the direction of movement until the platoon closes up. The scouting fire team leader awaits further word from the platoon commander before moving the team further into the woods.

When directed, the scouting fire team leader moves the team forward until they reach the far edge of the woods. The team is held at the edge of the woods and the fire team leader notifies the platoon commander of the situation. The platoon commander moves the platoon to a position where it can cover the scouting

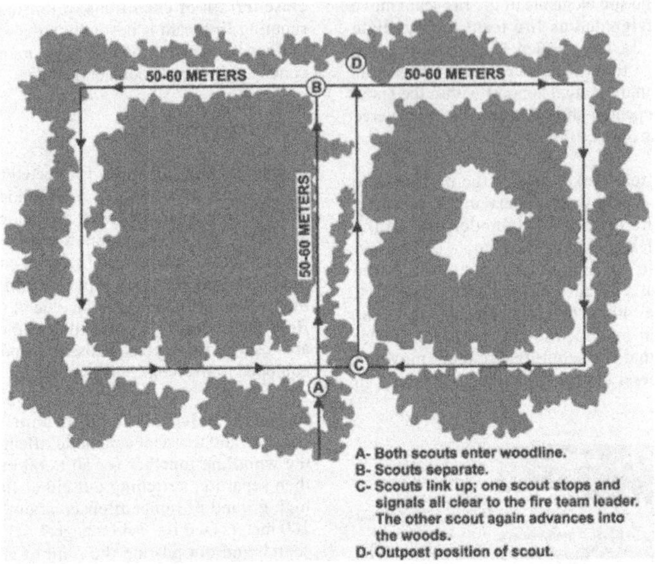

Figure 7-4. Searching Edge of a Woodline (Dense Underbrush/Poor Visibility).

fire team as it exits the woods and directs the team leader to move out and continue the scouting mission.

A scouting fire team passing through woods ahead of its unit maintains a distance allowing visual and oral communications. If an obstacle is encountered, reconnaissance to its front and flanks must be carried out. When advancing along a road or path, scouts precede the platoon to provide necessary protection and to prevent surprise fire on the platoon. When crossing a road or path, they reconnoiter well to the flanks before signaling "all clear" to the platoon.

The scouting fire team will not exit the woods until the arrival of the platoon commander, who will then be given an opportunity to alter the disposition or direction of march. The point where the platoon exits the woods is considered a danger area where the platoon is vulnerable to enemy fire. The scouting fire team is sent ahead to reconnoiter the danger area, as well as the next area to be occupied by the platoon. They signal back whether conditions require a halt, an advance or a quick rush across the open area. The scouting fire team leader must be continually on the lookout for signals from the rear.

Action Under Fire

When a scouting fire team is fired upon, they immediately take cover, locate targets, and return fire. The scouting fire team leader then determines—

- Location of enemy (range and reference points).
- Extent of position (location of flanks).
- Types of positions (obstacles, bunkders, fighting holes, etc.).
- Number of enemy.
- Enemy weapons (machine guns, mortars, tanks, etc.).

The platoon commander assesses the situation as quickly as possible based on the limited information obtained. Usually the platoon commander brings up the remaining squads, sets up a base of fire, and assaults the enemy position. Should the enemy position prove too strong for the platoon, the platoon remains engaged with the enemy as a base of fire until the remainder of the company is committed to clear the enemy resistance.

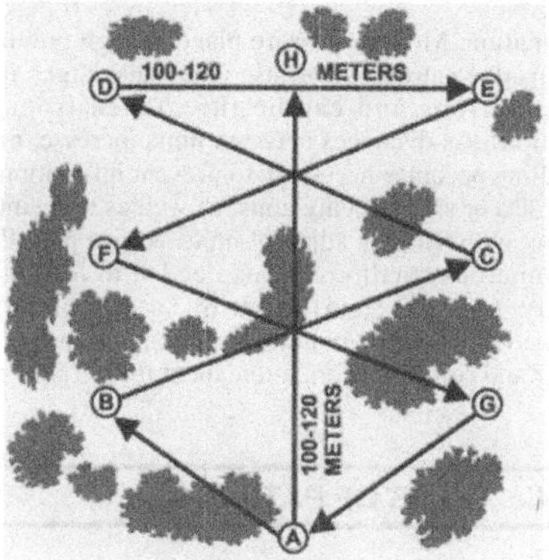

A- Both scouts enter woods. Assistant automatic rifleman remains here and covers forward movement of rifleman.

B through G - Rifleman moves to these positions, stopping and observing before moving to next position.

G to A - Rifleman reports results of his search to the assistant automatic rifleman.

A to H - Rifleman moves to out-post position.

Figure 7-5. Searching Edge of a Woodline (Light Underbrush/Good Visibility).

7004. ACTION WITH AN ENVELOPING UNIT

When a platoon is given the mission to envelop an enemy position, a scouting fire team is employed for protection and reconnaissance in the same way as when the platoon is advancing in the approach march.

Part 2. Infantry Patrolling

Chapter 8. Fundamentals of Infantry Patrolling

This chapter begins Part 2 and provides basic information about infantry patrols; specifically, their purpose, types, and missions. Infantry patrol training is also addressed and keys to successful patrolling are presented. Subsequent chapters of this part cover patrol organization, preparation, movement, and reconnaissance actions.

8001. DEFINITIONS

A patrol is a detachment of ground, sea or air forces sent out for the purpose of gathering information or carrying out a destructive, harassing, mopping-up or security mission (Joint Publication [JP] 1-02). The mission to conduct a patrol may be given to a fire team, squad, platoon or company.

8002. RELATION OF PATROLLING TO SCOUTING

Each patrol member must be knowledgeable in the principles of scouting and maintain membership of a larger team. To develop the teamwork skills required among the members of a patrol, additional training beyond the basic principles is necessary to become a well-trained scout. A patrol member must respond quickly to the decisions and orders of the patrol leader. There must be complete confidence among all members of the patrol and the confidence that they, as a team, will be successful in their mission.

8003. PURPOSE

A commander must have current information about the enemy and the terrain in order to employ the unit effectively. Patrols are an important means of gaining this information and are used to destroy enemy installations, capture enemy personnel, perform security missions or prevent the enemy from gaining information. Modern warfare places a high premium on effective patrolling because units have larger areas of operations and can be threatened from all directions. As distances between units increase, more patrolling becomes necessary to prevent infiltration by guerrillas or small enemy units, as well as to maintain contact with friendly adjacent units. Active patrolling by numerous small groups is needed to locate the enemy and gather information on the enemy's disposition, strength, morale, and weapons, as well as gather and confirm information about the terrain.

8004. TYPES OF PATROLS

Classification as to Mission

Reconnaissance Patrol

Reconnaissance patrols gather information about the enemy, terrain or resources. Relying on stealth rather than combat strength, they gather this information and fight only when necessary to complete the mission or to defend themselves. The distance covered by reconnaissance patrols varies based on the terrain and mission. The squad is ideally suited for reconnaissance patrol missions because of its relative small size and its experience of working together.

Combat Patrol

A combat patrol is a fighting patrol assigned missions that require engagement with the enemy in combat. Larger and more heavily armed than reconnaissance patrols, combat patrols have a mission to capture enemy documents, provide security, and capture or destroy enemy equipment and installations. Such action is ordinarily followed by a return to friendly positions. Regardless of the mission, the patrol reports any information concerning the enemy and terrain acquired during the accomplishment of the assigned mission. There are four types of combat patrols: raid, contact, ambush, and security (normally conducted by a Marine rifle platoon). A rifle platoon reinforced with crew-served weapons is normally considered the minimum size for contact, economy of force or ambush patrols. In some situations, such as the capture of a small enemy outpost, a rifle platoon could conduct a raid. However, a raid is a complex mission and, due to the organization of a raid force (command, reconnaissance, assault, support, security, and reserve elements), a rifle company is normally the smallest force assigned to a raid.

Classification as to Means of Movement

Foot Patrols

Movement by foot is the most common means; however, there are inherent disadvantages. Foot patrols travel slowly and carry limited quantity and types of equipment and supplies. Range and area coverage is relatively restricted. Foot patrols also have apparent advantages in that they have fewer restrictions as to terrain that can be covered; are more difficult for the enemy to detect; provide thorough coverage within limits of range; and are generally not inhibited by weather.

Motorized Patrols

Where terrain and road networks permit, a motorized patrol overcomes the inherent disadvantages of the foot patrol. Mechanized forces require patrolling units that can keep pace with them. However, motorized patrols are restricted to certain types of terrain, and tend to bypass areas that may be advantageous to and occupied by enemy infantry.

Waterborne Patrols

Waterborne patrols move over seas, lakes, rivers and streams, canals, and other inland waterways. The water is either used as a medium of entry to an objective area or is the actual patrol route. Waterborne patrols are limited by the location of water routes in the terrain and tend to bypass areas that may be advantageous to and occupied by the enemy.

Helicopterborne Patrols

Where terrain is extremely difficult or the enemy situation precludes the use of vehicle or motorized patrols, helicopterborne patrols are a method or means to conduct a patrol.

8005. TRAINING

Training is essential to successful patrolling. Premature and unordered actions by members of the patrol destroy coordination and control. Leaders are trained to issue their orders calmly to inspire confidence and discipline, and to avoid misunderstanding. Patrol members must work together and fight as a team. Training should develop the following skills:

- Expertise in handling individual and special weapons, and familiarity with enemy weapons that may be captured.
- Recognize camouflaged personnel, equipment, and defensive positions; ability to pick up fleeting targets fire the rifle from any firing position.
- Understand fire discipline and, after weapon firing, immediately change position (see fig. 8-1).
- Quick and accurate observation skills, and the ability to recall and transmit clearly and briefly, both orally and in writing.

Figure 8-1. Changing Position After Firing.

- Recognize and quickly respond to improvised signals, visually or by sound.
- Ability to swim with weapon and equipment.
- Use issued or improvised camouflage suits and garnish helmet in order to blend with the surroundings. Smudge face, hands, and any bright surfaces of weapons and equipment with some substance, such as mud or charcoal, to prevent the reflection of light.
- Silence self, equipment, and weapon.
- Use antimalarial and water purification tablets.
- Acclimation to temperature extremes.
- Develop a sense of direction and learn how to follow a course by compass, stars, sun, flow of streams, prominent terrain features, and by observing other natural phenomenon. Learn to determine the distance traveled from a known point and to keep a record of azimuths and the distance traveled on each azimuth (dead reckoning).
- Call for and adjust indirect fire assets.
- Familiarization with all communications assets and the use of field expedient antennae.

8006. KEYS TO SUCCESSFUL PATROLLING

Regardless of the category or means of conducting a patrol, the keys to successful patrolling are—

- Detailed planning. Every portion of the patrol must be planned, all possible contingencies considered.
- Productive, realistic rehearsals. Each phase of the patrol is rehearsed, beginning with actions in the objective area. Similar terrain and environmental conditions are used when conducting rehearsals.
- Thorough reconnaissance. Ideally, the patrol leader will physically conduct a reconnaissance of the route and objective. Photographs and/or maps will be used to supplement the reconnaissance.
- Positive control. The patrol leader must maintain positive control, this includes supervision during patrol preparations.
- All-around security. Security must be maintained at all times, particularly near the end of the patrol where there is a natural tendency to relax.

CHAPTER 9. PATROL ORGANIZATION

Organizing a patrol is a two-step process: the general organization of the entire patrol and the task organization of various patrol elements. Normally, the nature of patrolling does not permit long preparation periods and rehearsals to specifically build a unit for each mission. Accordingly, the patrol leader must combine unit integrity considerations with proven concepts of patrol organization.

9001. GENERAL ORGANIZATION

The patrol leader establishes a patrol headquarters and elements to accomplish the mission.

The headquarters is composed of the patrol leader and the personnel who provide support for the entire patrol, such as a forward observer, corpsman, and radio operator.

The major subdivisions of reconnaissance and combat patrols are elements. The existing infantry structure (squads and fire teams) is reinforced as required. For example, a reinforced platoon tasked to conduct a combat patrol that will raid an enemy outpost could be organized as follows:

- Platoon headquarters (command element):
  - Patrol leader (platoon leader).
  - Assistant patrol leader (platoon sergeant).
  - Navigator.
  - Radio operator (company tactical net).
  - Radio operator (patrol tactical net).
  - Corpsman.
- The first squad (security element) provides security en route to the objective area (the point, flank security, and rear security) and at the objective area (the flanks and the objective rally point).
- The second squad (support element) supports by fire for the attack, covering fire for the withdrawal, and supporting fires to cover the crossing of danger areas.
- The third squad (assault element) provides the assault force to attack and seize the objective; searchers to clear the objective; pacers, compass man, navigator, and the assistant patrol leader en route and back from the objective area.

Any attachments a patrol may have (i.e., demolition team, scout snipers, and machine gun squad) will be added to the element that supports its function. For example, the demolition personnel should go with the unit conducting the attack, and scout snipers and machine gunners should stay with the support squad.

9002. TASK ORGANIZATION

The preceding paragraphs described the elements necessary for a patrol to accomplish its mission. These elements reflect the internal functions or tasks required for the patrol to succeed. Depending on the METT-T, there are various methods of grouping these elements together. Task-organization is the further subdivision of patrol elements into teams that are required to perform essential tasks. In creating teams, unit integrity of infantry units should be maintained.

The patrol is organized so each individual, team, and element is assigned a specific task, but capable and prepared to perform other tasks. This may not be possible for certain specialist tasks requiring a trained technician.

CHAPTER 10. PATROL PREPARATION

For a patrol to succeed, all members must be well trained, briefed, and rehearsed. The patrol leader must have a complete understanding of the mission and a thorough understanding of the enemy and friendly situations. The patrol leader should make a complete reconnaissance of the terrain to be covered (either visual or map), and must issue an order to the patrol, supervise preparations, and conduct rehearsals.

10001. MISSION

The mission assigned to a patrol must be clear and oriented toward one objective with a specific task and purpose. More than one primary objective or indefinite missions invites confusion, casualties, and failure.

10002. FACTORS INFLUENCING PATROL SIZE

The circumstances under which patrolling by infantry units is conducted make it necessary for combat patrols to be able to fight offensively, security patrols to defend themselves, and reconnaissance patrols to move quickly and only fight if necessary to break contact or defend themselves. The size of a patrol depends on METT-T.

Generally, a patrol should contain the least number of members needed to accomplish the mission. Combat missions ordinarily require larger patrols than reconnaissance missions.

10003. COMMANDER'S DUTIES

Determining Patrol Requirements

The need for conducting patrols derives from the commander's stated mission (issued by higher level commander) and other specified and implied tasks (secondary or supporting efforts necessary to accomplish the stated mission). This requires analyzing the unit's mission and determining the necessary reconnaissance and/or combat tasks that must be performed. By considering the mission, along with time available, the commander develops the overall concept of operations to include the patrol plan and the specific tasks for each patrol.

Assigning Units

When assigning patrol missions, maintaining the integrity of the existing unit while considering the skills and experience of the unit and its leader are critical factors to the infantry commander. To provide operational depth and equitable apportionment of hazardous assignments, the commander ensures that each of the subordinate leaders and units develops the skills and experience necessary to conduct successful patrols. Assignment of patrol units must consider the commander's concept of operations as a whole and the plans for subsequent employment of assigned forces after completion of the assigned patrolling mission.

Providing Adequate Time

The commander must allow the patrol sufficient preparation time by completing the mission analysis, estimate, and preparation of orders. The commander should use warning orders to alert subordinates to possible requirements and afford them the opportunity for concurrent planning.

Providing the Patrol Leader Information

The following information should be provided:

- A simple, straightforward explanation of the mission, particularly for night patrols.
- General routes (defined by checkpoints) or exact routes (defined by avenues of approach or other terrain features) to follow.
- Enemy composition, disposition, and strength.
- Location and activities of friendly troops.
- Outposts or other security elements through which the patrol is to pass.
- Terrain conditions.
- Missions and routes of other patrols.

- Time patrol is to depart and return.
- Method of reporting information while on patrol (radio, messenger), place where messengers are to be sent, and place where the patrol leader is to report upon completion of the patrol.
- The challenge and password to be used for exiting and reentering friendly lines.
- Special instructions such as locations to be avoided and essential elements of information higher headquarters is seeking.
- Report barrier/obstacle plan of friendly units if known; support available from friendly unit (e.g., medical evacuation [MEDEVAC], react force).
- Fire support available.

Providing Special Skills and Equipment

There will be situations when the unit assigned to conduct a patrol does not have the necessary technical skills or equipment organic to it to successfully accomplish the patrol mission. In these cases, additions (attachments) are made to the unit, such as—

- Machine gun and/or shoulder-launched, multipurpose assault weapon (SMAW) team and/or squad.
- Forward observers (mortar/artillery).
- Radio operator(s).
- Combat engineers.
- Tracked vehicle crewman to assess the ability to traverse the terrain.
- Corpsman.
- Other personnel (snipers, translators) as required.

Providing Miscellaneous Support

The commander must ensure that the patrol leader is provided with the food, water, ammunition, radios and batteries, maps, special clothing, and any other items required by the unit (includin attachments) needed for the mission. Post-patrol support such as debriefings by intelligence personnel must also be planned.

Reviewing the Patrol Leader's Plan and Preparations

Once the patrol leader receives the mission, conducts visual and/or map reconnaissance, and develops the plan, the commander confirms the patrol leader's understanding of the mission and plan for accomplishing it. This discussion between the patrol leader and commander ensures that the patrol leader understands the commander's intent and is properly prepared to provide the patrol leader an opportunity to ask for clarification or additional support, if required.

Debriefing the Patrol

Upon return of the patrol, the commander receives the patrol report at a debriefing attended by the patrol leader and all patrol members. The debriefing should be conducted as soon as possible following the patrol's return, while information is still fresh in the minds of the patrol members. A patrol report, based on the information collected during the debrief, is generated and forwarded to the next higher com-mander.

10004. PATROL LEADER DUTIES

The patrol leader organizes and prepares the patrol by using the six troop-leading steps to make the best use of resources available. These steps are to—

1. Begin planning.
2. Arrange for reconnaissance and coordination.
3. Make reconnaissance.
4. Complete the plan.
5. Issue the order.
6. Supervise.

Begin Planning

The patrol leader begins by evaluating all factors affecting the mission. He looks for possible courses of action that lead to a decision, and then transforms this decision into an order. The first step includes making an initial assessment and decision on using available time, issuing a warning order and initial preparatory tasks, and initiating his estimate.

The initial planning effort assesses the time, assistance, and information available, and plans the proper use of each. Time allowances include reconnaissance; completion of the estimate and order; troop preparation; and such briefings, rehearsals, and inspections as required before beginning the patrol.

The patrol leader reviews the mission and the attachments and/or support available and decides what preparatory efforts must begin immediately. Proper use of subordinates to manage these initial tasks during this period reduces preparation time and frees the patrol leader for proper planning and reconnaissance. A warning order is issued using a modified five-paragraph order format (situation, mission, execution, administration and logistics, and command and signal [SMEAC]) as a checklist. (Refer to app. A for specific information that a patrol warning order should contain within each checklist item.

Once the initial preparations are set in motion, the patrol leader begins estimating by analyzing the mission; considering the friendly, enemy, and operating environments; considering each course of action available against what the enemy might do; comparing the courses of action in terms of mission accomplishment, capabilities, and probable casualties; and choosing one that becomes the basis for concept and order. While the patrol leader should use and organize notes, estimates must be done quickly and accurately, particularly for immediate situations. To organize thinking, the patrol leader uses METT-T, which consists of:

- Mission—the mission assigned to the patrol and how it relates to the mission of the commander who is sending the patrol.
- Enemy—what is known or suspected of enemy presence and capabilities, habits and characteristics, and fighting techniques.
- Terrain and weather—including ground, vegetation, drainage, weather, and visibility.
- Troops and support available—friendly situation and support available.
- Time available—the constraints and impact of time on preparation and mission accomplishment.

The estimate begins with mission analysis, which is the most important part of the entire planning process. Here, the patrol leader considers the specified tasks of the mission assigned and identifies other significant actions (specified and implied tasks) that must be undertaken to accomplish the stated mission. These tasks are arranged in sequence of accomplishment. The tasks and sequence create the framework for developing courses of action for the patrol concept of operations.

Arrange for Reconnaissance and Coordination

The patrol leader arranges a personal reconnaissance to observe as far forward as possible and also coordinates with the appropriate commanders for the patrol's "passage of lines" (see paragraph 11001 for complete definition) and supporting fires. The patrol leader also coordinates with other patrol leaders who may be operating in the same or adjacent areas and requests that the commander assigning him the patrol mission coordinate the patrol action with adjacent commanders, local security, and night defensive fires, as appropriate. The patrol leader may delegate any or all of these arrangements to the assistant patrol leader if the patrol leader requires the time for planning.

Make Reconnaissance and Complete the Estimate

The patrol leader uses personal reconnaissance to answer questions that arise from the map reconnaissance and METT-T evaluation. Specific points include passage points, lanes through obstacles, locations of friendly listening posts and observation posts, possible approach and return routes, enemy positions (if any), and intermediate observation points on the way to the objective.

In selecting approach and return routes, the patrol leader chooses routes that best use concealment and avoid opposition and obstacles. To lessen the chances of ambush by the enemy, the return trip is planned along a different route. In addition to personal reconnaissance and review of the map and aerial photographs, the advice of other patrol leaders who already are familiar with the terrain and the objective area should be considered.

After compiling information about the situation and possible time constraint, the patrol leader completes an estimate. The first step is developing courses of action, each of which will provide for movement to the objective area, mission accomplishment, and the return, based on the tasks and their sequencing identified in mission analysis. While the eventual concept of operations is presented in order of occurrence, the patrol leader must develop the courses of action by either backward or forward planning. In situations where the objective is well defined and there is sufficient information to plan the action for mission accomplishment (reconnaissance or combat), the patrol leader begins the scheme for accomplishing the mission at the objective and then, planning backwards, considers the options for getting there and back.

The following sections on movement to and return from the objective area, reconnaissance missions, and combat missions discuss methods and options available to the patrol leader in developing the courses of action. The principal variables between courses of action will be who, where, and how in the following:

- Patrol task organization.
- Routes to the objective area.
- Observation point(s) (reconnaissance patrol), ambush site, form of maneuver, type of ambush (combat patrol), fire support plan.
- Return routes.

The patrol leader then mentally considers the progress of each course of action (a map or simple sketch is a useful aid) against expected and unexpected enemy action. By comparing the options against each other and prospective enemy opposition, the patrol leader chooses the course of action that has the best chance of success. Includes in this mental preview process is the time to determine the patrol plan for unexpected contingencies (enemy attack/counterattack, casualty handling). These contingency actions, together with the selected course of action, become the patrol concept of operations.

Once the patrol leader determines the scheme of maneuver, the fire support required to accomplish the mission is addressed including the fire power organic to the patrol and what additional indirect fire support will have to be provided by other units. When planning for indirect fire support, the patrol leader considers the following questions:

- Will artillery, mortar, or close air support be required at the objective area (combat patrols)?
- What artillery and mortar targets exist along the routes to and from the objective area that can be employed by the patrol if it encounters the enemy during movement (reconnaissance patrols and combat patrols)?
- What additional fire support will be required to cover the patrol's movement from the objective area back to the friendly area once the enemy is aware of the patrol's actions at the objective area (combat patrols)?

The effect that casualties have upon the patrol depends upon many factors. Generally, more casualties can be expected in a combat patrol than in a reconnaissance patrol. A patrol may continue on to the objective carrying its casualties, send them back with a detail of Marines, abort the mission and return the entire patrol with the casualties, or call their parent unit for assistance.

Some factors that determine what action the patrol leader takes are: patrol's mission; unit's standing operating procedure for handling wounded; number of casualties and nature of their injuries; availability of aid, helicopters or other means of casualty evacuation. Helicopter evacuation should only be used for the most serious casualties. For infantry units conducting patrols in proximity to the enemy, helicopter evacuation of casualties may compromise the patrol's mission and force the patrol to return to friendly positions before the mission is completed.

The patrol leader determines the requirement for nuclear, biological, and chemical (NBC) defense equipment. Gas masks should always be carried due to the availability of riot control agents (RCAs) to the enemy. If chemical or biological agents have been employed in the area that the patrol must pass through, protective garments will have to be worn by patrol members for part of or the entire patrol. Wearing extra clothing and carrying extra equipment affects the speed of the patrol's movement. A contingency plan for post patrol decontamination must be developed.

Complete the Plan

At this point, the patrol leader has completed the basic thinking necessary for accomplishing the assigned mission. The patrol leader prepares the patrol order to spell out the details, assign tasks to subordinates, and explain the entire endeavor for ease of understanding by the other members of the patrol.

Prepare the Order

The patrol leader's order contains more detailed information than discussed in the warning order. Orders follow the prescribed five-paragraph order format but contain greater detail. The patrol order is a modified 5-paragraph order; the major modifications are to paragraphs 3a and 3c. The format for the order is contained in appendix B.

Issue the Order

The patrol leader asks for a status report on the initial preparatory tasks assigned to subordinate leaders and specialists when the warning order was issued. When the patrol leader has completed planning and initial preparations have progressed to the point where the patrol order may be issued, the members of the patrol are assembled. Roll call is taken to ensure all patrol members are present, then the prepared order is issued.

This will be the only opportunity for the patrol leader to issue detailed instructions. The mission, in particular, must be unmistakably clear so that once the patrol is committed, all subordinate leaders can act with unity of purpose.

Whenever possible, the patrol leader should have a Marine, such as the navigator, build a terrain model using dirt, sand, twigs, etc., explaining the concept of operations for movement to the objective area, actions at the objective area, and the return. Terrain models provide patrol members with a clear and simple layout of the area of operations and key terrain.

Supervise

Inspections and rehearsals, vital to proper preparation, are conducted even though the patrol leader and patrol members are well experienced in patrolling. Inspections determine the patrol's state of physical and mental readiness.

The patrol leader inspects before rehearsals to ensure completeness and correctness of uniform and equipment. The following areas are checked:

- Camouflage.
- Identification tags, Geneva Convention cards.
- Prescribed equipment, weapons, and ammunition are available and serviceable.
- Tape and other items are used to "silence" equipment (prevent noise produced during movement).
- Items that could provide information to the enemy (e.g., letters and papers) remain behind.
- Unnecessary equipment and excess weight remain behind.

The patrol leader questions each patrol member to ensure the following is known:

- The mission, planned routes (primary and alternate), and the fire support plan of the patrol.
- The individual's role: what to do and when to do it.
- What others are to do and how their actions impact.
- Challenges and passwords, codes, reporting times, radio call signs, frequencies, and any other pertinent details.

There is usually a period of time between final rehearsal and departure. The patrol leader reinspects just before departure to ensure all equipment is still in working order and the unit is ready to embark on the mission.

Rehearsals ensure the operational proficiency of the patrol. Plans are checked and needed changes are made. The patrol leader verifies the suitability of equipment. It is through rehearsals that patrol members become thoroughly familiar with the actions to take during the patrol.

If the patrol will operate at night, both day and night rehearsals are conducted. Terrain similar to that over which the patrol will operate is used. All actions are rehearsed. If time is limited, the most critical phases are rehearsed. Action at the objective is the most critical phase of the patrol and is always rehearsed.

An effective method is to talk the patrol through each phase, describing the actions and having each member perform individual duties. When satisfied, the patrol leader walks the patrol through all phases of the patrol using only the signals and commands to be used during the actual conduct of the patrol. Rehearsals continue until the patrol is thoroughly familiar with the plan. The rehearsal is also used to test the soundness of the patrol order and patrol organization.

After the rehearsal, the patrol leader makes final adjustments to the plan and patrol organization based on what was learned during the rehearsal and from other sources, such as the S-2 and adjacent patrols. When this is completed, the patrol leader issues final instructions to subordinate leaders noting any changes made in the patrol organization or plan. While the subordinate leaders are briefing the remainder of the patrol members, the patrol leader reports to the commander stating that the patrol is ready to begin the mission. The patrol leader also coordinates the location and time that the patrol can test fire all weapons prior to departure.

CHAPTER 11. MOVEMENT TO AND RETURN FROM THE OBJECTIVE AREA

This chapter provides guidance to patrol leaders for movement to and return from the objective area. Action in the objective area depends on whether the patrol is assigned a reconnaissance mission, combat mission or security mission. Chapters 12 and 13 provide detailed guidance on these types of missions.

11001. PASSAGE OF LINES

A passage of lines is an operation in which a force moves forward or rearward through another force's combat positions with the intention of moving into or out of contact with the enemy. (JP 1-02)

During the initial preparation for the patrol, the patrol leader selects a patrol assembly area and reconnoiters the area of passage designated by the commander. In coordination with the unit commander responsible for the area of passage, the patrol leader identifies gaps or lanes in minefields and wire obstacles and locates local security elements through which the patrol will pass. The patrol leader also checks the route from the patrol assembly area to the passage point or contact point where the patrol will depart friendly lines. If possible, both the route to the passage point and the route through the frontlines should be concealed from the enemy's view.

The patrol leader also reconnoiters the area for return passage of lines and coordinates with the unit commander responsible for the area of passage for passage points and lanes as necessary. The patrol leader observes these points from the direction that the patrol will use upon return to friendly lines, if possible, to aid in recognition upon return. The patrol leader provides the forward unit with information about the size of the patrol, general route, and expected time of return. The manner of challenge and recognition of the returning patrol should be coordinated in detail.

Upon return to friendly forward local security squads and/or frontlines, the patrol leader leaves the patrol in a covered position and moves forward with a radio operator and at least one Marine for security to make contact with the friendly unit in the manner agreed. After contact is made and recognized, the patrol leader rejoins the patrol and takes them to the passage point, personally checking in each member.

11002. ORGANIZATION FOR MOVEMENT

The patrol's task organization establishes the elements and teams needed to accomplish the mission in the objective area and to and from the objective. The patrol leader determines the formation(s) in which the patrol moves to the objective area.

Formations

The proper use of patrol formations is critical to the patrol's success. The squad and fire team formations described in FMFM 6-5 (proposed MCWP 3-11.2), *Marine Rifle Squad*, also apply to infantry patrolling formations. Because the movement of the patrol must be concealed from the enemy, the patrol normally moves through terrain that provides concealment. Control of the patrol in this type of terrain is difficult; thus the column formation, which is easily controlled, is normally used. However, as various types of terrain are encountered, the patrol leader uses the same considerations in determining the appropriate formation used in other infantry operations.

The standard squad and fire team formations are adaptable to a patrol. The patrol may change formations en route to match the situation and terrain. The patrol leader may have to sacrifice some control for better dispersion or give up some speed for greater stealth and more security. Other considerations include—

- Visibility, weather, terrain, and vegetation will influence dispersion and control of individuals and units. These factors may also affect the enemy; if visibility is good for the patrol, it is also good for the enemy. Two pieces of luminous tape worn on the back of the collar will aid in control and movement on dark nights. The collar is turned down when near the enemy. The tape can also be worn on the back of the cap, but should be covered or removed when near the enemy.

- Preserving the integrity of fire units (fire teams and, if attached, machine-gun teams) is of primary importance. If team members are detached from a machine-gun team, the unit integrity is lost and effectiveness is reduced. The patrol leader must also position fire units so as not to mask their fires.
- All-around defense of the patrol must not be sacrificed under any circumstances. The conventional squad and platoon formations provide adequate firepower in any direction required. When attachments are made to the patrol, the attachments are positioned within the formation to enhance the firepower of the patrol. If a fire unit, such as a machine-gun team or squad, is attached, it is incorporated into the all-around defense of the patrol by modifying the conventional formation (see fig. 11-1). It is permissible to employ machine-gun teams individually during movement. However, assault rocket squads and teams should be employed primarily as rifle fire units for movement. Firing the SMAW or AT-4 from within a patrol formation can present a serious back-blast danger to patrol members and should only be fired on the patrol leader's direction.
- Time allotted for mission accomplishment is also a major consideration. In selecting the formations, the patrol leader must consider the speed of the movement required to meet the time constraints (if any) imposed on the patrol. If required to meet a time schedule, a formation that permits rapid movement should be used. Speed, however, must never be permitted to force the patrol leader to make rash tactical decisions.

Exercise of Control

The patrol leader is positioned for best control over the patrol. The assistant patrol leader moves at or near the rear of the patrol. Other subordinate leaders move with their elements. All patrol members assist by staying alert and passing on signals and orders. A signal to halt may be given by any patrol member, but the signal to resume movement is given only by the patrol leader.

Arm-and-hand signals are the primary means of communication within a patrol and should be used exclusively when near the enemy. All members must know the standard infantry signals (refer to FMFM 6-5), as well as any special signals required, and be alert to receive and pass them to other members.

The patrol leader should speak just loudly enough to be heard. At night, or when close to the enemy, the

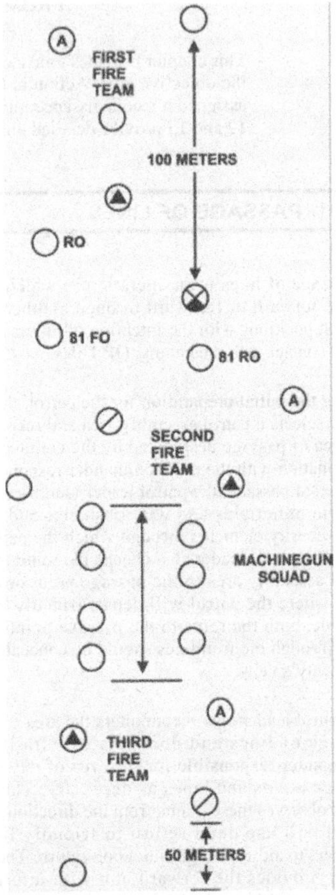

Figure 11-1. Squad Patrol with Attached Mortar Forward Observer Team and Machine Gun Squad.

patrol leader halts the patrol and has subordinate leaders come forward. They speak in a low voice and then pass the information to their subordinates by moving from member to member.

Radios provide a means of positive control within a large patrol; however, radios should be used only when arm-and-hand signals or face-to-face contact between the patrol leader and subordinate leaders is impractical. When close to the enemy, words are spoken into the microphone with hands cupped over it in a low voice.

Other sound signals may be used if the patrol leader is sure they serve the purpose intended. Planned sound signals are rehearsed before starting on the patrol. Sound signals used must be simple, natural sounds that are few in number and easily understood. Bird and animal calls are seldom satisfactory.

Infrared equipment, such as the sniper scope and infrared filters for the flashlight, may be used as means of sending and receiving signals and maintaining control at night.

Luminous tape may be used to assist in control at night. Small strips on the back of the cap or collar of patrol members aid in keeping visual contact with the front member. However, the luminous tape must be covered when near the enemy.

An important aspect of control is accounting of personnel, especially after crossing danger areas, enemy contacts, halts, and exiting and re-entering friendly lines. The patrol leader may arrange for the last member to send up the count automatically after crossing danger areas, enemy contact, and halts. In large patrols or when moving in a formation other than a column, subordinate leaders check members and report the count to the patrol leader by the quickest method appropriate to the circumstances.

Navigation

One or more members are assigned as navigators for the patrol. Their function is to assist the patrol leader in maintaining direction by the use of the compass.

The patrol leader assigns at least two members as pacers to keep track of the distance from point to point. The average of their count is used to approximate the distance traveled. Pacers are separated so they do not influence each other's count. Pacers must know how to convert their own paces to meters.

The route is divided into increments, with each increment starting at a recognizable point on the ground. The pacers begin their counts from zero at the beginning of each leg. This makes the pace count easier to keep and checks for accuracy.

The pace count is sent forward when the patrol leader turns to the member behind and in a low voice says, "send up the pace" or uses the arm-and-hand signal of tapping his boot. This signal is passed to both pacers, who in turn send up the pace count in meters; for example, "two-hundred" or "one-seven-five."

Patrol members must understand that the counts of both pacers are sent forward. The patrol leader must know the counts of both men in order to check them.

Security

The patrol is dispersed consistent with control, visibility, cover, and concealment. Scouts are employed to the front, flanks, and rear of the patrol to provide security. For the patrol members traveling in the main body of the patrol, areas of responsibility are assigned to the front, flanks, and rear. Scouts are the eyes and ears of the patrol leader. They move when and where directed by the patrol leader and maintain contact with the patrol leader at all times, except when momentarily obscured by vegetation or other terrain features.

Front

Small patrols (squad size) may employ from one scout up to a fire team as the point, depending on the enemy situation, terrain, and route being followed by the patrol. Normally, squad-size patrols will use two scouts as the point. The point is responsible for investigating the route of advance immediately to the front of the patrol.

The point moves as far ahead of the patrol as visibility and terrain permit. When visibility is good, the point may precede the main body by as much as 100 meters. The point travels right and left ahead of the patrol, searching the area over which the patrol will pass.

The point maintains direction by knowledge of the general route to be followed and visual contact with the patrol leader. The patrol leader or the navigator ensures that the point is proceeding correctly.

The point, which stays far enough ahead of the patrol to provide security, is not a trail breaker for the patrol. If the point loses contact with the patrol, the point waits for the main body to catch up or moves rearward if contact is not quickly regained.

One of the navigators may be positioned with the point. One or more members works as the point while the other is the navigator.

Flank

Flank security for a patrol of squad size or less may be provided by using one or two members on either flank. If two scouts are assigned to a flank, one is positioned to observe the patrol leader and the other works farther out from the patrol. The scout who must observe the patrol leader remains within a maximum distance of 100 meters. The scout farther out remains in sight of the inside scout but normally does not move more than 20 or 25 meters away and remains prepared to relieve flank security regularly. Moving through dense woods or jungle may render the use of flank security impractical because of reduced visibility. In such cases, it moves with the patrol itself, but maintains observation to its assigned flanks.

Rear

A small patrol normally has only one rifleman assigned as rear security. An interval between the member assigned as rear security and the last Marine of the patrol is maintained at the limit of visibility, up to 50 meters. This member maintains rear security for the patrol by constantly observing to the rear.

Halting

Speed of movement is slower at night than in day patrols and reduces the danger of a Marine becoming separated from the patrol. The patrol occasionally halts to observe and listen for enemy activity; this is called a security halt. Upon signal, when reaching a danger area and periodically throughout movement en route, every member freezes in place, remains quiet, observes, and listens. It may be necessary to call a security halt just after departing friendly areas and just before entering friendly areas.

The patrol may halt briefly to send a message, eat, rest, check direction, or make a reconnaissance. The area selected should provide adequate concealment and cover, as well as favor the defense. All-round security is established and the patrol leader ensures all members move out when the patrol resumes movement. (For extended halts, see paragraph 11005.)

Infiltration

The disposition of enemy forces may sometimes prevent a patrol from entering the enemy occupied area as a unit; however, pairs of scouts or fire teams may slip through without being discovered. (Refer to FMFM 6-5 for infiltration techniques and procedures).

11003. CONTROL MEASURES FOR MOVEMENT

Checkpoints

A checkpoint is a predetermined point on the surface of the Earth used as a means of controlling movement, a registration target for fire adjustment or reference for location (JP 1-02). Checkpoints are a means of control between the parent unit and the patrol. These locations are decided upon and coordinated before the patrol leaves, so that both the patrol members and parent unit will know the patrol's location when it reports in. The parent unit can follow the progress of the patrol without transmitting coordinates over the radio that the enemy could monitor.

Rally Points

A rally point is an easily identifiable point on the ground where units can reassemble and reorganize if they become dispersed (MCRP 5-12C). It should provide cover and concealment, be defensible for at least a short time, and be easily recognized and known to all patrol members. All rally points are considered and identified as tentative rally points until they are occupied, found to be suitable, and designated as rally points by the patrol leader.

Initial Rally Point

The initial rally point is within friendly lines where the patrol can rally if it becomes separated before departing the friendly area or before reaching the first en route rally point. It may be the assembly area where the patrol waits while the patrol leader contacts the last friendly position through which the patrol will pass. The location of the initial rally point must be coordinated with the forward unit commander.

En route Rally Points

En route rally points are between the initial rally point and the objective rally point; and from the objective rally point back to the point where the patrol reenters friendly lines. They are determined as the patrol passes through a likely area that is suitable for a rally point.

Objective Rally Point

The objective rally point is located nearest the objective where the patrol makes final preparations prior to approaching the objective. It also serves as a location where the patrol reassembles after completing actions on the objective. The objective rally point must be suitable to accommodate those activities accomplished prior to actions on the objective. This position must provide the patrol concealment from enemy observation and, if possible, cover from enemy fires. It may be located short of, to a flank or beyond the objective. It should be out of sight, sound, and small-arms range of the objective area. The patrol leader's reconnaissance of the objective is made from this position; it is the release point from which patrol elements and teams move to the objective to accomplish the mission.

Rally Point Selection

The patrol leader selects likely locations for tentative rally points during reconnaissance or map study. A tentative rally point must be confirmed and announced after examination proves suitability. A tentative initial rally point and a tentative objective rally point are always selected and identified in the patrol order. If necessary, the patrol leader selects additional rally points en route as suitable locations are reached. When the patrol reaches a danger area that cannot be bypassed, such as an open meadow or stream, the patrol leader selects a rally point on both the near and far side. If good locations are not available, the patrol leader designates the rally points in relation to the danger area. The patrol leader will say, for example, "50 meters this side of the trail" or "50 meters beyond the stream."

Rally Point Use

The initial rally point and en route rally points are designated to enable the patrol to reassemble if it is unavoidably separated or dispersed. Identifying features are pointed out. The patrol leader ensures that the information is passed to all patrol members. When crossing a danger area, a near side rally point and a far side rendezvous point are designated. These rally points should only be used when all other methods of retaining control of the patrol have failed. The success of the patrol may be jeopardized if it is dispersed and cannot rally expeditiously.

If the patrol has left the friendly area and becomes dispersed, patrol members return to the last designated rally point (the initial or an en route rally point) unless the patrol leader gives other instructions.

As previously noted, the patrol leader selects two rally points at the near and far sides of danger areas that cannot be bypassed. If the patrol becomes separated or dispersed at a danger area, and there has been no enemy contact, the patrol should reassemble at the rally point on the far side of it. If, however, the patrol is separated or dispersed at a danger area as a result of enemy contact, members who have already crossed the danger area assemble and reorganize at the rally point on the far side; members who have not crossed assemble and reorganize at the rally point on the near side. In the absence of the patrol leader and unless directed otherwise, the senior member at the rally point on the near side takes charge, attempts to move the rallied patrol members to the rally point on the far side, and rejoins the remainder of the patrol.

Rally Point Actions

The patrol leader plans the actions to be taken at rally points and instructs the patrol accordingly in the patrol order. Planned actions at the initial rally point and en route rally points must provide for the continuation of the patrol as long as there is a reasonable chance to accomplish the mission. Plans for actions at rally points should provide for—

- Recognition signals for assembly at rally points.
- Minimum number of members and maximum amount of waiting time required before the senior member at the rally point moves the rallied patrol members onward toward the objective or returns to friendly lines.
- Instructions for patrol members who find themselves alone at a rally point.

11004. PRECAUTIONS AT DANGER AREAS

A danger area is any place where the patrol is vulnerable to enemy observation or fire (open areas, roads, trails, and obstacles such as barbed wire, minefields, rivers and streams, and lakes). Any known or suspected enemy position the patrol must pass is also a danger area. The patrol leader plans for crossing each danger area and includes these plans in the order.

The patrol reconnoiters the near side of a danger area first, then the patrol leader sends scouts to reconnoiter the far side. Once the scouts report that the far side is clear of the enemy, the remainder of the patrol crosses the danger area. As each individual or group crosses the danger area, they are covered by those remaining and by those who have successfully crossed. Enemy obstacles are avoided since they are usually covered by fire.

In crossing a river, the near bank is reconnoitered first; then the patrol is positioned to cover the far bank. Scouts are sent across to the far bank. After the far bank has been reconnoitered and the scouts report that it is clear of the enemy, the patrol crosses as rapidly as possible. This may be done individually or in pairs. If crossing the river requires swimming, the patrol uses improvised rafts to float equipment, weapons, and ammunition across. (Refer to MCRP 3-02C, *Water Survival Handbook.)*

A road or trail is crossed at or near a bend or where the road is narrow. Observation is restricted and, if the enemy is present, the patrol is exposed as short a time as possible. The near side is reconnoitered first, then scouts are sent across to reconnoiter the far side. This includes reconnoitering the tentative rally point on the far side. Once the scouts report "all clear," the remainder of the patrol crosses rapidly and quietly.

If the patrol must pass close to an enemy position, it takes advantage of battlefield noises to cover the sounds of movement. If supporting fires are available, the patrol leader can call for them to divert the enemy's attention as the patrol passes.

11005. HIDE

When a patrol is required to halt for an extended period in an area not protected by friendly troops, the patrol moves into a location which, by the nature of the surrounding terrain, provides passive security from enemy detection. Such an assembly area is termed a *hide*. To establishment a hide—

- Cease all movement during daylight hours to avoid detection.
- Hide the patrol for an extended period while the patrol leader conducts a detailed reconnaissance of the objective area.
- Rest and reorganize after extended movement.
- Reorganize after a patrol has infiltrated the enemy area in small groups.

The patrol leader's plan must include tentative hide locations when the patrol's mission dictates an extended halt within enemy areas. These tentative locations must be confirmed by actual ground reconnaissance prior to occupation by the patrol. The plan for a hide includes both passive and active security measures.

Passive security measures are—

- Avoid built-up areas.
- Select an area remote from all human habitation.
- Avoid known or suspected enemy positions.
- Avoid ridge lines, topographic crests, valleys, lakes, and streams.
- Avoid roads and trails.
- Avoid open woods and clearings.
- Select areas offering dense vegetation, preferably bushes and trees that spread out close to the ground.

Active security measures—

- Establish security covering all likely avenues of approach into the site.
- Establish communications (wire, radio, signal, runner) with posted security to provide early warning of enemy approach.
- Select an alternate area for occupation if the original hide is compromised or found unsuitable.
- Plan for withdrawal in the event of discovery.
- Establish an alert plan with a certain percent of the personnel awake at all times.
- Organize the elements of the patrol so necessary activities can take place with a minimum amount of movement.

The size of the area physically occupied by a patrol in a hide and the number of security posts required are governed by the terrain, quantity and quality of cover and concealment, and size of the patrol.

If the situation permits, a hide can also be used as the final preparation position and/or objective rally point.

11006. IMMEDIATE ACTIONS UPON ENEMY CONTACT

A patrol may make contact with the enemy at any time. Contact may be through observation, a meeting engagement or ambush. Contact may be visual, in which the patrol sights the enemy but is not itself detected. When this is the case, the patrol leader can decide whether to make or avoid physical contact, basing his decision on the patrol's assigned mission and capability to successfully engage the enemy unit.

When a patrol's assigned mission prohibits physical contact (except that necessary to accomplish the mission), its actions are defensive in nature. Physical contact, if unavoidable, is broken as quickly as possible and the patrol, if still capable, continues its mission.

When a patrol's assigned mission permits or requires it to seek or exploit opportunities for contact (as in the case of a combat patrol), its actions are offensive in nature, immediate, and positive.

In patrolling, contacts (visual or physical) are often unexpected at very close ranges, and short in duration. Effective enemy fire often provides leaders little or no time to fully evaluate situations and issue orders. In these situations, immediate action provides a means for swiftly initiating positive offensive or defensive action, as appropriate.

Two types of physical contact with the enemy are meeting engagement and ambush. Meeting engagement is a combat action that occurs when a moving force, incompletely deployed for battle, engages an enemy at an unexpected time and place. It is an accidental meeting where neither the enemy nor the patrol expect contact and are not specifically prepared to deal with it. An ambush is a surprise attack from a concealed position.

Immediate Actions

Immediate actions are designed to provide swift and positive small unit reaction to visual or physical contact with the enemy. They are simple courses of action in which all Marines are well trained. Minimal signals or commands are required and they are developed as needed for the combat situation. The signals can, in many cases, be initiated by any member of the unit. It is not feasible to attempt to design an immediate action drill to cover every possible situation. It is better to know the immediate action drill for each of a limited number of situations that may occur during a patrol. Arm-and-hand signals associated with immediate actions—such as FREEZE, ENEMY IN SIGHT, and HASTY AMBUSH RIGHT or LEFT—are contained in FMFM 6-5 (proposed MCWP 3-11.2).

Immediate Halt

When the patrol detects the enemy but is not itself detected, the situation requires the immediate, in-place halt of the patrol. The first member visually detecting the enemy gives the silent signal for FREEZE. Every member halts in place, weapon at the ready, and remains absolutely motionless and quiet until further signals or orders are given.

Air Observation and/or Attack

These actions are designed to reduce the danger of detection from aircraft and casualties from air attack.

When an enemy or unidentified aircraft that may detect the patrol is heard or observed, the appropriate immediate action drill is FREEZE. The first member hearing or sighting an aircraft that may be a threat signals FREEZE. Every member halts in place until the patrol leader identifies the aircraft and gives further signals or orders. Members of the patrol must not look up at the aircraft as sunlight can reflect off their faces even when camouflaged.

When an aircraft detects a patrol and makes a low level attack, the immediate action drill air attack is used. The first member sighting an attacking aircraft shouts, "AIRCRAFT," followed by the direction of the incoming attack: FRONT, LEFT, REAR or RIGHT. The patrol moves quickly into line formation, well spread out, at right angles to the aircraft's direction of travel. As each member comes on line, the member hits the ground, using available cover, then positions the body perpendicular to the aircraft's direction of travel, to present the shallowest target possible (see fig. 11-2 on page 11-8). Between attacks (if the aircraft returns or if more than one aircraft attacks), patrol members seek better cover. Attacking aircraft are fired upon only on command of the patrol leader.

Meeting Engagement

Hasty Ambush. This immediate action is used to avoid contact and to prepare to initiate an unplanned ambush on the enemy. It may often be a subsequent action after the command freeze. When the signal HASTY AMBUSH is given (by the point member, patrol leader or another authorized patrol member), the entire patrol moves quickly to the right or left of the line of movement, as indicated by the signal, and takes up the best available concealed firing positions (see fig. 11-3 on page 11-9). The patrol leader initiates the ambush by opening fire and shouting, "FIRE"; thus ensuring initiation of the ambush if the weapon misfires. If the patrol is detected before this, the first member aware of detection initiates the ambush by firing and shouting. The patrol leader may decide not to initiate the ambush in order to avoid contact unless the patrol is detected. When used as an offensive

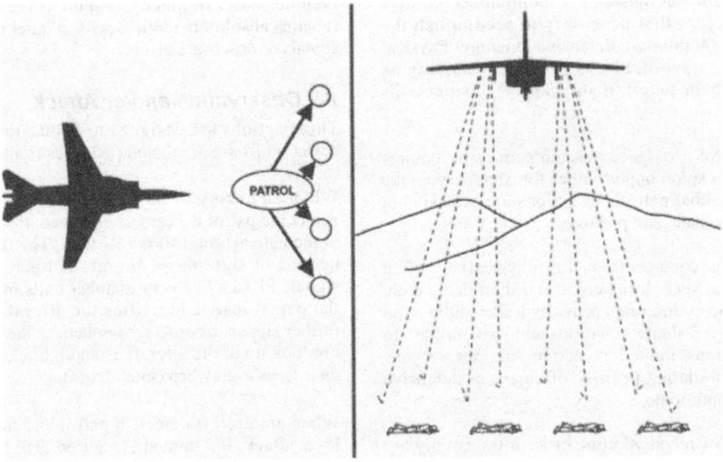

Figure 11-2. Immediate Action, Air Attack.

measure, the enemy is allowed to advance until he is in the most vulnerable position before the ambush is initiated. An alternate means for initiating the ambush is to designate an individual (for example, point or last member) to open fire when a certain portion of the enemy unit reaches or passes that member.

Immediate Assault. This immediate action drill is used defensively to make and quickly break undesired but unavoidable contact (including ambush) and offensively to decisively engage the enemy (including ambush). When used in a meeting engagement, members nearest the enemy open fire and shout, "CONTACT," followed by the direction of the incoming attack: FRONT, LEFT, REAR or RIGHT. The patrol moves swiftly into line formation and assaults (see fig. 11-4).

Defensive Measures

When used defensively, the assault is stopped if the enemy withdraws and contact is broken quickly. If the enemy stands fast, the assault is carried through the enemy positions and movement is continued until contact is broken.

Offensive Measures

When used offensively, the enemy is decisively engaged. Escapees are pursued and destroyed until orders to break contact are given by the patrol leader. If the patrol is fired upon from beyond 50 meters, the patrol must break contact as quickly as possible and continue the mission. If it engages the enemy any longer than necessary to break contact, it may put the mission in jeopardy.

Fire and Maneuver

Fire and maneuver is one means to break contact. One portion of the patrol returns the enemy fire while another portion moves by bounds away from the enemy. Each portion of the patrol covers the other by fire until contact is broken by all.

Clock System

The clock system is another means to break contact. Twelve o'clock is the direction of movement of the patrol. The patrol leader shouts a direction and a distance. For example: "TEN O'CLOCK-TWO HUNDRED," means the patrol should move in the direction of ten o'clock for 200 meters. Patrol members keep their same relative positions as they move so the original formation is not disrupted. Subordinate leaders must be alert to ensure that the members of their elements and teams receive the correct order and move as directed.

Counter Ambush

When a patrol is ambushed, the immediate action drill used depends on whether the ambush is a near ambush

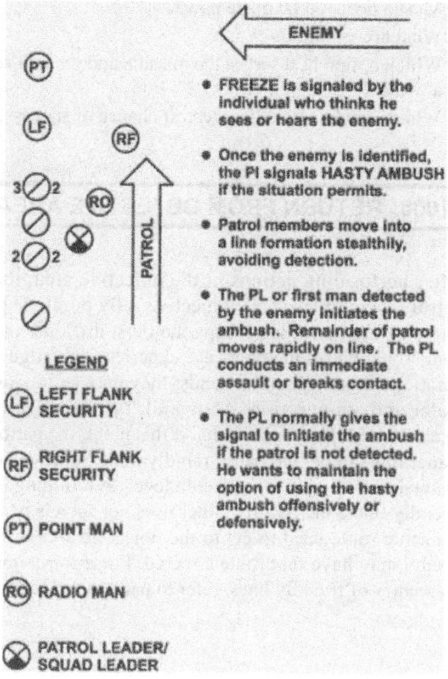

Figure 11-3. Immediate Action, Hasty Ambush.

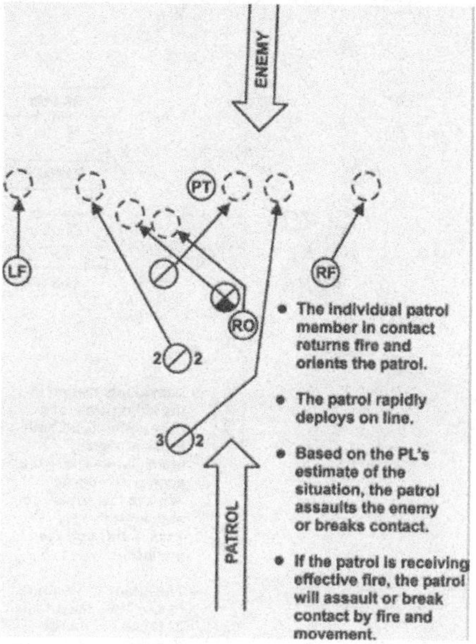

Figure 11-4. Immediate Action, Immediate Assault.

(the enemy is within 50 meters of the patrol) or a far ambush (the enemy is beyond 50 meters of the patrol). Fifty meters is considered the limit from which the ambushed patrol can effectively launch an assault against the enemy.

Near Ambush. In a near ambush, the killing zone is under very heavy, highly concentrated, close range fires. There is little time or space for members to maneu-ver or seek cover. The longer they remain in the killing zone, the greater the chance they will become casualties. Therefore, if members in the killing zone are attacked by a near ambush, they immediately assault without order or signal directly into the ambush position, occupy it, and continue the assault or break contact, as directed. This action moves them out of the killing zone, prevents other elements of the ambush from firing on them without firing on their own members, and provides positions from which other actions may be taken (see fig. 11-5 on page 11-10). Members not in the killing zone maneuver against the enemy as directed. The assault continues until all patrol members are outside of the killing zone.

Far Ambush. In a far ambush, the killing zone is also under very heavy, highly concentrated fires, but from a greater range. This greater range provides members in the killing zone maneuver space and some opportunity to seek cover at a lesser risk of becoming a casualty. If attacked by a far ambush, members in the killing zone, without order or signal, immediately return fire, take the best available positions, and continue firing until directed otherwise. Members not in the killing zone maneuver against the ambush force, as directed (see fig. 11-6 on page 11-10). The assault is continued against the enemy or until the order to break contact is given.

In each situation, the success of the counter ambush employed depends on the members being well trained in quickly recognizing the distance from which an ambush is initiated and well rehearsed in the proper reaction.

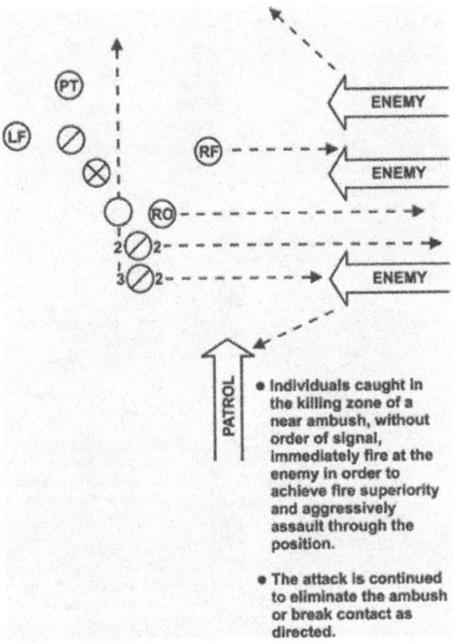

Figure 11-5. Near Ambush.

11007. PATROL LEADER'S ACTION IN A DEVELOPING SITUATION

While good patrolling depends on good planning, the patrol leader's plan must be flexible. Every combat situation develops differently than expected, and the patrol leader must be ready to quickly adapt to the situation as it develops. One of the most difficult tasks in battle is to recognize the correct moment for making a decision. Generally, it is more difficult to determine the moment for making a decision than it is to formulate the decision itself. When the situation demands, decisions must be made promptly without waiting for more information. In a developing situation, the patrol leader should use the following questions as a guide for battlefield decisionmaking:

- How has the situation changed?
- How does the change affect mission accomplishment and the immediate superior's mission?
- Must a decision be made now?
- What are the options?
- Which option best serves the mission and the unit as a whole?
- Which option offers the greatest chance of success?

11008. RETURN FROM OBJECTIVE AREA

After performing actions in the objective area, the patrol reassembles at the objective rally point. This phase of the patrol is perhaps the most difficult and dangerous. Patrol members are experiencing fatigue, emotional letdown, and wounds; they may be low on water and ammunition. Above all, the enemy was likely alerted if not in pursuit. At this point, the patrol leader must move the patrol rapidly but carefully and maintain patrol security at a high level. In returning to friendly lines, the patrol neither uses nor travels near the same route used to get to the objective area. The enemy may have that route covered. For information on reentry of friendly lines, refer to paragraph 11001.

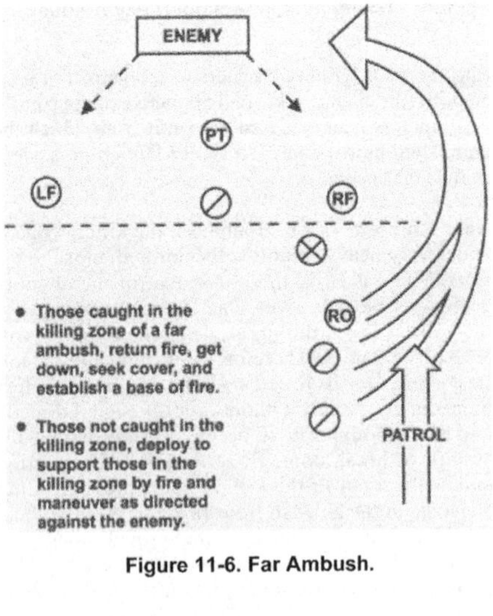

Figure 11-6. Far Ambush.

CHAPTER 12. RECONNAISSANCE PATROLS

The commander needs accurate, timely information about the enemy and the terrain the enemy controls to assist in making tactical decisions. Reconnaissance patrols are one of the most reliable means for obtaining this information. These patrols engage the enemy only when necessary to accomplish their mission or for protection. They generally avoid combat, accomplish the mission by stealth, and do not maintain contact with the unit that sent them out. A reconnaissance patrol is capable of carrying the search for information into the area occupied by enemy forces—usually beyond the range of vision of friendly ground observation posts—and examining objects and events at close range.

12001. GENERAL MISSIONS

General missions for reconnaissance patrols include gathering information about location and characteristics of friendly or hostile positions and installations, terrain (routes, stream crossings), and obstacles.

12002. SPECIFIC MISSIONS

Locate the Enemy

Reconnaissance patrols try to determine the—

- Location of enemy forces, installations, and equipment.
- Identification of enemy units and equipment.
- Strength of enemy forces.
- Disposition of enemy forces.
- Movement of enemy personnel and equipment.
- New or special types of weapons.
- Presence of mechanized units.
- Unusual enemy activity.
- Presence of NBC equipment.

Reconnaissance of Enemy Wire Obstacles

A patrol with the mission of investigating enemy-emplaced wire obstacles employs a formation providing for all-around security and takes precautions against being observed by the enemy. The patrol leader and one Marine inspect each gap and establish its location by means of compass bearings to prominent objects in the rear of enemy or friendly positions or through the use of global positioning satellite receivers.

Reconnaissance of Contaminated Areas

A patrol with the mission of investigating a contaminated area reconnoiters and marks the area's boundaries. Patrol members wear protective clothing and field protective masks. The patrol leader reports the extent of the area, the ttype of agent used, the terrain and vegetation and the method of marking the contaminated area. A sketch of the contaminated area, prepared by the patrol leader, should be included in the patrol report.

Reconnaissance of Enemy Minefields

Patrols assigned to reconnoiter enemy minefields are often composed of specially trained personnel. The type of mine is identified and the possibility of bypassing the area determined.

Reconnaissance of Terrain

The reconnaissance patrol is particularly suited for gathering information about the terrain within their area of operations. (Refer to MCRP 2-15.3B, *Reconnaissance Reports Guide.*) Commanders must know the location and condition of bridges, streams, and roads in order to make proper plans for the movement of troops. The following information should be obtained by a reconnaissance patrol assigned such missions:

- Bridges:
  - Maximum load capacity.
  - Material used in construction (wood, stone, concrete or steel).

- Material used for abutments (dirt, wood, masonry, concrete).
- Type and number of supporting members of the bridge.
- Condition and dimensions of bridge flooring.
- Presence or absence of mines and demolition charges on the bridge and at entrances/exits.
- Terrain crossed (ravine, stream, road).
- Primary use (rail, auto, foot).
- Location and type of an alternate crossing if bridge is unusable.

- Streams and fords:
  - Width and depth of stream.
  - Composition of the bottom (mud, sand, gravel, rock).
  - Speed of current in miles per hour.
  - Composition and gradient of banks.
  - Surrounding terrain and vegetation cover.
- Roads:
  - Width.
  - Composition of surface.
  - Condition.
  - Road blocks.
  - Gradient/degrees of slopes.
  - Curves (location and length).
  - Pitch and culvert locations.
  - Primary use and frequency of use.

12003. TYPES OF RECONNAISSANCE

Route

Route reconnaissance is a directed effort to obtain detailed information of a specified route and all terrain from which the enemy could influence movement along that route. (Refer to MCRP 5-12C.) It is also described as a form of reconnaissance focused along a specific line of communications—such as a road, railway, or waterway—to provide new or updated information on route conditions and activities along the route. (Refer to MCRP 5-12A, *Operational Terms and Graphics*.)

A route reconnaissance normally precedes the movement of forces. It provides detailed information about a specific route and the surrounding terrain that could be used to influence movement along that route. Considerations include trafficablility, danger areas, critical points, vehicle weight and size limitations, and locations for friendly obstacle emplacement.

The objective area is normally defined by a line of departure, a route, and a limit of advance.

Area

An area reconnaissance is a directed effort to obtain detailed information concerning the terrain or enemy activity within a prescribed area such as a town, ridge line woods or other features critical to operations. (Refer to MCRP 5-12C.) An area reconnaissance could also be made of a single point, such as a bridge or installation. (Refer to MCRP 5-12A.)

Emphasis is placed on reaching the area without being detected. Enemy situations encountered en route are developed only enough to allow the reconnoitering unit to report and bypass.

Zone

A zone reconnaissance is a directed effort to obtain detailed information concerning all routes, obstacles (to include chemical or radiological contamination), terrain, and enemy forces within a zone defined by boundaries. A zone reconnaissance is normally assigned when the enemy situation is vague or when information concerning cross-country trafficability is desired. (Refer to MCRP 5-12C.)

The commander specifies specific routes or areas of interest within the zone. The zone to be reconnoitered usually is defined by a line of departure, lateral boundaries, and a limit of advance.

12004. TASK ORGANIZATION

A reconnaissance patrol is organized around the current structure of the Marine rifle squad with one or more of the squad's fire teams assigned as the reconnaissance element to reconnoiter or maintain surveillance over the objective. At least one fire team acts as a security element whose functions are to—

- Secure the objective rally point (see para. 11003).
- Give early warning of enemy approach.
- Protect the reconnaissance unit.

Reinforcing the squad is based on METT-T. If two squads are required, then the mission should be assigned to a platoon.

A company may send a reconnaissance patrol to specifically reconnoiter a given area (a riverbed or a bridge), or to maintain surveillance on a particular streambed for the next four nights. Reconnaissance patrols perform three basic functions en route to and from the objective:

- Provide control.
- Provide security while moving and in the objective area.
- Conduct reconnaissance or surveillance.

Depending on the size of the reconnaissance patrol, an element must be tasked to reconnoiter the area or zone. In the case of a surveillance mission, two teams are recommended to share the same task (so one could lay up and rest, while the other lays low and hides to maintain strict silence and no movement while surveillance is being maintained on the objective). Whatever the case, each unit that has a surveillance mission or reconnaissance mission must be thoroughly briefed as to what essential elements of information are to be collected as well as their location in the objective area.

The patrol should be organized with one or two fire teams to actually conduct the reconnaissance mission and the remaining fire team to provide security. Each fire team should be prepared to assume either mission. A small area reconnaissance patrol needs only one fire team for the assigned mission. A patrol with a wider area reconnaissance mission should use two fire teams to physically conduct the mission and one fire team for cover and/or security.

The security element for a reconnaissance patrol should be organized to cover the likely avenue of approach into the objective area, to protect the units conducting the reconnaissance, and to cover the objective rally point.

The variations of the special organization for reconnaissance patrols illustrate the patrol leader's flexibility in organizing the patrol to meet mission requirements. The patrol leader is not limited to the variations reflected herein but can choose any special organization as long as that task organization accommodates the requirement for command and control, reconnaissance, and security.

12005. SIZE OF RECONNAISSANCE PATROLS

A reconnaissance patrol should be kept to the minimum number of personnel required to accomplish the mission. A mission requiring a patrol to remain away from its unit for a considerable period of time, or one requiring a patrol to send back information by messenger, increases the size of the patrol. Reconnaissance patrols seldom exceed a squad in strength. Unit integrity should be preserved whenever possible. Intelligence personnel, interpreters, and other specialists, such as radio operators or engineers, are assigned to a patrol if the particular mission requires.

12006. RECONNAISSANCE EQUIPMENT

Patrol members are armed and equipped as necessary for accomplishing the mission. The automatic rifle in each fire team provides a degree of sustained firepower in case of enemy contact. The patrol should have at least two: pairs of binoculars, pairs of wire cutters, maps, compasses, and watches. Night observation equipment may be used. Pencils and small notebooks are carried so notes and sketches can be made. A message book with message blanks and overlay paper is mandatory.

12007. RECONNAISSANCE PATROL ACTIONS AT THE OBJECTIVE AREA

Route

The patrol leader halts and conceals the patrol near the objective area in the final preparation position. The patrol leader then conducts leader's reconnaissance to confirm the plan for positioning the security teams and employing units assigned to the reconnaissance mission. The patrol leader returns to the patrol and positions the security to provide early warning of enemy approach and secure the objective rally point. The reconnaissance unit(s) then reconnoiters the objective area (route). The reconnaissance unit may move to several positions, along or adjacent to the specific route, in order to conduct a thorough reconnaissance. After completing the reconnaissance, each reconnaissance team moves to the objective rally point and reports to the patrol leader. The patrol then returns to friendly lines and the leader makes a full report.

Area

The patrol leader halts and conceals the patrol near the objective area in the final preparation position. The patrol leader then conducts leader's reconnaissance to pinpoint the objective and confirm the plan for positioning the security teams and employing units assigned the reconnaissance mission. The patrol leader returns to the patrol and positions security to provide early warning of enemy approach and secure the objective rally point. The reconnaissance unit(s) then reconnoiters the objective area. The reconnaissance unit may move to several positions, perhaps making a circle around the objective area, in order to conduct a thorough reconnaissance. When the reconnaissance is completed, the patrol leader assembles the patrol and tells members what has been observed and heard. Other patrol members contribute observations. The patrol then returns to friendly lines and the patrol leader makes a full report.

Zone

The patrol leader halts the patrol at the final preparation position, confirms the plan, and conducts leader's reconnaissance. The patrol leader positions the security team and sends out the reconnaissance team. When the entire patrol is used to reconnoiter the zone, it provides its own security. After completing the reconnaissance, each reconnaissance team moves to the objective rally point and reports to the patrol leader. The patrol then returns to friendly lines.

CHAPTER 13. COMBAT PATROLS

Combat patrols are assigned missions that usually include engaging the enemy. They are fighting patrols. Every combat patrol has a secondary mission: gaining information about the enemy and terrain. Combat patrols are employed in both offensive and defensive combat operations and they assist the parent unit in accomplishing its mission by inflicting damage on the enemy; establishing and/or maintaining contact with friendly and enemy forces; denying the enemy access to key terrain; and probing enemy positions to determine the nature and extent of enemy presence.

13001. TASK ORGANIZATION

A combat patrol is organized around the current structure of the Marine rifle platoon. A combat patrol leader should use the unit's normal organization (fire team, squad, and platoon) in assigning functions, patrol missions, and chain of command. Combat patrols must be able to perform the following four basic functions en route to and from the objective:

- Provide control.
- Provide security.
- Provide support by fire.
- Attack or assault the objective.

A rifle platoon could task-organize as a combat patrol as follows:

- Platoon headquarters (patrol headquarters).
- First squad (security).
- Second squad (support).
- Third squad (assault).

Every combat patrol must—

- Provide a control mechanism in the form of a headquarters.
- Designate a unit (a fire team or squad) to provide security while moving en route to the objective or while at the objective. At the objective area, this unit isolates the objective area, secures the objective rally point, and covers the withdrawal.
- Designate a unit to act as support. This unit provides the base of fire in the attack or covers withdrawals or advances.
- Designate a unit(s) to conduct the attack or assault. This unit(s) engages the enemy at the objective area by fire and maneuver or movement. It also operates immediately in the objective area (searching, demolition, prisoners of war, etc.). Paragraph 9001 outlines the general organization of combat patrols. As in the case with reconnaissance patrols, the task organization of a combat patrol depends on the specific mission assigned. If any special requirements are generated because of the specific mission, the patrol is task-organized to fit the needs of the mission.

13002. EQUIPMENT

Combat patrols are armed and equipped as necessary for accomplishing the mission. In addition to binoculars, wire cutters, compasses, and other equipment generally common to all patrols, it usually carries a high proportion of automatic weapons and grenades. Communications with higher headquarters is important as success of the mission may depend on being able to call for supporting fires. Also, internal radio communications with the units and teams may be useful. However, the patrol must not be so overburdened with equipment as to impede movement or mission accomplishment.

13003. RAID PATROLS

A raid is a surprise attack on an enemy force or installation with the attacking force withdrawing after accomplishing its mission. Raids destroy or capture enemy personnel or equipment, destroy installations, or free friendly personnel who have been captured by the enemy. Patrolling techniques are used in planning and when moveing to and from the objective. (Refer to MCWP 3-41.2, *Raids*.) Surprise, firepower, and violence of action are the keys to a successful raid. Patrols achieve surprise by attacking—

- When the enemy is least prepared (e.g., during periods of poor visibility such as darkness, rain, fog, or snow).
- From an unexpected direction. (This might be accomplished by approaching through a swamp or other seemingly impassable terrain.)
- With concentration of firepower at critical points within the objective.

Planning

A successful raid requires detailed planning. The leader of a combat patrol engaged in raiding must anticipate probable situations and decide upon definite courses of action to meet them. Rehearsals are imperative.

A raid patrol conducts such missions as destroying an enemy outpost or seizing prisoners from an observation post or lightly defended position.

While preparing for the mission, the patrol leader requests fire support required for the accomplishment of the mission. If practical, artillery and mortars should be employed to isolate the objective to prevent movement of enemy reinforcements into the area.

Execution

The leader's plan must be detailed and complete. All of the considerations outlined in chapter 11 must be covered. Patrol formations must provide for ease of control and all-around security while moving to and from the objective area and provide for rapid and coordinated deployment of the various units once the objective area is reached. The leader's plan usually includes the encirclement of the hostile position—either physically or by fire—in order to isolate it during the assault.

The final simultaneous assault against the objective develops when enemy defensive fires at the objective are suppressed by either friendly fire superiority or surprise. The assault is covered by the fire of the unit assigned the function of support by fire.

Grenades, SMAWs, and demolitions are most effective for clearing bunkers.

Security units are posted to isolate the objective. The patrol leader signals them when the withdrawal begins. As a minimum, security is on each flank and to the rear (at the objective rally point).

Actions at the Objective

The patrol leader halts the patrol near the objective at the final preparation position. Security is established and the leader's reconnaissance is made with appropriate subordinate leaders. When the leaders return to the patrol, they confirm previous plans or announce any changes. Movements are arranged so all units reach their positions simultaneously. This improves the patrol's capability for decisive action, if prematurely detected by the enemy.

The teams of the security element move to positions to secure the objective rally point, give early warning of enemy approach, block avenues of approach into—and prevent enemy escape from—the objective area. As the assault element moves into position, the security element informs the patrol leader of all enemy activity, firing only if detected or on the patrol leader's order. Once the assault element commences action, the security element prevents enemy entry into or escape from the objective area. The security element covers the withdrawal of the assault element (and support element, if employed) to the objective rally point, withdrawing only upon order or prearranged signal.

As the assault element approaches the objective, it deploys early enough to permit immediate assault if detected by the enemy. Each team uses stealth while moving into proper position. On command, or if one or more of the assault element is detected and fired upon by the enemy, the support element opens fire to neutralize the objective, then ceases or shifts fire according to prearranged plans and signals. As supporting fires cease or shift, the assault element assaults the objective. Demolition, search, and other teams are protected by the assault element while they work. On order, the assault element withdraws to the objective rally point.

If a support element is employed, its leader deploys teams to provide fire support for the assault element. Each member of the support element must know the scheme of maneuver to be used by the assault element, specific targets or areas to be neutralized by fire, and the signals that will be employed to commence, shift, and cease fires. The support element withdraws on order of the patrol leader. At the objective rally point, the patrol leader quickly reorganizes the patrol and begins the return movement to friendly lines.

13004. CONTACT PATROLS

General

Contact patrols establish and/or maintain contact to the front, flanks or rear by—

- Establishing contact with an enemy force when the definite location of the force is unknown.
- Maintaining contact with enemy forces through direct and/or indirect fires, or observation.
- Avoiding decisive engagement with the enemy.

Task Organization and Equipment

Task organization and equipment depend on the known enemy situation and anticipated enemy contact. A patrol sent out to establish contact with an enemy force is organized, armed, and equipped to overcome resistance of light screening forces in order to gain contact with the main enemy force. It is not organized and equipped to engage the main enemy forces in combat. Communication is paramount; radios must be reliable over the entire distance covered.

Actions at the Objective

The patrol leader selects a series of objectives. Once an objective is reached, the patrol leader initiates a planned set of actions in order to establish and maintain contact with the enemy. These plans and actions are guided by the missions to establish or maintain contact—not to engage in decisive combat. Contact with the enemy is maintained for the purposes of surveillance, applying pressure, and preventing seizure of the initiative. If the contact patrol becomes decisively engaged with the enemy, many of the tasks originally assigned to the patrol cannot be accomplished, since the enemy has seized the initiative and friendly forces are not forced to react.

13005. AMBUSH PATROLS

General

An ambush is a surprise attack from a concealed position upon a moving or temporarily halted target. It is one of the oldest and most effective types of military actions. Ambush patrols conduct ambushes of enemy patrols, resupply columns, and convoys. The intent of an ambush is to place the enemy in a dilemma where staying in the kill zone or attempting to move out of it prove equally lethal. The ambush may include an assault to close with and decisively engage the enemy, or the attack may be by fire only.

Purpose of Ambushes

Ambushes are executed for the general purpose of reducing the enemy's over-all combat effectiveness and for the specific purpose of destroying its units. The cumulative effect of many small ambushes on enemy units lowers enemy troop morale and harasses the enemy force as a whole.

Destruction is the primary purpose of an ambush because loss of men killed or captured, and loss of equipment and supplies destroyed or captured, reduces the overall combat effectiveness of the enemy.

Harassment, though less apparent than physical damage, is a secondary purpose of ambushes. Frequent ambushes force the enemy to divert men from other missions to guard convoys, troop movements, and carrying parties. When enemy patrols fail to accomplish their mission because they are ambushed, the enemy is deprived of valuable information. A series of successful ambushes causes the enemy to be less aggressive and more defensive minded. His men become apprehensive, overly cautious, reluctant to go on patrols, seek to avoid night operations, are more subject to confusion and panic if ambushed, and in general, decline in effectiveness.

Classification of Ambushes

A *deliberate ambush* is one in which prior information about the enemy permits detailed planning before the patrol departs for the ambush site. Information needed to plan a deliberate ambush includes the size, composition, and organization of the force to be ambushed; how the force operates; and the time it will pass certain points or areas. A deliberate ambush may be planned for such targets as—

- Any force if sufficient prior information is known.
- Enemy patrols that establish patterns by frequent use of the same routes or habitually depart and reenter their own areas at the same point.
- Logistic columns.
- Troop movements.

An *ambush of opportunity* is conducted when available information does not permit detailed planning before the patrol departs. This is the type of ambush that an infantry unit normally conducts. An ambush of opportunity should not be confused with a hasty ambush. An ambush of opportunity is a planned ambush; a hasty ambush is an immediate action. In planning for an ambush of opportunity, the patrol must be prepared to execute any of several courses of action based on the types of targets that may be ambushed and must rehearse prior to departure. The course of action taken is determined when the opportunity for ambush arises.

The patrol leader may be directed to reconnoiter an area for a suitable ambush site, set up at the site selected, and execute an ambush against the first profitable target that appears.

The patrol may depart just after dark, move to a specific point, observe until a designated time, ambush the first profitable target after that time, and return before daylight.

A *hasty ambush* is an immediate action where the patrol makes visual contact with an enemy force and has time to establish an ambush without being detected. The actions for a hasty ambush must be well rehearsed and accomplished through the use of hand and arm signals given from the patrol leader.

Types of Ambushes

There are two types of ambushes: point and area. The *point ambush* is one where forces are deployed to attack along a single killing zone. The *area ambush* is one where forces are deployed as multiple related point ambushes.

A point ambush, whether independent or part of an area ambush, is positioned along the enemy's expected route of approach. Formation of the forces conducting the ambush is important because, to a great extent, the formation determines whether a point ambush is able to deliver the heavy volume of highly concentrated fire necessary to isolate, trap, and destroy the enemy.

The ambush formation to be used is determined by careful consideration of possible formations and the advantages and disadvantages of each in relation to—

- Terrain.
- Visibility.
- Forces.
- Weapons and equipment.
- Ease or difficulty of control.
- Target to be attacked.
- Combat situation.

For a detailed discussion of ambush formation, see appendix D.

Ambush Operation Terms

The *ambush site* is the location where an ambush is established.

The *killing zone* is that portion of an ambush site where fires are concentrated to trap, isolate, and destroy the target. On little-traveled roads, an obstacle placed in a defile, in the woods, on a bridge or on a steep upgrade can be used effectively to force vehicles to halt, and thus render the occupants vulnerable to attack. Antitank mines may be emplaced and the occupants of the wrecked vehicle killed or captured while still dazed by the explosion.

A *near ambush* is a point ambush where the attacking force is located within reasonable assaulting distance of the killing zone (50 meters is a guide). A near ambush is most often conducted in close terrain, such as a jungle or heavy woods.

A *far ambush* is a point ambush where the attack force is located beyond reasonable assaulting distance of the killing zone (beyond 50 meters is a guide). A far ambush may be more appropriate in open terrain offering good fields of fire or when the target will be attacked by fire only.

Factors for a Successful Ambush Patrol

There are many factors that give the ambush its best chance of success. The ideal situation would be to position the ambush on favorable terrain and have detailed planning completed beforehand.

The patrol leader of an ambush looks for the most favorable terrain in which the enemy is canalized between two obstacles with limited opportunity to attack or escape. Suitable areas include defiles, small clearings, bends in trails, and steep grades. Dense undergrowth adjacent to the ambush site permits observation from concealed positions. The ambush patrol should have maximum cover and concealment, not only for the firing positions, but for the routes of withdrawal. The enemy should be in an area offering as little protection from fire as possible. Favorable fields of fire include stretches of road, trail or open ground of at least 100 meters for machine guns and 15 meters for rifle fire and grenades. The ambush site can be improved by constructing obstacles—such as felled trees, wire, land mines, or booby traps—to impede the enemy.

Planning

A deliberate ambush or an ambush of opportunity requires thorough planning.

A deliberate ambush plan is based on extensive knowledge of the enemy and terrain, and is planned and rehearsed in great detail. A physical reconnaissance of the ambush site is made during the preparation phase and information gained is incorporated into the plan. All likely immediate actions of the enemy when ambushed are examined. Planned counteractions are developed and rehearsed.

In planning an ambush of opportunity, any available information on the enemy and terrain is used. A tentative plan for the ambush that incorporates all anticipated actions is developed and rehearsed. However, the bulk of planning is done concurrently during the patrol leader's reconnaissance of the prospective ambush site. In a rapidly developing situation, hasty ambush immediate action is employed.

The route and ambush site considerations apply to both deliberate ambushes and ambushes of opportunity. A primary route that allows the patrol to enter the ambush site from the rear is planned. Entering the prospective killing zone is avoided. If the killing zone must be entered to place mines or explosives, care is taken to remove any tracks and signs that might alert the enemy and compromise the ambush. If mines or explosives are to be placed on the far side of the ambush site, or if the appearance of the site from the enemy's viewpoint is to be checked, a wide detour is made around the killing zone. Care is taken to remove any tracks that might reveal the ambush. An alternate route from the ambush site to the objective rally point, as in other patrols, is planned.

Maps and aerial photographs are used to carefully analyze the terrain. When possible, an on-the-ground reconnaissance of the ambush site is made prior to occupation. Obvious ambush sites are avoided as the element of surprise is even more difficult to achieve in these areas. An ambush site must provide for—

- Favorable fields of fire.
- Occupation and preparation of concealed positions.
- Canalization of the target into the killing zone. (An ideal killing zone restricts the enemy on all sides, confining him to an area where he can be quickly and completely destroyed. Natural obstacles, such as cliffs, streams, embankments, or steep grades, are used whenever possible to force vehicles to slow down. Man-made obstacles, such as barbed wire, mines, and craters in the roads, are used to supplement natural obstacles.)
- Covered routes of withdrawal that enable the ambush force to break contact.
- Avoidance of enemy pursuit by fire.

Occupation of Ambush Site

The surrounding area is searched for enemy patrols prior to occupation of the ambush site. Ambush formations are used to physically deploy the patrol in a manner to inflict maximum destruction upon the enemy and to provide maximum security to the patrol. Ambush formations are contained in appendix D.

Positions

The patrol is moved into the ambush site from the objective rally point. Security is positioned first to prevent surprise while the ambush is being established. Automatic weapons are then positioned so each can fire along the entire killing zone. If this is not possible, overlapping sectors of fire are provided to cover the entire killing zone. The patrol leader then selects his position, located so he can tell when to initiate the ambush. Riflemen and grenadiers are positioned and sectors of fire are assigned to cover any dead space left by the automatic weapons. The patrol leader sets a time by which positions are to be prepared. Patrol members clear fields of fire and prepare positions in that order, with attention to camouflage for both.

Suitable Objective Rally Point

An easily located objective rally point is selected and made known to all patrol members. The objective rally point is located far enough from the ambush site so that it will not be overrun if the enemy assaults the ambush. Routes of withdrawal to the objective rally point are reconnoitered. Situation permitting, each man walks the route he is to use and picks out checkpoints. When the ambush is to be executed at night, each man must be able to follow his route in the dark. After the ambush has been executed, and the search of the killing zone completed, the patrol is withdrawn quickly but quietly, on signal, to the objective rally point where it reorganizes for the return march. If the ambush was not successful and the patrol is pursued, withdrawal may be by bounds. The last group may arm mines, previously placed along the withdrawal route, to further delay pursuit.

Local Security

Security must be maintained. Security elements do not usually participate in the initial attack, but protect the rear and flanks, and cover the withdrawal.

Patience

The Marines of the ambush force must control themselves so that the ambush is not compromised. Patience and self-discipline are exercised by remaining still and quiet while waiting for the target to appear, particularly if the patrol occupies the ambush site well ahead of the arrival of the enemy. Patience is necessary so as not to alert the enemy to the presence of the ambush.

Surprise

Surprise must be achieved, or the attack is not an ambush. If complete surprise cannot be achieved, it must be so nearly complete that the target is not aware of the ambush until too late for effective reaction. Surprise is achieved by careful planning, preparation, and execution so that targets are attacked when, where, and in a way for which they are least prepared.

Coordinated Fires

Properly timed and delivered fires contribute heavily to the achievement of surprise, as well as to destruction of the enemy. The lifting or shifting of fires must be equally precise; otherwise, the assault is delayed and the enemy has an opportunity to recover and react. All weapons, mines, and demolitions are positioned and all fires, including those of available artillery and mortars, are coordinated to achieve the following results:

- Isolation of the killing zone to prevent the enemy's escape or reinforcement.
- Surprise delivery of a large volume of highly concentrated fires into the killing zone.

Control

Close control of the patrol is maintained during movement to, occupation of, and withdrawal from the ambush site. This is best achieved through rehearsals and establishment and maintenance of good communications. When the enemy approaches, the temptation to open fire before the signal is given is resisted. The patrol leader must effectively control all elements of the ambush force. Control is most critical at the time the enemy approaches the killing zone. Control measures must provide for—

- Early warning of enemy approach.
- Fire control. Withhold fire until the enemy has moved into the killing zone, then open fire at the proper time.
- Initiation of appropriate action, if the ambush is prematurely detected. Individual patrol members must be prepared to react if detected by the enemy prior to the initiation of the ambush.
- Timely and orderly withdrawal of the ambush force from the ambush site and movement to the objective rally point.

It is important to remember that an ambush patrol should have four distinct signals: one to open fire (with an alternate signal to open fire to be used at the same time as the primary); a signal to cease fire or shift fire; a signal to assault or search the killing zone; and a signal to withdraw. The signal to open fire should meet two criteria: first, it should be the firing of a weapon that will kill the enemy; secondly, it should be a weapon reliable in any weather condition. A good primary signal is a Claymore mine, and an alternate signal would be a closed bolt weapon (M16A2). Open bolt weapons (M240G, M249) should not be relied upon to initiate an ambush.

Execution of an Ambush

The manner in which the patrol executes an ambush depends primarily on whether the ambush's purpose is harassment or destruction. To a lesser degree, the execution of the ambush is determined by whether the ambush is deliberate or an ambush of opportunity.

When the primary purpose is harassment, the patrol seals off the area with security teams to prevent enemy reinforcement and escape. Maximum damage is inflicted with demolitions and automatic weapons fire. The patrol delivers a very heavy volume of fire for a short time and withdraws quickly and quietly. The patrol avoids being seen by the enemy.

When the primary purpose of the patrol is destruction, the area is sealed off with security units. Maximum damage is inflicted with demolitions, antitank weapons, and automatic weapons fire from the support team or element. When these fires cease or shift, an assault is launched into the killing zone with heavy fire and violence to complete destruction. The assault unit provides security, while designated teams search and/or capture personnel and destroy vehicles and equipment. On the patrol leader's command, or by prearranged signal, all units withdraw to the objective rally point and move out quickly.

When the patrol's primary purpose is to obtain supplies or capture equipment, security units seal off the area. Demolitions and weapons are used to disable vehicles. The assault unit must use care to ensure its fire does not damage the desired supplies or equipment. Designated teams secure the desired items; other teams then destroy enemy vehicles and equipment.

The most successful ambush is one where the attacker is deployed and concealed in such a way that the enemy will unknowingly be surrounded by fire. The usual method is for the attackers to deploy themselves along a trail or route the enemy will travel. The enemy is permitted to pass by the center of the attacker's force so that the attack can be made from the front, flank, and/or rear. One or two men are posted well forward and to the rear along the route to prevent any enemy from escaping. All fires should be delivered simultaneously on a prearranged signal.

An effective method of luring the enemy is for an ambushing patrol to cut communication or electrical wire. The patrol then deploys and ambushes the line repair crew when it arrives. Since the line crew may be protected by riflemen, the attackers must be careful to engage the entire party.

Vehicles and foot personnel moving on well-established transportation routes can sometimes be captured by altering or moving directional signs so as to divert the enemy into an area where he can be more readily attacked. The attack can best be accomplished at an obstacle, such as a stream or gully, that forces the enemy to stop or slow down.

After the enemy has been ambushed and destroyed, the unit quickly withdraws over a prearranged route to the objective rally point. Speed is very important, since the noise of the ambush could alert other nearby enemy units.

13006. SECURITY PATROLS

General

Security patrols are assigned missions that may or may not require them to engage the enemy. They are used in proximity to defensive positions, on the flanks of advancing units or in rear areas. Purposes of security patrols are to detect infiltration by the enemy, destroy infiltrators, and protect against surprise and ambush.

In any situation where there is a threat of attack, such as a rear area threatened by guerrillas or a facility that is under threat of a terrorist attack, all Marines, not just the infantry, must know how to conduct a security patrol.

In just the offensive operations, infantry units provide security patrols to screen their flanks, areas, and routes. Whereas, in defensive operations, security patrols are used to prevent the enemy from infiltrating an area, detect and destroy infiltrators, and prevent surprise attacks. In rear areas, particularly when there is guerrilla or terrorist threat, the requirement to conduct security patrols increases for all Marine air-ground task force (MAGTF) units ashore, particularly aviation and combat service support units.

Task Organization and Equipment

Generally, a Marine rifle squad or similar sized organization is considered ideal for security patrols. Communications are important to higher headquarters so that they receive information from the patrol; and communications are important to the patrol to request fire support, etc. The radio the patrol carries must have the range necessary for higher headquarters to be able to receive transmissions from anywhere along the patrol route, and the patrol must have a secondary means of reporting (i.e., flare signals upon contact).

Patrol Procedures

All of the procedures presented in previous sections are to be used in security patrols.

Patrol Planning

Security patrol planning includes—

- Rehearsing prior to departing friendly lines.
- Maintaining communications.
- Support by organic weapons.
- Reinforced if necessary.
- Using varied routes and never establishing a routine pattern.
- Staying within proximity of friendly units.

Patrol Techniques

Within rear areas, an irregular pattern of patrol is established and changed daily. Outside of friendly lines it would be prudent to establish a definite preplanned route for the patrol, of which all adjacent units know the route. The parent unit commander establishes frequent checkpoints for control. If checkpoints are designated, the patrol leader treats them as individual objectives to be searched and cleared.

The patrol has a definite plan as to what to do if contact with enemy is made, how to break contact, how to defend itself, and how to call for supporting fires. It is imperative that patrol members know what to do if they become split or separated; i.e., location of rally points and how to be recovered.

13007. URBAN PATROLS

General

As national strategy continues to focus on regional vice global conflicts, the Marine Corps will continue to conduct urban operations in various operational environments. Cities and towns are often the center of economic and political power and are therefore extremely vulnerable to urban insurgent activities and violence. The lessons learned from recent operations in Somalia, as well as experiences gained by British forces in Northern Ireland, provide the foundations of urban patrolling.

Enemy Forces

Enemy forces in urbanized areas range from organized military forces to low intensity engagements with insurgents, such as terrorists or local gangs. For Marine Corps doctrine and tactics, techniques, and procedures (TTP) for dealing with the higher intensity threat in urbanized areas, refer to MCWP 3-35.3, *Military Operations on Urbanized Terrain.*

Common Threat Tactics

Cities provide cover and concealment for both friendly forces and enemy forces. However, enemy forces generally find active support only in certain areas of the town or city. The urban-based insurgent or terrorist usually lives in a friendly community or in one where the people are too frightened to withhold support or inform anyone about the situation. The insurgent or terrorist normally maintains close contact with leaders and others friendly to the cause. The enemy will often have efficient communication and intelligence systems, sometimes involving women and children to provide cover for its activities.

The urban insurgent normally cannot, like his rural counterpart, establish bases and recruit large military units. He is generally an individual or a member of a relatively small group. He relies on the cover afforded by the city's populace and terrorizing them to coerce loyalty or support.

Urbanized areas tend to give the insurgent and/or terrorist many opportunities to initiate action and gain advantage. The normal presence of large numbers of people in cities provides the foe an opportunity to mass crowds quickly and manipulate demonstrations easily. The presence of women and children during mass demonstrations may restrict the courses of action available to friendly forces. Major incidents stemming from overreaction or excessive use of force by friendly forces may provide the insurgent with propaganda material. Publicity is easily gained in an urbanized area because major incidents can't be completely concealed from the local population. Insurgent successes can be exploited to discredit the ability of host nation police, friendly forces, and the civil government, and gain recognition for the insurgents' cause.

The urban insurgent or terrorist can usually be expected to operate more boldly than his rural counterpart. This is reflected in the enemy's tactics. A single sniper or bomber may be the norm in the urbanized area, whereas the rural threat is generally the more conventional ambush. In urbanized areas, explosive devices can be easily emplaced and used effectively against large groups or select individuals. Enemy forces may be expected to employ the following tactics in urbanized areas:

- Using local communications, such as radio and newspapers, for propaganda purposes.
- Disrupting industry and public services through strikes and sabotage.
- Generating widespread disturbances designed to stretch the resources of the security force.
- Creating incidents or massing crowds in order to lure the patrol or reaction force into a trap.
- Provoking security forces in the hope that they may react improperly, therefore discrediting the security force by means of propaganda.
- Sniping at roadblocks, outposts, sentries, and patrols.
- Attacking friendly bases with rockets and mortars.
- Planting explosive devices, either against specific targets or indiscriminately, to cause confusion and destruction, and lower public morale and confidence.
- Using ambush patrols.
- Firing on friendly helicopters.

Principles of Urban Patrolling

Patrolling in an urban environment often presents conditions considerably different and often more complex than those encountered in rural and less inhabited areas. While the principles of patrolling are still relevant in an urban situation, the nature of urban patrolling has led to the development of six specific urban patrolling principles. They are—

1. **Depth.** The restrictive, canalizing nature of urbanized terrain usually limits a patrol's ability to disperse laterally. To prevent the patrol from bunching up, patrols normally maintain dispersion along the length of a patrol formation.

2. **Mutual Support.** The positioning of units in-depth within the patrol enables one unit to cover another unit's movement and facilitates immediate action during various situations. Aircraft, vehicles, and snipers also provide good mutual support.

3. **Deception and Pattern Avoidance.** Deception and pattern avoidance are normally a planning consideration of the headquarters directing the entire patrolling plan. By varying patrol routes, durations, and departure times, hostile actions commonly used against urban patrols—such as enemy ambushes, and roadblocks—are more difficult to plan and may be preempted.

4. **Intra-patrol Communication.** Elements of an urban patrol must have the means to communicate with each other. Ideally, each element will possess a radio enabling it to remain in continuous communication that facilitates rapid response and reporting to higher headquarters, reaction force coordination, and coordination of actions with other patrols or fire support agencies.

5. **Establishment of a Reaction Force.** The nature of urbanized terrain (its compartmentalization) makes urban patrols more vulnerable to a wider range of hostile actions. The requirement for immediate, coordinated reinforcement of a patrol is best satisfied by employment of an established reaction force. The reaction force requires superior mobility (relative to the enemy's) and fire support to be effective for this task.

6. **Three-Dimensional Threat.** Patrolling in an urban environment requires constant attention to its three-dimensional aspect; hostile actions can originate from rooftops, streets, subsurface levels or combinations of all levels at once.

Classification of Urban Patrols

Mission

The vast majority of urban patrols are overt in nature, with their presence readily apparent to the local populace. Most urban patrols are combat vice reconnaissance patrols. The vast majority of urban patrols are security patrols. However, units may be assigned secondary tasks of reconnoitering specific or general areas along the patrol routes.

Raids normally involve a swift penetration of an objective to secure information, confuse the enemy or destroy installations. Raids include a planned withdrawal upon completion of the assigned mission.

Movement

Means of movement are as follows:

- Dismounted: movement on foot.
- Vehicular: movement by motorized, mechanized or armored vehicle.
- Helicopterborne: movement by helicopter, however, helicopterborne patrols will usually involve dismounted or vehicular movement after patrol insertion.
- Combination: movement using a combination of methods.

Task-Organization

Units task-organize to meet the specific requirements of the mission and situation. The Marine rifle squad is ideally suited for urban patrolling and can easily integrate attached specialists required by specific missions. Specialists who may accompany urban patrols include—

- Interrogator-translator team (ITT) and counterintelligence team (CIT) Marines.
- Explosive ordnance disposal (EOD) personnel.
- Members of host nation or allied military forces.
- Interpreters.
- Local community leaders.
- Local law enforcement officers.
- Public affairs personnel escorting media representatives.

Dismounted Patrol Organization

Patrols should maintain unit integrity (fire team and/or squad) in organizing elements for an urban patrol. Similar to reconnaissance patrols, the vulnerability of urban patrols necessitates that all elements must provide for their own security in addition to the combat aspect of the mission. Combined assault and security teams are an effective method to organize for all-around security.

Urban Patrol Base Operations

Urban patrols may operate from an established patrol base that may be located within the unit's assigned area of operations or an area designated for a patrol. The patrol base should be located in a building used exclusively for this purpose. Patrol bases may also be located within a larger site that houses other agencies, such as a higher headquarters' command echelon.

If located within a larger site, the patrol base will be included in the overall facility security plan. If isolated, the patrol base must consider the following security factors:

- External security:
  - Barrier plan.
  - Sentry posts.
  - Local security patrols.
- Internal security:
  - Covered positions for all Marines.
  - Contingency plan for hostile actions against patrol base.

The Reaction Force

The high probability of an urban patrol becoming involved in a hostile or volatile incident requires the establishment of a dedicated reaction force for rapid reinforcement, support or extraction of the patrol. Ideally, the reaction force is—

- Large enough and task-organized in a manner that it can meet and quickly defeat the expected threat. Reaction forces are normally tiered with a lead element (normally one-third of the unit's size) and a main body force (the other remaining two-thirds of the force). Reaction force response times routinely are determined in advance by higher headquarters.
- Ready to respond immediately.
- Motorized or mechanized and supported by close air support and other fire support.
- Familiar with the area of operations.
- Briefed on the patrol's plans and monitors the status of patrols in progress.
- Task-organized to be multimission capable.
- Able to communicate with the higher headquarters, fire support assets, patrol base, and the patrol.
- Controlled by higher headquarters, once employed.

Patrol Preparation

Planning

Higher headquarters will—

- Designate the area for patrol.
- Provide intelligence briefs and updates.
- Ensure liaison with allied forces and the civilian populace.
- Provide special equipment and personnel required for the mission (scout snipers, public affairs officer, interpreters, etc.)
- Provide urban maps, photos, terrain models as required.
- Consider deception and pattern avoidance when issuing mission.
- Prescribe rules of engagement (ROE).

Intelligence Brief

An intelligence brief is conducted by the S-2 officer or representative prior to a patrol conducting its mission. The brief adresses the situation relevant to the specific patrol (e.g., routes, areas, updated enemy situations).

Coordination

Higher headquarters will effect liaison with adjacent and allied forces, as well as civil authorities and other agencies, having a possible effect on the patrol. The patrolling unit generally follows the same procedures as those used during patrol planning and execution in a jungle or forest environment.

The Urban Patrol Order or Warning Order

The Urban Patrol or Warning Orders use the same format and considerations as noted in this publication for patrolling. They rely heavily on a detailed terrain model, photographs, and subterranean construction to ensure complete understanding of the plan.

Rehearsals

The limited size of the patrol base usually precludes the need for full-scale rehearsals. Immediate action drills, such as crossing danger areas, are rehearsed in as much detail as possible, despite the limited available space.

Inspections

Initial and final inspections are conducted in the same manner as other patrols. Attached personnel must be fully integrated into the patrol and familiar with the plan and unit standing operating procedures (SOPs).

Conducting an Urban Patrol

Movement

Individual and unit movement considerations are generally the same as those for other patrols. However, urban environments require consideration of additional factors. Because of these factors, an urban patrol leader should—

- Ensure that each movement within a patrol takes place under the observation or cover of another individual or element of the patrol.
- Know where cover can be taken in the event of a hostile incident or action.
- Be prepared for contact with civilians, especially children, during the patrol and be aware that they may intentionally attempt to distract patrol members.
- Expect the presence of vehicles (both moving and stationary) along the patrol route.
- Expect members of the patrol to be approached by dogs and what action to take if threatened.

Patrol Formations

Squad-sized Patrols. The need for immediate fire power outweighs the dangers of becoming canalized. In contrast to other types of patrols, the headquarters element of an urban patrol will normally locate at the lead of the patrol column. This allows the patrol leader greater flexibility through control of two combined assault and security (A&S) teams. The leaders of these elements tactically stagger their members on each side of the road (see fig. 13-1 on page 13-12).

A&S teams follow in trace of the headquarters element and maintain unit integrity on separate sides of the street. One unit will remain slightly to the rear to create a staggered interval between Marines on either side of the street. This allows A&S teams to take lateral routes in support of headquarters element without having to cross a street to do so.

Platoon-sized Patrols. Squads will generally travel abreast of each other, moving along parallel routes. The interval between squad-sized units and/or teams is situation-dependent, but is usually between 100 and 150 meters (roughly two city blocks; this often prevents visual contact between the units). The intent is to create less of a target to an aggressor, yet still allow the patrol to quickly react to an incident. Individuals within units or teams will move in a staggered column as in a squad-sized patrol. See figure 13-2, which is located on page 13-13).

Night

Night patrols will generally be at least squad-sized and will generally use the same formation as that for day patrols. At night, it may be necessary to close distances between individuals or elements to maintain control. Consideration should be given to the use of night vision devices and thermal weapons sights.

Navigation, Control, and Security Measures

The patrol leader is ultimately responsible for the navigation. The headquarters element normally functions as the base unit during movement. The designated navigator is normally assigned from within the headquarters element. City maps are often inaccurate or outdated; however, when used with aerial photographs and other navigational aids, they can be effective for urban navigation.

Checkpoints and phase lines should be related to major streets (alleys, buildings, bridges) for easy identification. Arrival at checkpoints and crossing of phase

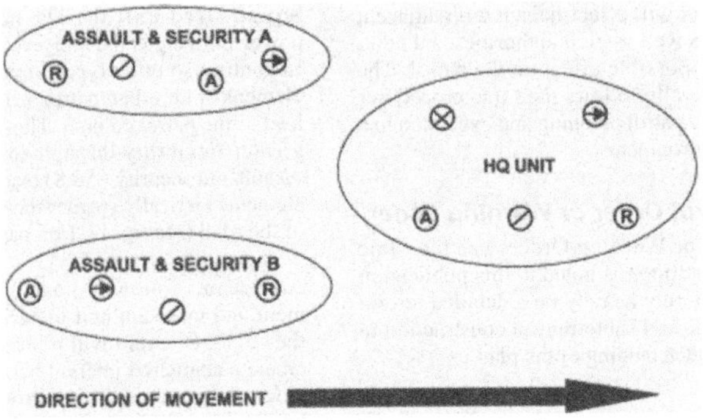

Figure 13-1. Squad-sized Dismounted Urban Patrol Formation.

lines should be relayed to higher headquarters using established brevity or codewords. A detailed patrol overlay is issued to both the reaction force and higher headquarters to keep them advised of routes and control measures used. A terrain model should be maintained at higher headquarters to aid in supervision and control of the patrol.

The use of camouflage should be limited to avoid frightening and confusing the local populace. The often overt nature of urban patrols may negate the need for camouflage. Patrols are sometimes deployed to show force presence and usually move on the urban street in plain view.

During daylight, patrols will routinely vary their rate of movement ranging from short halts to brief periods of double-timing. The British term for this urban patrolling technique is *hard-targeting*, meaning it makes the patrol harder for an enemy to target. Altering the rate of movement is intended to frustrate the enemy's ability to coordinate an attack or ambush against a targeted patrol.

Patrols should use short security halts, with Marines taking up mutually supporting firing positions. Marines must always work in pairs, ensuring mutual support. The last Marine in the element will provide rear security, but stays in his buddy's sight.

Individual Tasks

Individuals may be assigned collateral tasks performed throughout the patrol that may increase the patrol's situational awareness. Individual tasks may include—

- Vehicle spotter: looks for suspicious or known insurgent vehicles.
- Personnel spotter: observes and attempts to recognize previously identified enemy in crowds.
- Talker: attempts to gain information from casual conversations with the local populace. (Talkers are usually subordinate leaders or Marines with foreign language skills.)
- Searcher: conducts physical searches of vehicles and personnel while other patrol members provide cover and security.
- Marksman: engages point targets when the tactical situation does not permit massed or high volume fire. Other patrol members provide security to cover the marksman's engagement.

Departure of Friendly Lines

Urban patrols must vary their departure times to prevent being ambushed while exiting the patrol base.

Individual elements will usually depart exits at staggered times and at different movement rates, especially where sentries cannot provide cover. (This technique is used when the environment adjacent to the patrol base is dangerous.)

Once an element has exited friendly lines, a short halt is conducted in a predetermined, covered initial rally point 50 to 100 meters from the base. This ensures all elements are in position before the patrol continues.

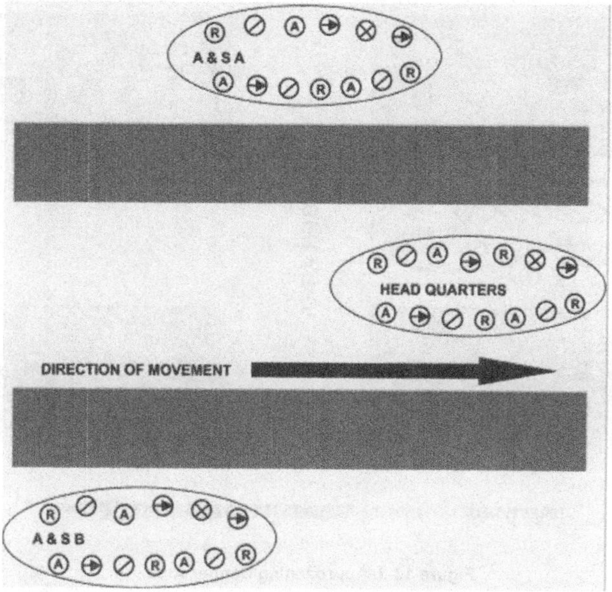

Figure 13-2. Platoon-sized Dismounted Urban Patrol Formations.

Exit points, routes from the base, departure techniques, and locations of IRPs should be varied constantly to avoid setting a pattern. This pattern avoidance may also include using vehicles to insert patrols away from the urban patrol base and employing empty vehicles as part of a deception plan.

Danger Areas

Urban patrols may encounter hundreds of danger areas during a single patrol. The three-dimensional threat requires keen situation awareness by every patrol member. Many danger areas can be dealt with simply by avoidance, while others require an adjustment of patrol formation, movement rate, etc. In the urban environment, places to be treated as danger areas are points that pose a major threat to the patrol, such as local political and religious headquarters, weapons containment areas, roads and routes that canalize movement and direct fire, and any area with a history of repeated contact.

Near and far side rally points are designated and briefed during the issuance of the patrol order. Squad-sized formations may use the A&S teams to provide flank security for the headquarters element and for each other. The headquarters element identifies the danger area and takes up a position on the near side of intersection. Individuals provide all-around security (see fig. 13-3 on page 13-14).

Two Marines (one from each A&S team) are designated to move through the headquarters element and establish respective firing positions on the near side of the danger area covering the patrol's near side flanks. They are followed by a second pair (again, one Marine from each A&S team) that moves across to the far side of the danger area and establishes respective firing positions covering the patrol's far side flanks (see fig. 13-4 on page 13-14).

Once near and far side flank security is established, the headquarters element moves across to the far side of danger area (see fig. 13-5 on page 13-15).

The remaining A&S team members then cross the danger area and join the headquarters element on the far side (see fig. 13-6 on page 13-15).

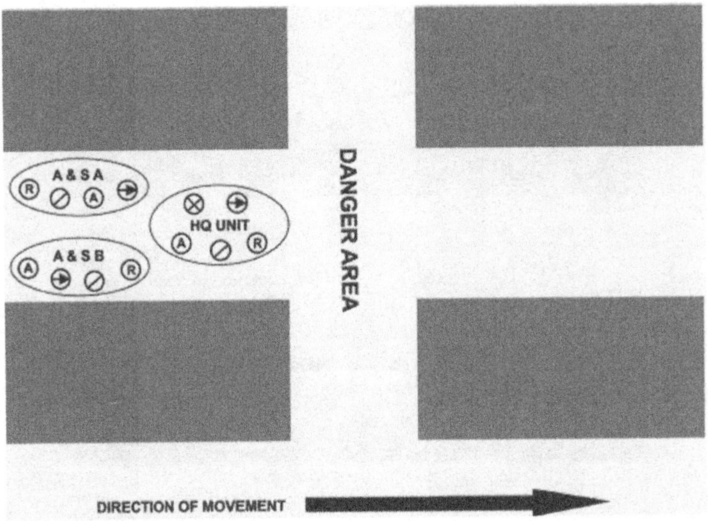

Figure 13-3. Approaching Danger Area.

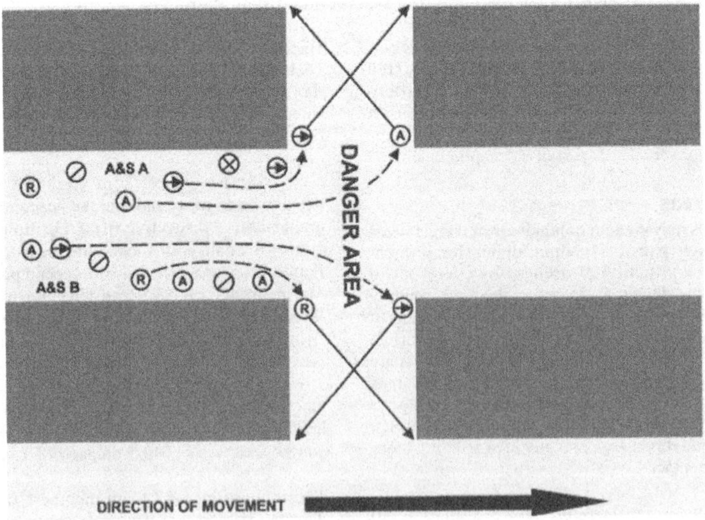

Figure 13-4. Securing Flanks.

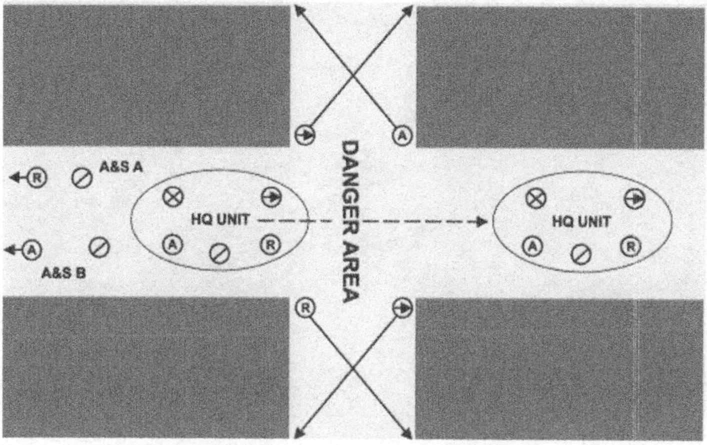

Figure 13-5. HQ Element Moves Through.

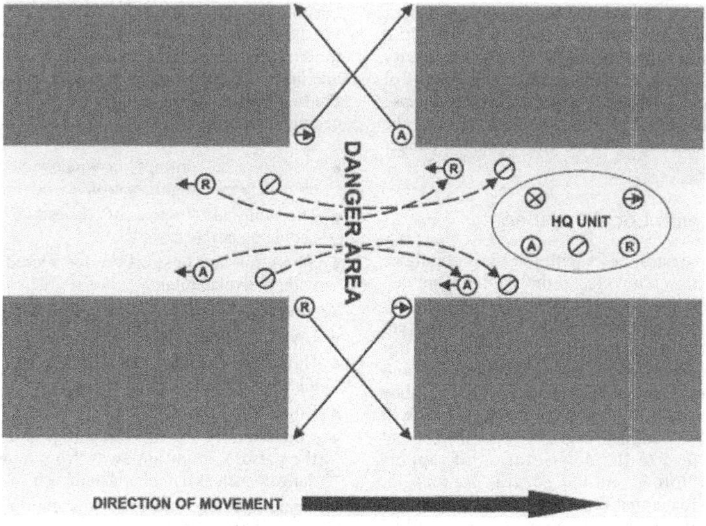

Figure 13-6. A&S Elements Move Through.

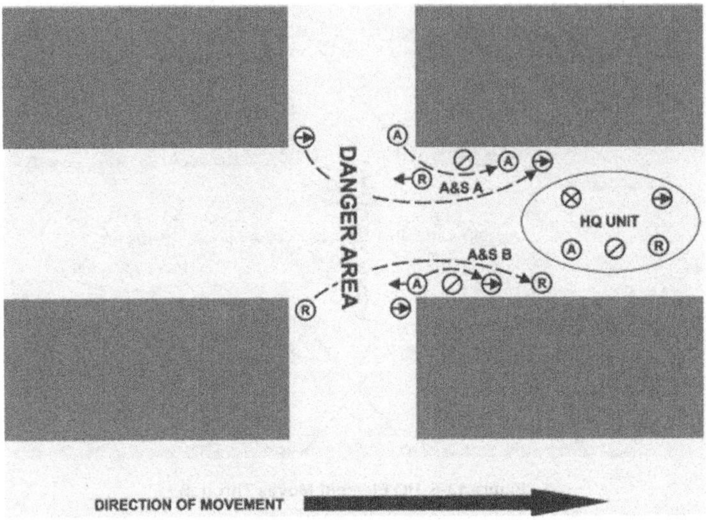

Figure 13-7. Patrol Resumes Movement Along Intended Route.

Once the trail A&S team members cross the danger area, they take up rear security and cover the movements of the near and far side flank security teams as they return to their positions in the patrol formation. The near side security team should collapse back first, followed by their far side counterparts. The patrol then resumes its advance away from the danger area (see fig. 13-7).

Interaction with Local Populace

Urban patrols must interact with the local populace. Patrols are at first a novelty to the civilians but can quickly become an unwelcome intrusion. The movement of the patrol must be fast enough to prevent the enemy from massing their fires upon it, but deliberate enough to ensure adequate security and mutual support. Patrol members must realize that they are usually the only Marines the local populace will encounter and that an inappropriate gesture, comment or act could lead to the deterioration of rapport between U.S. forces and the general population. Marines must remember that the vast majority of the individuals with whom they come in contact will be noncombatants attempting to survive in trying political, economic, and social situations.

Hostile incidents often seem to occur spontaneously, but there are usually indications that can alert Marines to imminent danger. The most obvious are the sudden alteration of normal routines, patterns, and attitudes of the local populace or other unusual activity. Some examples include—

- Observers on rooftops, in windows, etc., who are obviously tracking the patrol.
- The unusual absence of pedestrian traffic and people on porches.
- Stores, markets or street vendors closed suddenly or without explanation.
- Changes in civilian attitude toward patrol members.
- Unknown individuals or vehicles in the patrol area.
- Unfamiliar vehicles parked in the patrol area (possible car bomb).
- Roadblocks.
- Children throwing rocks at patrols to possibly draw the patrol's attention away from a more serious danger, such as a deliberate ambush.
- Vehicles riding unusually low due to overloading (possibly ferrying people, weapons, explosives).
- Agitators trying to provoke an incident with patrol members.

- Absence of the usual stray dogs (dogs are adept at sensing danger and avoiding it).
- Anti-American graffiti suddenly appearing in the patrol area.
- Pictures of enemy leaders and martyrs posted in the patrol area.
- Civilian workers failing to appear at U.S. or friendly bases.
- Normal deliveries and pick-ups conducted late or early without reason.
- Sudden change of civilian sentiment in newspaper articles, radio broadcasts or other media.
- Women and children leaving to live elsewhere.

Immediate Actions Upon Enemy Contact

Reaction to Sniping

Snipings are often executed from a single firing point, but coordinated snipings delievered from multiple points are not uncommon.

Patrol element or team leaders should constantly try to identify likely firing points and anticipate their own reactions to a possible shooting. Normally, snipers in an urban environment have a detailed withdrawal plan. Once a patrol comes into contact with a sniper, the patrol leader must immediately assess the situation and maneuver his patrol accordingly. The patrol's mission, location, size, ROE, and location of the threat often determine whether the patrol will attempt to neutralize the targeted sniper. If the patrol leader decides to kill or capture the sniper, he uses planned and rehearsed immediate actions to maneuver and counter the sniper's assault. The goal is to kill the sniper or cut off his escape and capture him. There are three immediate reactions to neutralizing a sniper: initial contact, immediate follow-up, and subsequent follow-up.

Initial Contact. The initial contact is made when the sniper fires the first shot. The patrol must react immediately and positively to get behind the firing position in order to kill or capture the gunman. The period of contact ends when the gunman is killed or captured, or the patrol element or team leader on the scene ends it. The following technique is the same for both squad- and platoon-sized patrols:

- The element or team in contact attempts to identify the firing position and maneuvers designated marksman into position to return well-aimed and controlled fire. Other members of the patrol take up positions to cover the marksman's engagement.
- The patrol element or team leader in contact sends initial contact report to the patrol leader, who notifies higher headquarters.
- The element or team leader in contact determines appropriate cut-off positions and relays them to flanking elements or teams.
- The patrol element or team leader in contact continues to observe the firing point, but does not enter it due to the possibility of booby traps. Flank elements or teams set up along likely escape routes.
- The incident ends when either the sniper ceases fire or is neutralized.

Immediate Follow-Up. Regardless of the fate of the gunman, isolation of the firing point is necessary to prevent reinforcement and preserve forensic evidence (scent, spent casings, etc.). If not under fire, members of the patrol element or team cordon off the area surrounding the firing point. Flanking elements or teams maintain their positions and prevent civilians from entering the area. The patrol leader moves to link-up with the element or team in contact (if not his own), and makes an estimate of the situation. The patrol leader sends a SPOTREP to higher headquarters.

Subsequent Follow-Up. The aim of the subsequent follow-up is to use follow-on forces to clear the building of remaining resistance or to obtain evidence that can be used to capture the gunman. The patrol leader establishes a position where he can brief arriving units (reaction force commander, S-2 representative, EOD personnel, etc.). Once the arriving units have been briefed, recommendations are made to higher headquarters via radio. No one is allowed into the cordon without the patrol leader's approval.

Reaction to Becoming Decisively Engaged

If a patrol becomes decisively engaged from numerous firing positions, the following immediate action should be taken:

- All patrol members move to available cover and return accurate fire on identified firing points.
- The patrol leader assesses the situation and makes a decision to either request the reaction force or break contact.
- If the reaction force is requested, the patrol will maintain its position until the reaction force arrives. The patrol should use fire and maneuver to gain better tactical positioning and support the arrival of the reaction force. When the reaction force arrives,

its commander may decide to either clear occupied buildings or cover the patrol during its extract.

Reaction to Bomb Threat or Discovery

The use of command-detonated explosive devices is a common ambush tactic employed by a terrorist or insurgent in an urban environment. The appropriate response to a reported threat or an actual discovery is generally involves four steps (known as the four Cs)—

1. The patrol leader **CONFIRMS** the presence of the suspicious item.

2. Without touching or moving anything suspicious, patrol elements or teams **CLEAR** the immediate danger area to a minimum of 100 meters. The area is cleared from the suspected device outward, inform civilians as to the reason for evacuation.

3. A&S teams establish a **CORDON** to secure the cleared area. Avenues of approach are cordoned off to keep people out and to protect EOD or engineer personnel clearing the device. The assistant patrol leader acts as the cordon commander and informs the patrol leader when the cordon is secure. An effective cordon technique is to tape off the area with engineer tape, creating both a physical and psychological boundary.

4. **CONTROL** of the area is maintained throughout the bomb clearing operation by the patrol leader. The patrol leader sends a report to a superior concerning details of the device (if known) and the area affected. The patrol leader coordinates with arriving personnel (EOD, engineers, etc.). The patrol leader maintains communication with the assistant patrol leader and keeps the Marines informed of the progress of the clearing operation.

Reaction to a Bomb Detonation

Bombs may be used by an insurgent as a means of initiating an ambush on mounted or dismounted patrols, in which case the actions for decisive engagement apply. Immediate action in response to an isolated explosion is similar to that used in reaction to a sniping and breaks down into the same three phases:

Initial Contact. The patrol leader attempts to identify the likely initiation point and sends an initial contact report to higher headquarters. If the bomb was command-detonated, the patrol leader sends his A&S teams deep to cut off the bombers' escape routes. Any casualties are moved a minimum of 100 meters from explosion and out of the line of sight to it.

Immediate Follow-Up. The A&S teams may need to maneuver to positions behind the likely initiation point to cut off escape. Once in position, personnel checks are conducted and any suspects are detained. The patrol leader coordinates requests for required support (MEDEVAC, reaction force, etc.)

Subsequent Follow-Up. Due to the possibility of secondary detonations, the four Cs (confirm, clear, cordon, and control) can be conducted as in reaction to a bomb discovery or bomb threat.

Civil Disturbances

Urban patrols must prepare to react to spontaneous aggression by the local populace. In many cases, civil disturbances are organized by the enemy to draw dismounted patrols into a targeted area, or to distract them from enemy activity occurring elsewhere. Civil disturbances are generally divided into two categories: minor aggressive actions, and full-scale rioting.

Minor aggressive actions are activities characterized by rock-throwing or use of devices such as Molotov cocktails and may either be directed at the patrol or take place between different ethnic factions of the population. Minor aggressive actions are normally spontaneous in nature and may have minimal or limited objectives for the insurgents.

Full-scale rioting events are usually in response to another major event or incident that may enflame the populace. Full-scale riots are well-planned and orchestrated, with clear objectives or targets in mind. At times, patrols will need to attempt to maintain control of a civil disturbance situation; however, dismounted and mobile small unit patrols should generally avoid potential flashpoints. Procedures to handle civil disturbances are as follows:

- The patrol leader reports the incident to headquarters and attempts to diffuse the crowd by talking to crowd leaders.
- If the patrol leader determines the size of the disturbance is too large for the force to handle, the patrol should move away from the disturbance to a safer, more remote covered area and occupy positions to observe and report the situation to higher headquarters. To prevent the patrol from being pursued by the crowd, the patrol should move quickly and change direction, often at road junctions, to gain distance from it.

Patrol members should maintain dispersion to create a more difficult target. They should face the crowd at all times to see and avoid any projectiles thrown. Individual self-discipline must be maintained throughout the disturbance. Marines charging into the crowd or throwing objects back at the crowd will only worsen the situation. If pursued or trapped, the patrol leader may consider using riot control agents (combat support, pepper gas, etc.) to disperse or slow a crowd's movement. The use of riot control agents must be authorized under the established ROE. If the patrol leader believes the situation is deteriorating beyond the patrol's ability to control it, the patrol leader should request the reaction force, which may be better equipped to handle a large riot or mob.

Break Contact

As with patrols in rural areas, the patrol leader may be forced to break contact as a result of decisive engagement with the enemy. On the basis of his estimate of the situation, the patrol leader will normally break contact in one of the following ways:

- As a patrol, with elements providing cover for movement as defined by clock direction and distance.
- As individual units/teams taking separate routes out of the area, then linking up at a designated rally point a safe distance away from the engagement.

As in any contact with enemy forces, smoke may be employed to screen movement. Fire support agencies can be utilized to suppress targets; riot control agents can be employed to disrupt enemy movement.

Reentry of Friendly Lines

The reentry of a dismounted patrol into an urban patrol base is no different from that of a patrol conducted in a rural area. The same planning considerations and control methods apply.

Missions Related to Urban Patrolling

House Calls

House calls missions are usually part of a coordinated effort to collect information within the area of operations. They involve obtaining up-to-date information on particular houses and occupants. When possible, local police should accompany patrols to do the actual talking to the occupants. If this is not possible, a technique that may be used by an urban patrol is—

- A&S elements move to provide cover around the target house.
- The headquarters element provides security just outside the house.
- The patrol leader and one Marine for security enter the house, if invited. If not invited, they talk to the occupants on the doorstep. Patience and tact are required in requesting information. An interpreter should be present when language differences exist.

Vehicle Checkpoints

Urban terrorists or insurgents commonly use vehicles to transport personnel, weapons, explosives, and equipment. Civilian vehicles are often used for these purposes, creating the requirement to check as many vehicles as possible. While permanent, fortified checkpoints may be conducted along approaches into an urbanized area, dismounted patrols can be employed to establish hasty vehicle checkpoints to stop vehicles and to keep the enemy off-guard. The two common types of vehicle checkpoints patrols establish are hasty and deliberate.

Hasty checkpoints are deployed anywhere based upon the decision of the patrol leader. Patrols must not set patterns through the frequent use of the same sites.

Deliberate checkpoints are tasked by higher headquarters to achieve a specific purpose. Time and locations are carefully considered to avoid setting patterns.

The general layout for a squad-sized, two-way dismounted checkpoint is depicted in figure 13-8 on page 13-20. The technique is as follows:

- The A&S teams are positioned stealthily in blocking positions on both sides of the road.
- Both the patrol leader and assistant patrol leader act as "talkers" for each direction of traffic (with local police or interrogator or translator Marines acting as interpreters) while a two-man team from the headquarters element physically searches the vehicles.
- Obstacles or parked vehicles may be employed to create a staggered roadblock in center of the checkpoint to slow approaching vehicles.
- The checkpoint location should be sited so that approaching vehicles cannot see it until they have passed a security team, and they have no escape route then available.
- Signs announcing the checkpoint should be displayed a safe distance from the search area for safety to both drivers and Marines.
- Normally higher headquarters will issue criteria that determines which vehicles are searched, but random checks of cars should normally be made as

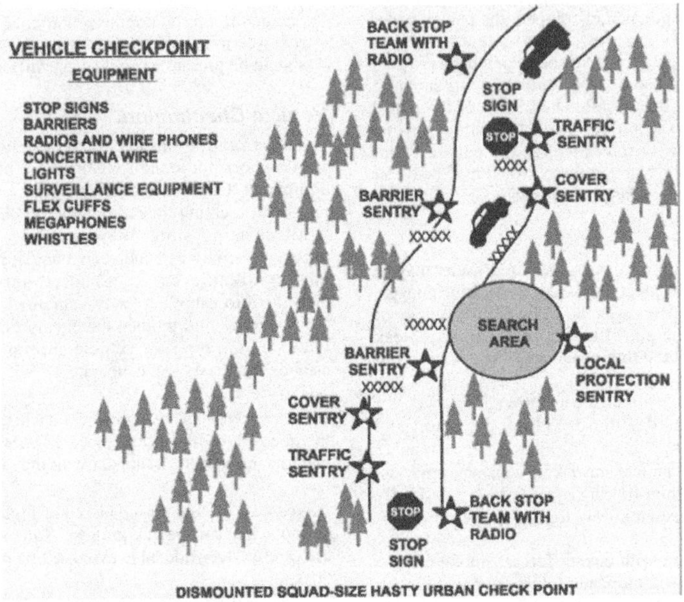

Figure 13-8. Dismounted Squad-sized Hasty Urban Vehicle Checkpoint.

well. While the vehicle is being searched the driver should accompany the searcher around the vehicle.

- Vehicle occupants should be made to exit the vehicle and then searched. Whenever possible, women should be searched by female police or female Marines.
- All patrol members should conduct themselves with courtesy and politeness. If nothing is found, an apology for the inconvenience is recommended.
- A hasty vehicle checkpoint should not be conducted any longer than 30 minutes for security reasons.
- The ROE should dictate whether or not action should be taken against vehicles that fail to stop at the checkpoint. Failure of a vehicle to stop does not automatically give authorization to fire.

Observation Posts

Urban observation posts are established to provide extended security, not only for patrol bases but also for patrols operating within the observation post's sector of observation. Observation posts can be established in conjunction with sniper operations and for providing overwatch for patrols operating within their sector of observation. Observation posts are nor-mally positioned on dominating terrain or in buildings outside the patrol base itself.

Insertion to the observation posts and conduct of observation may be either overt or clandestine in nature. Overt observation posts usually will be hardened positions to increase security. A patrol provides cover while the observation post is being inserted. A clandestine observation post relies on stealth of insertion and occupation for protection. It is normally positioned in abandoned buildings to cover sectors of observation that overt observation posts cannot. Because of their nature, clandestine observation posts are difficult to successfully establish and should not be manned for an extended period of time. Orders establishing observation posts (and patrols) must address the method of extraction as well as actions upon compromise/attack.

Cordon and Search

The cordon and search mission involves isolating a predesignated area by cordoning it off and systematically searching for enemy personnel, weapons, supplies, explosives or communications equipment. While large-scale cordon and search operations are planned and rehearsed in advance and normally entail extensive coordination with local law enforcement agencies, a squad-sized urban patrol may often conduct a cordon and search of a point target—searching one house or building identified by intelligence as a possible weapons cache.

The basic principle of a search of a populated area is to conduct it with limited inconvenience to the population. The populace may be inconvenienced to the point where they will discourage urban guerillas or insurgents from remaining in the area, but not to the point that they will assist the enemy as a result of the search.

Upon receiving intelligence that warrants the searching of a building or a specific tasking from higher headquarters, A&S elements of the patrol move to establish an inner cordon around the target building to seal it off, with the primary intent of preventing movement out of the targeted building.

On order, the designated reaction force deploys to establish an outer cordon, oriented outward some distance from the inner cordon and covering routes leading into the area in order to prevent outside interference/reinforcement. The reaction force maintains a reserve to reinforce either cordon or react to unfolding events (civil disturbance in response to the operation).

Once the cordons have been established, the patrol leader, with the assistance of local police or interpreters, informs the local populace that a building is about to be searched, that a house curfew is in effect (if permitted by higher headquarters), and that all occupants should remain indoors. Occupants of the target house are instructed to gather at a central location to stay out of the way of the search party.

The headquarters element, having linked up with any required assistance (explosive ordnance disposal (EOD), ITT, etc.) now acts as the search party and accompanies local police. A female searcher should be included in the party, if necessary.

Occupants are searched and screened first for possible enemy personnel. Apprehended persons are evacuated as soon as possible.

The head of the household should accompany the search party throughout the operation to be able to counter incriminating evidence and possible accusations of theft and looting against Marines. If possible, a prominent member of the local community should act as a witness.

Buildings are best searched from top to bottom. Ideally, the search is conducted with the assistance of combat engineers using mine detectors to locate hidden arms and ammunition.

If the targeted building is empty or the occupant refuses entry, it may be necessary to forcefully enter the premises to conduct a search of the dwelling. If an unoccupied house containing property is searched, arrangements should be made with the local community to secure it until its occupants return. Unnecessary force and damage to property should be avoided during the search.

Motorized Urban Patrols

The advantages of a motorized urban patrols is their ability to capitalize on the speed, mobility, and protection offered by various vehicles. They may be motorized, mechanized or armored vehicles or a combination. Generally, motorized urban patrols possess greater combat power than dismounted patrols and can cover larger areas faster than dismounted patrols.

The disadvantages of motorized urban patrols is that they are restricted to roads and are vulnerable to ambush by the enemy. They are also restricted in their ability to interact with the local populace.

Motorized patrols are generally organized in the same manner as dismounted patrols (see fig. 13-9). Unit integrity is maintained when assigning personnel to specific vehicles.

The urban patrolling principles apply to motorized patrols in much the same manner as dismounted patrols. Mutual support and depth are achieved by maintaining constant observation between vehicles and coordinating support with any dismounted patrols in the area. All-around security is achieved through the use of constant observation as well as the vehicle's mobility and firepower. Positive communications between units or teams are maintained through vehicle radios. Patrol routes and speeds are varied to promote deception or pattern avoidance.

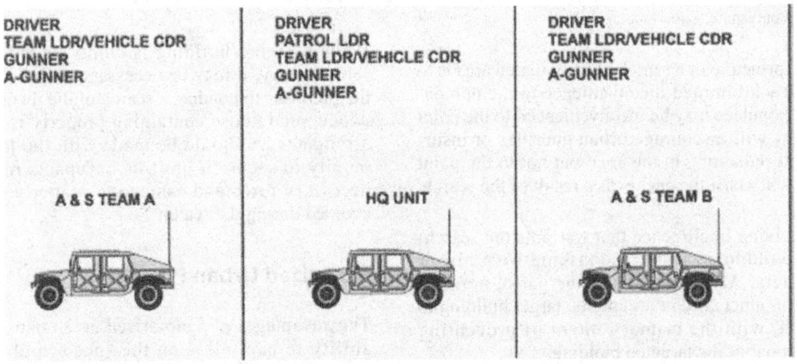

Figure 13-9. Organization of a Squad-sized Motorized Patrol.

The canalizing nature of streets and alleys force vehicular patrols to use a traveling overwatch movement technique to reduce vulnerability to ambushes. All vehicles travel at a moderate rate of speed with the lead vehicle stopping only to investigate potential danger areas. If vehicles must stop in danger areas, designated crew members' will dismount to provide security. The gunner will remain at the ready and in the turret while the driver remains in the driver's seat with the vehicle running.

Vehicles should move at a high rate of speed only when responding to an incident. At all other times, vehicle speed should be between 15 to 20 mph to allow for quick reaction and good observation. Distances between vehicles should be approximately 50 meters (one half to one city block) or such that visual contact and mutual support are ensured. Particular care is taken at major road junctions and other danger areas to ensure individual vehicles do not become isolated.

Vehicles with doors removed generally enhance observation and overall security, yet expose Marines to thrown objects, theft and concealment.

CHAPTER 14. INFORMATION AND REPORTS

It is necessary that patrol leaders and all patrol members be trained in observing and reporting their observations accurately. The leader of a patrol should have all members of the patrol immediately signal or report any information obtained. These reports should not be restricted to information about the enemy, but should also include information about the terrain, such as newly discovered roads, trails, swamps, and streams. The leader includes all information in the report to the officer dispatching the patrol. (Refer to MCRP 2-15.3, *Reconnaissance Reports Guide,* for detailed report formats.)

14001. REPORTING

The officer dispatching the patrol instructs the patrol leader on whether and when messages are to be sent back during the patrol and what communication means to use. Messages may be oral or written. They must be accurate, clear, and complete. Every message should answer the question what, where, and when. For a detailed discussion on reporting, see paragraph 6002.

Verbal Messages

A patrol leader sending a verbal message should make it simple, brief, and avoid using numbers and names. The messenger should accurately repeat the message back to the patrol leader before leaving.

Written Messages

In preparing written messages, the patrol leader must distinguish between fact and opinion. Information about the enemy should include: strength; armament and equipment; actions; location and direction of movement; unit destination, if known; time enemy was observed; and the patrol's location when the observation was made. Use of an overlay or sketch may often simplify the message.

Messengers

A messenger team is given exact instructions as to where to deliver the message and the route to take. Any information obtained along the route should be reported at the time the message is delivered. Messengers must be given all practical assistance. If in danger of capture, the messenger immediately destroys the message.

Radio and Signals

If the patrol is provided with a radio, a definite radio schedule for checking in must be arranged before departure of the patrol. The patrol leader takes every precaution to ensure that codes and copies of messages are not captured by the enemy. If a close reconnaissance of enemy lines is required, the radio should be left in a concealed location at a safe distance from the enemy. Once a report is sent by radio, the patrol should immediately leave the area to avoid the possibility of detection by enemy locating devices. Pyrotechnics (flares, colored smoke, grenades) and air-ground panels may also be used by patrols for reporting information by a prearranged signal.

SALUTE Report

Information must be reported as quickly, accurately, and as completely as possible. An established method to remember how and what to report about the enemy is to use the acronym SALUTE:

Size
Activity
Location
Unit
Time
Equipment

An example of such a report is: "Seven enemy soldiers, unit unknown, traveling SW, crossed road junction on BLACK RIDGE at 211300 August carrying one machine gun and one rocket launcher."

14002. CAPTURED ITEMS

Every patrol should make a practice of searching enemy casualties, prisoners, and installations first for booby traps, then for equipment, papers, maps, messages, orders, diaries, and codes. Search techniques should be practiced often in order to reduce the time exposed to potentially dangerous situations. Items found are collected by the patrol leader and turned in with the patrol report. The items found are marked as to time and place of capture. When possible, captured items should be linked to a specific prisoner who possessed the items or to the place where the items were found. When this is done, the enemy prisoner of war tag and item tag are marked accordingly. The patrol leader must impress upon the members of the patrol the importance of turning in all documents and equipment. Furthermore, the patrol leader must ensure all information gathered by the patrol is quickly disseminated.

14003. PRISONERS

A patrol normally does not capture prisoners unless required by the mission. If prisoners are taken, the "5S and T" rule applies. This memory aid stands for: search, segregate, silence, speed, safeguard, and tag.

Search. Prisoners are body-searched thoroughly for weapons and documents as soon as they have been captured. This search must include the helmet, body armor, and gas mask. These items are left with the prisoner for protection until the patrol is completed. Weapons, equipment, and documents are tagged and immediately sent to the patrol leader.

Segregate. Prisoners are segregated into isolated groups: officers, noncommissioned officers (NCOs), privates, deserters, and civilians. By segregating prisoners, it makes it more difficult for leaders to organize escapes and issue orders to subordinates.

Silence. Silence is essential. Do not allow prisoners to talk to each other.

Speed. Speed is required in getting prisoners to the commander who dispatched the patrol. Timely information secured from prisoners is essential.

Safeguard. Prisoners are safeguarded as they are moved. They are restrained, but not abused. If the patrol will soon reach friendly positions, prisoners are not given cigarettes, food or water until they have been questioned by interrogators. If the patrol will not return to a friendly position for a lengthy time, the prisoners are given food, water and medical aid.

Tag. Equipment and personal effects recovered from prisoners are tagged so that they may later be matched back to specific individuals.

14004. PATROL REPORT

Every patrol leader makes a report when the patrol returns. Unless otherwise directed, the report is made to the person ordering the patrol. If the situation permits, the report is written and supported by overlays and/or sketches. The patrol leader's report should be a complete account of everything of military importance observed or encountered by the patrol while on the assigned mission. It should include the following information:

- Size and composition of patrol.
- Tasks and purpose (mission).
- Time of departure.
- Time of return.
- Routes, out and back (show by sketch, azimuth, trace on map).
- Terrain (general description to include any man-made or natural obstacles and critical terrain features which, if occupied by either enemy or friendly forces, would allow them to control the surrounding area).
- Enemy (size, activity, location, unit, time, equipment).
- Any map corrections (show on map).
- Miscellaneous information not covered elsewhere in report.
- Results of enemy encounters.
- Condition of patrol, including disposition of any dead or wounded.
- Conclusion and recommendations.

14005. PATROL CRITIQUE

After the patrol has rested and eaten, the patrol leader should hold a critique. Constructive criticism is made. It is an excellent time to prepare for future patrols by going over lessons learned as a result of the patrol.

Appendix A. Patrol Warning Order

The warning order is issued as soon as practical with all available information included to assist patrol members in preparation.

1. Situation. Friendly and enemy situation information necessary for initial preparation.

2. Mission. Statement of what the patrol is to accomplish, and the purpose for accomplishing it. When, how, and where will be discussed in the patrol order.

3. Execution

a. Task Organization. General patrol organization and assignment of responsibilities if known; otherwise, promulgate in the patrol order.

b. Tasks. Alert subordinate leaders to patrol tasks (e.g., stream crossing, helicopter rappelling, demolitions) requiring preparation prior to departure.

c. Coordinating Instructions

(1) Time schedule and location for individual preparation, rest, briefings, in-spection, rehearsal, and departure.

(2) Time, place, uniform, and equipment for receiving the patrol order.

(3) Tasks for subordinate leaders to direct and supervise the initial preparation which may include drawing ammunition, rations, and special equipment; conducting immediate action drills or other necessary individual or unit training; meeting and briefing attachment personnel; reconnoitering the area for passage of lines; and coordinating with the necessary unit leaders.

(4) Preliminary guidance to specialists and key individuals regarding their roles and organization within the patrol.

4. Administration and Logistics

a. Individual uniform, equipment, weapons, and prescribed load of rations, water, and ammunition.

b. Crew-served weapons (if required) and guidance regarding distribution of weapons and ammunition during movement.

c. Special equipment requirements (wire cutters, demolitions, radios, flashlights, infrared equipment, mines, binoculars) and their distribution during movement.

d. Restricted or prohibited items.

5. Command and Signal

a. Designation of assistant patrol leader and his role in preparation.

b. Designation of navigators and radio operators (if required).

c. Brief outline of patrol leader's schedule for preparation, and where he can be reached.

APPENDIX B. PATROL ORDER

A patrol order follows a warning order. Any of the following subjects that have been addressed in the warning order may be omitted with the exception of the mission statement. A patrol order is more detailed than a 5-paragraph squad order in that a great deal of attention is given to individual duties.

The patrol order also provides orientation information, both enemy and friendly, that impacts the patrol and enemy forces. This information can include, but is not limited to, weather, terrain, visibility, NBC considerations, local population situation, terrain model and/or map orientation, and behavior.

1. Situation

a. Enemy Forces

(1) Composition, disposition, and strength are based on size, activity, location, unit, time, and equipment (SALUTE).

(2) Capabilities and limitations to defend, reinforce, attack, withdraw, and delay (DRAW-D).

(3) Enemy's most probable course of action.

b. Friendly Forces

(1) Mission of next higher unit (task and commander's intent).

(2) Adjacent unit missions (task and intent). Also identify left, front, right, and rear.

c. Attachments and Detachments. (date and time effective).

2. Mission. The mission is the task to be accomplished, and its purpose (who, what, where, when, and why). For patrols, specify if the mission or time has priority.

3. Execution

a. Commander's Intent and Concept of Operations

(1) Commander's intent.

(2) The concept of operations tells the where, how, and who and lays out the patrol leader's general scheme of maneuver and fire support plan. It outlines the following:

(a) Task organization of the patrol.

(b) Movement to the objective area, to include navigation method.

(c) Actions in the objective area.

(d) The return movement, to include navigation method.

(e) Use of supporting forces (including illumination, if required).

b. Subordinate Element Missions. Subordinate element missions (task and purpose) are assigned to elements, teams, and individuals, as required.

c. Coordinating Instructions. This paragraph contains instructions common to two or more elements, coordinating details, and control measures applicable to the patrol as a whole. At a minimum, it includes—

(1) Time of assembly in the assembly area.

(2) Time of inspections and rehearsals (if not already conducted).

(3) Time of departure and estimated time of return.

(4) Location of departure and reentry of friendly lines and the actions associated with departure and reentry.

(5) Details on the primary and alternate routes to and from the objective area.

(6) Details on formations and order of movement.

(7) Rally points and actions at rally points.

(8) Final preparation position and actions at this position.

(9) Objective rally point and actions at this point.

(10) Actions at danger areas.

(11) Actions in the event of enemy contact.

(12) Details on actions in the objective area not covered elsewhere.

(13) Estimated time of patrol debriefing upon return.

4. Administration and Logistics

a. Changes/additions to uniform, equipment, and prescribed loads from that given in the warning order.

b. Instructions for handling wounded and prisoners.

5. Command and Signal

a. Command Relationships. Identify key leaders and chain of command.

b. Signal. Challenge and password, arm and hand signals, special signals, and radio frequencies and call signs.

APPENDIX C. PATROL EVALUATION CHECKLIST

This appendix provides a comprehensive checklist of critical patrolling steps, techniques, and procedures to aid unit leaders to critique patrol performance during training. Unit leaders may use the list as they observe the performance of a patrol to provide the patrol leader with a detailed analysis of the performance.

Warning Order	Yes	No
Did the PL ensure all patrol members were present before issuing the warning order?		
Did the PL issue a brief statement on the enemy situation?		
Did the PL issue a brief statement on the friendly situation?		
Did the PL state the mission in a clear, positive manner?		
Did the PL list all members of the patrol including attachments?		
Was the chain of command for the patrol covered fully?		
Were all members of the patrol asigned positions and duties in particular squads and teams?		
Were all the necessary individual duties assigned?		
Did the PL follow established principles in organizing the patrol into squads and teams?		
Was each patrol member assigned a particular weapon to carry on patrol?		
Did the PL list all special equipment needed to accomplish the mission?		
Was required special equipment assigned to the proper element to carry?		
Did the PL select uniform and equiment common to all based on METT-T?		
Coordination with Adjacent Units		
Did the PL coordinate with other patrols operating to the right and left?		
Was the route out and back coordinated?		
Was the time of departure and return cocordinated?		
Were call signs and frequencies coordinated?		
Was a signal for the FPF coordinated so as not to approach friendly lines during this time?		
Coordination With Front Line Units		
Did the coordinator pass the size of the patrol?		
Was the time of departure and return coordinated?		
Did the coordinator give a general area of oeprations for the patrol?		
Did the coordinator ask for information on known or suspected enemy positions and/or obstacles?		
Did the coordinator ask about information on the latest enemy activity?		
Did the coordinator ask for detailed information on friendly fire support available and the unit's barrier plan?		
Was the location of the IRP established and coordinated?		
Did the coordinator ask the forward unit to monitor their patrol frequency?		
Was the current challenge and password confirmed?		
Did the coordinator request that all information coordinated be passed on to any relieving unti?		
Patrol Order		
General		
Did the PL check to ensure all patrol members were present before issuing the patrol order?		

	Yes	No
Did the PL issue the patrol order in a forceful, confident manner?		
Did the PL make maximum use of available visual aids in issuing the patrol order (i.e., terrain model, sand table, map board, chalkboard)?		
Did the PL issue the patrol order in correct sequence?		
Did the PL issue the entire patrol order without allowing interruptions by patrol members?		
Did the PL adequately answer all questions asked by patrol members		
Did paragraph 1a (Enemy Situation) include—		
A weather forecast for the period of operation?		
A description of the terrain over which the patrol was to operate?		
Identification or description of enemy units known to bein the area of operations?		
Known locations of enemy units?		
Recent activity of enemy units?		
Stength of enemy units in the operating area?		
Did paragraph 1b (Friendly Situation) include—		
Mission of the next higher unit?		
Locations, missions, and planned actions of units on right and left?		
Fire support available to support the patrol (as per the coordination)?		
Missions and routes of other patrols operating in the immediate area?		
Did paragraph 1c (Commander's Intent) include—		
All views and ides of what the commander wants to be accomplished?		
The final result desired by the commander?		
Did paragraph 1c (Attachments and Detachments) include—		
All attachments to patrol and effective time of attachment?		
All detachments from patrol and effective time of detachment?		
Did paragraph 2 (Mission) include, at a minimum—		
Who was to conduct the patrol?		
What the patrol was tasked to do (i.e., conduct a point recon patrol)?		
Where the action was to take place?		
Why the action was to take place (i.e., the purpose)?		
Did paragraph 3 a (Commander's Intent and Concept of Operations) include—		
A complete concept of operations?		
A detailed description of the mission of elements?		
A detailed description of the mission of teams?		
A detailed description of the duties of specific individuals (i.e., navigator, compass man, pace m an, corpsman, APL)?		
Time of departure and time of return?		

	Yes	No
Type of formations and order of movement to be used?		
Description of the route and alternate routes(s) to include azimuths and distances between checkpoints as a minimum?		
Techniques to be used in the departure from friendly areas?		
Techniques to be used for the reentry into friendly areas?		
Location of the IRP and a tentative ORP, to include grid coordinates and recognizable terrain features?		
Method for designating and passing rally points?		
Actions to be taken at rally points if their use became necessary?		
Actions to be taken in the event of enemy contact?		
Actions to be taken at danger areas?		
Complete, detailed description of actions to be taken at the objective?		
Times and locations for rehearsals and the order of priority for rehearsals?		
Times and lcoations for inspections and the methods of conduct (to include uniform and equipment to be worn and/or carried)?		
Where, when, and by whom the debriefing is to be conducted?		
Did paragraph 4 (Administration and Logistics) include—		
Rations to be carried or reference to the warning order if there is no change?		
Arms and ammunition to be carried or reference to the warning order if there is no change?		
Uniform and equipment to be worn and carried or reference to the warning order if there is no change?		
Method for handling wounded or dead?		
Method for handling prisoners?		
Did paragraph 5a (Signal) include—		
Complete description of all signals, code words, frequencies, and call signs to be used within the patrol?		
Were signals adequate?		
Radio call signs to be used?		
Primary and alternate frequencies to be used?		
Call signs and frequencies to be used with other units in the area of oeprations (if applicable)?		
Required reports to be transmitted to higher headquarters, along with occasions for reporting?		
Code words and brevity codes to be used between the patrol and higher headquarters?		
Current challenge and password to be used in friendly held areas?		
Current challenge and password to be used forward of FEBA?		
Did paragraph 5b (Command) include—		
Chain of command or reference to the warning order if there is not change?		
Location of the PL during all stages of operation?		

	Yes	No
Location of the APL during all stages of operation?		
Inspection Rehearsal		
General		
Did the P L adequately use the allotted inspection time?		
Was the inspection conducted in an orderly manner?		
During the inspection, did the PL check the following:		
Completeness and correctness of uniform?		
Completeness of all equipment necessary to accomplish the mission?		
Operational condition of equipment?		
Did the PL question members to ensure they knew—		
The mission of the patrol?		
The concept of operations?		
Their individual duties and responsibilities?		
Chain of command and succession?		
Duties and responsibilities of key personnel?		
Did the PL utilize the rehearsal area to conform as much as possible to area of operations?		
Were the following major actions rehearsed:		
Actions at the objective?		
Actions at danger areas?		
Actions on enemy contacts?		
Departure and reentry of friendly lines?		
Did the PL critique each action after it had been rehearsed?		
Were interpatrol communications and control measures checked at rehearsal?		
Did the PL maintain control of the patrol during the rehearsal?		
Movement		
Did the PL use proper formations for movement?		
Did the PL adequately control the point team?		
Did the PL check the compass men?		
Was the PL aware of his position at all times?		
Was light and noise discipline enforced?		
Was the rate of movement appropriate for the mission?		
Did the PL make full use of pace?		
Did the PL make full use of count?		
Was proper security maintained during movement?		
Did the PL select and pass proper rallying points?		
Did the PL ensure that all members knew location of rallying points?		

	Yes	No
Did the PL properly use arm and hand signals?		
Did the PL recognize and halt the patrol a safe distance from a danger area?		
If the PL conducted a map check, did he conduct it properly (light discipline, security, etc.)		
Did designated individuals reconnoiter the far side of a danger area properly?		
Were support teams emplaced properly?		
Did the PL use a proper formation in crossing the danger area?		
Were reports made on enemy contact and at checkpoints?		
Actions on Enemy Contact		
Did the PL halt patrol a safe distance from a tentative ORP?		
Did the PL issue an adequate frag order before departing to look for a tentative ORP?		
Did the PL take appropriate personnel to look for the ORP?		
Did the PL select a suitable ORP?		
Did the PL secure the site adequately?		
Did the patrol occupy the ORP as stated in the patrol order or subsequent frag order?		
Did the patrol occupy the ORP in an orderly manner?		
Did the PL maintain control of the patrol during occupation of the ORP?		
Did the PL issue a satisfactory contingency plan to the APL before departing on a leader's recon?		
Did the PL take appropriate personnel on a leader's recon?		
Did the PL maintain or ensure that the patrol leader had communications with the patrol?		
Did the patrol avoid being detected by the enemy during the leader's recon?		
If contact was made, did the PL take appropriate action?		
Was security maintained during the leader's recon?		
Did the PL ensure the objective was kept under surveillance?		
Were appropriate orders given to surveillance teams?		
If a leader recon proved a tentative ORP unsuitable, did the PL move the ORP?		
Was a satisfactory leader's recon conducted?		
Was the ORP move conducted in a satisfactory and orderly manner?		
On return to the ORP, did the PL issue appropriate frag order and allow enough time for dissemination?		
Actions at Objective		
Did the PL issue a frag order for action at the objective?		
Was security emplaced prior to the teams departing from the ORP?		
Did the PL employ the terrain at the objective to the best advantage?		
Was stealth maintained while moving into the objective?		

	Yes	No
Did the Pl exercise positive control of squads, teams, and individuals at the objective?		
Were actions at the objective in accordance with the details outlined in the patrol order?		
Did the PL make use of supporting arms at the objective?		
Was the action at the objective successful?		
Was withdrawal from the objective accomplished quickly and orderly?		
Did units withdraw according to the details reflected in the patrol order?		
Was reorganization at the ORP completed in an expeditious and orderly manner?		
Did the patrol withdraw from the ORP quickly and quietly?		

APPENDIX D. AMBUSH FORMATIONS

Formations discussed here are identified by names which correspond to the general pattern formed on the ground by deployment of the assault element.

Line Formation

The assault element is deployed generally parallel to the route of movement of the enemy. This positions the assault element parallel to the long axis of the killing zone and subjects the target to heavy flanking fire. The size of the force which can be trapped in the killing zone is limited by the area the assault element can effectively cover. The enemy is trapped in the killing zone by natural obstacles, mines, demolitions, and direct fires. (See fig. D-l.)

A disadvantage of the line formation is the chance that lateral dispersion of the target may be too great for

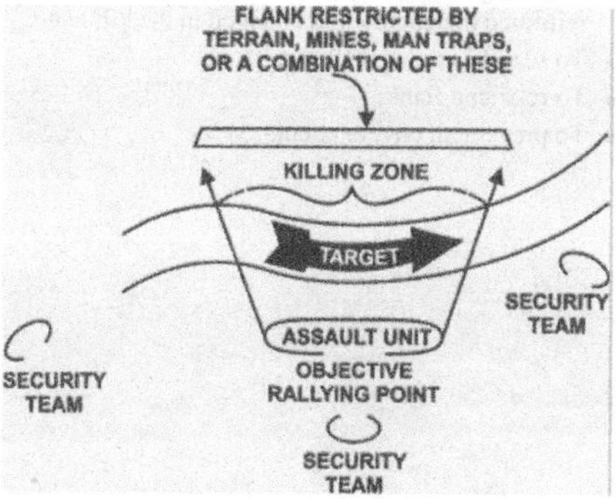

Figure D-1. Line Formation (Harassing or Destruction Ambush).

effective coverage. The line formation is appropriate in close terrain that restricts enemy maneuver, and in open terrain where one flank is protected by natural obstacles or can be protected by mines and demolitions. Similar obstacles can be placed between the attack force and the killing zone to provide protection from possible enemy counter attack. When an ambush is deployed in this manner, access lanes are left so that the enemy can be assaulted. (See fig. D-2.)

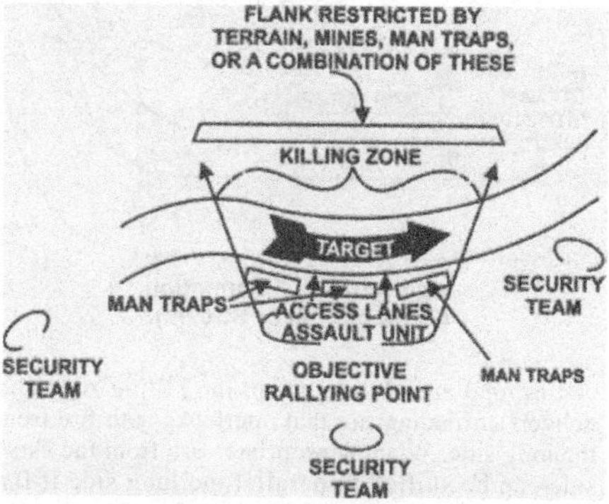

Figure D-2. Line Formation (Access Lanes for Assault of Target).

The main advantage of the line formation is its relative ease of control under all conditions of visibility.

L Formation

The "L" shaped formation is a variation of the line formation. This formation is very flexible because it can be established on a straight stretch of a trail or road (see fig. D-3) or at a sharp bend in a trail or a road (See fig. D-4 on page D-2).The long side of the assault element is parallel to the killing zone and delivers flanking fire. The short side of the attack force is at the

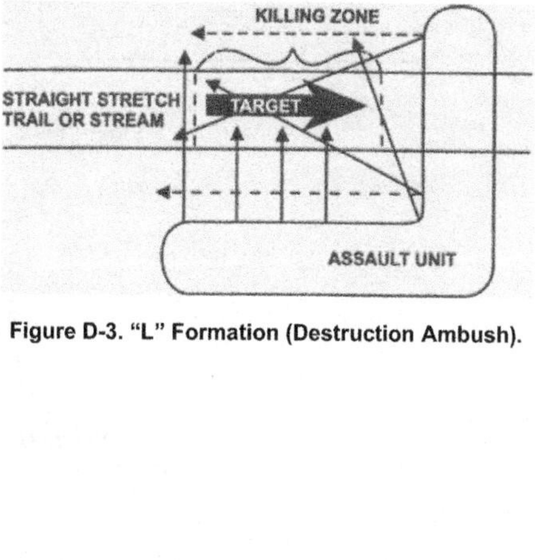

Figure D-3. "L" Formation (Destruction Ambush).

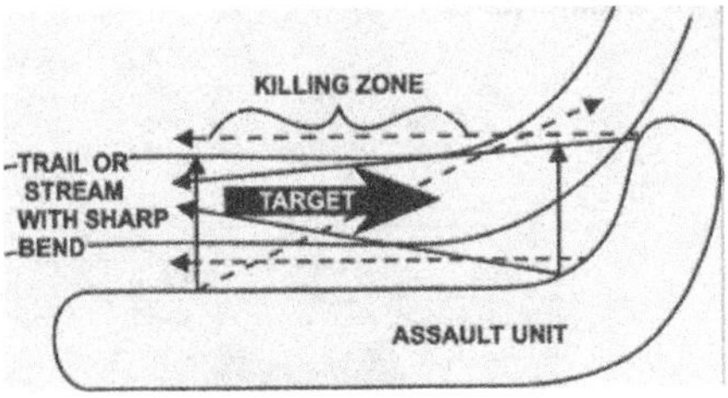

Figure D-4. "L" Formation (Bend of Trail or Stream).

end of, and at right angles to, the killing zone and delivers enfilading fire that interlocks with fire from the long side. When appropriate, fire from the short side can be shifted to parallel the long side if the enemy attempts to assault or escape in the opposite direction. In addition, the short side prevents escape and reinforcement. (See fig. D-5.)

"Z" Formation

The "Z" shaped formation is another variation of the line formation.

The assault element is deployed as in the "L" formation, but with an additional side so that the formation resembles a "Z". (See fig. D-6).

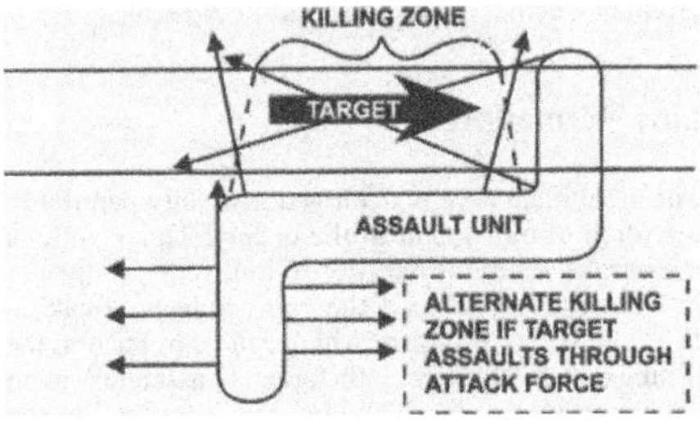

Figure D-6. "Z" Formation.

The additional wing may serve any of the following purposes:

- To engage an enemy force attempting to relieve or reinforce the enemy unit engaged in the kill zone.
- To seal the end of the killing zone.
- To restrict a flank.
- To prevent an envelopment.

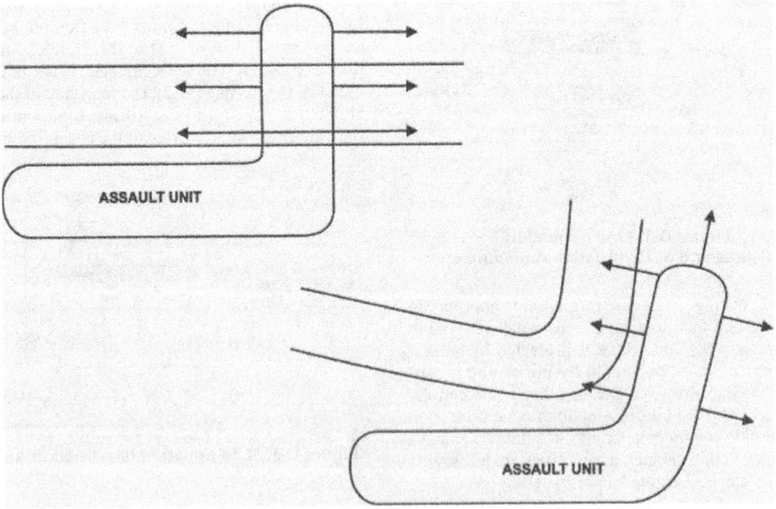

Figure D-5. "L" Formation.

"T" Formation

In the "T" shaped formation, the assault element is deployed across (perpendicular to) the enemy's route of movement so that its position forms the crossing of a "T" at the top. (See fig. D-7.)

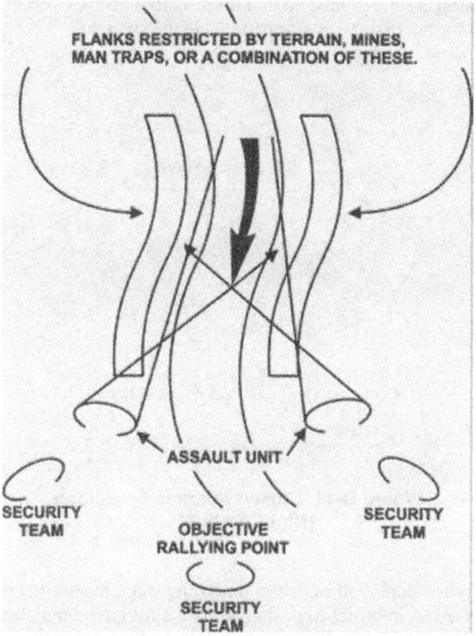

Figure D-7. "T" Formation.

This formation can be used day or night to establish an ambush to interdict movement through open areas that are hard to seal off.

A small force can use the "T" formation to harass, slow, and disorganize a larger force. When the lead elements of the enemy are engaged, they will normally attempt to maneuver right or left to close with the ambush. Mines and other obstacles placed to the flanks of the killing zones slow the enemy's movements and permit the ambush force to deliver heavy fire and withdraw without becoming decisively engaged.

The "T" formation can be used to interdict small groups attempting night movement across open areas. For example, the assault element is deployed along an avenue of approach with every second man facing the opposite direction. The attack of the enemy approaching from either direction requires only that every second man may shift to the opposite side of the formation. Each man fires only to his front and only when the enemy is at a very close range. Attack is by fire only and each man keeps the enemy under fire as long as it remains to his front.

If the enemy attempts to escape in either direction along the killing zone, each Marine takes the enemy under fire as the enemy comes into the Marine's sector of fire. The "T" formation is very effective at halting infiltration. But it has one chief disadvantage; there is a possibility that the ambush will engage a superior force at night while spread out. (See fig. D-8.)

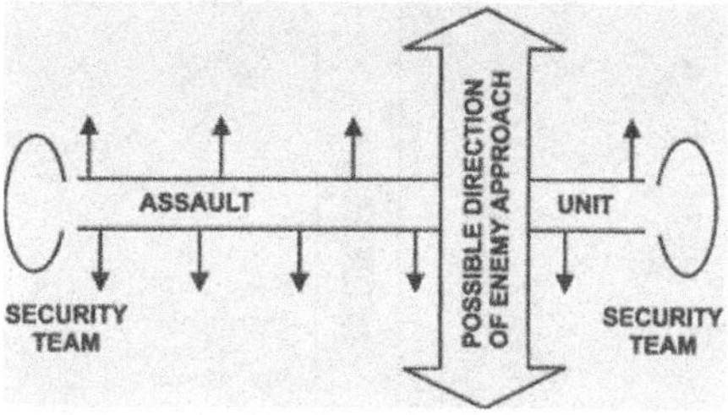

Figure D-8. "T" Formation (Target Approaching from Either Direction).

The "V" Formation

The "V" shaped formation is deployed along both sides of the enemy's route of movement so that it forms a "V"; care is taken to ensure that neither group (within the "V") fires into the other.

This formation subjects the enemy to both enfilading and interlocking fire. The "V" formation is best suited for fairly open terrain but can also be used in close terrain. When established in close terrain, the legs of the "V" close in as the lead element of the enemy force approaches the apex of the "V", and opens fire at a close range.

Here, even more than in open terrain, all movement and fire must be carefully coordinated and controlled to ensure that the fire of one wing does not endanger the other wing. The wider separation of forces makes this formation difficult to control, and there are few sites that favor its use. Its main advantage is that it is difficult for the enemy to detect the ambush until it is well into the killing zone. (See figs. D-9 and D-10.)

Triangle Formation

The triangle is a variation of the "V" formation and can be varied in three ways:

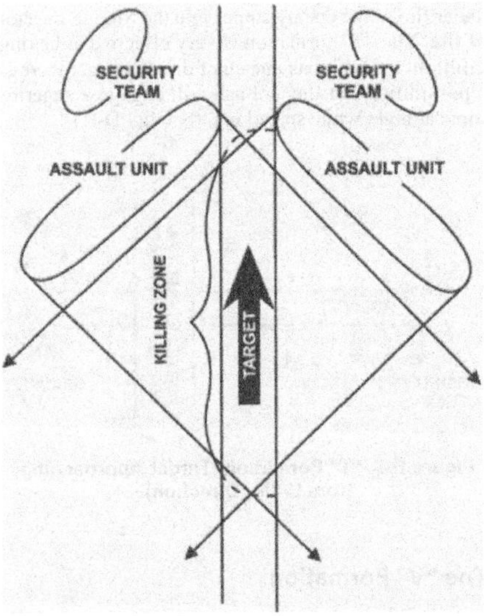

Figure D-9. "V" Formation (Open Terrain).

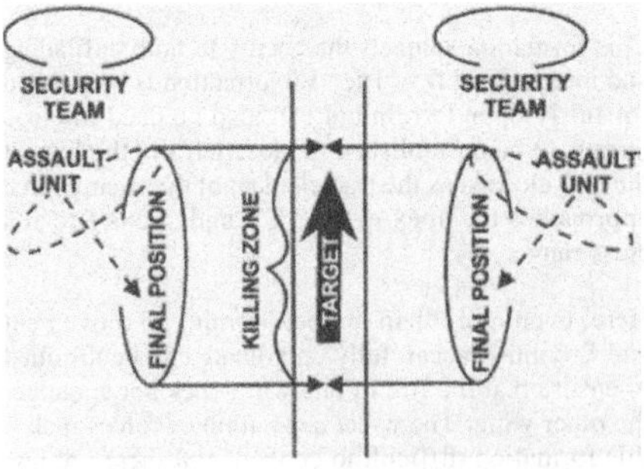

Figure D-10. "V" Formation (Close Terrain).

Closed Triangle Formation

(See fig. D-11.) The assault element is deployed in three teams, positioned so that they form a triangle (or closed "V"). An automatic weapon is placed at each point of the triangle and positioned so that it can be shifted quickly to interlock with either of the others. Men are positioned so that their sectors of fire overlap. Mortars may be positioned inside the triangle.

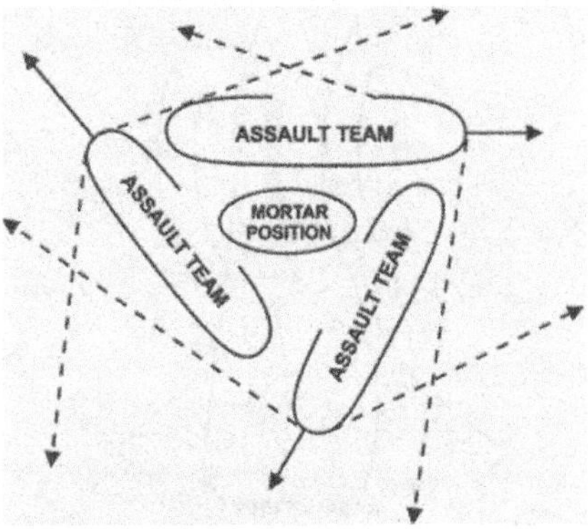

Figure D-11. Closed Triangle Formation (Night Ambush).

When deployed in this manner, the triangle ambush is used to interdict night movement through open areas. When enemy approach is likely to be from any direction, this formation provides all-around security, and security forces are deployed only when they can be positioned so that if detected by an approaching enemy, they will not compromise the ambush. Attack is by fire only, and the enemy is allowed to approach within close range before fire is initiated.

The advantages of the closed triangle formation are ease of control, all-around security, and the enemy can be brought under the fire of at least two automatic weapons, regardless of the direction they approach.

Disadvantages are that it requires a force of platoon size or larger to reduce the danger of being overrun by an unexpectedly large force and that one or more sides of the triangle may come under enfilade fire. The lack of dispersion, particularly at the points, increases the danger from enemy mortar fire.

Open Triangle Harassing Formation

This variation of the triangle ambush enables a small force to harass, slow, and inflict heavy casualties upon a large force without itself being decisively engaged. The assault element is deployed in three teams, positioned so that each team becomes a corner of a triangle containing the killing zone. (See fig. D-12.)

When the enemy enters the killing zone, the team to the enemy's front opens fire on the leading element. When the enemy counterattacks, the group withdraws and the team on the enemy's flank opens fire. When this team is attacked, the team to the opposite flank opens fire. This process is repeated until the enemy is pulled apart. Each team reoccupies its position, if possible, and continues to inflict the maximum damage possible without becoming decisively engaged.

Open Triangle Destruction Formation

The assault element is again deployed in three teams, positioned so that each team is a point of the triangle, 200 to 300 meters apart. The killing zone is the area within the triangle. The enemy is allowed to enter the killing zone; the nearest team attacks by fire. As the enemy attempts to maneuver or withdraw, the other teams open fire. One or more teams, as directed, assault or maneuver to envelop or destroy the enemy. (See fig. D-13 on page D-6.)

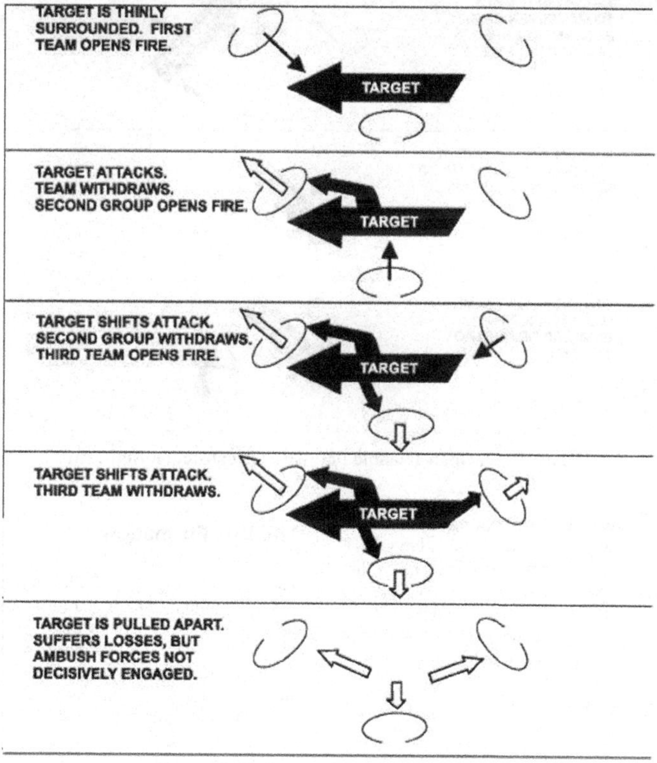

Figure D-12. Open Triangle Formation.

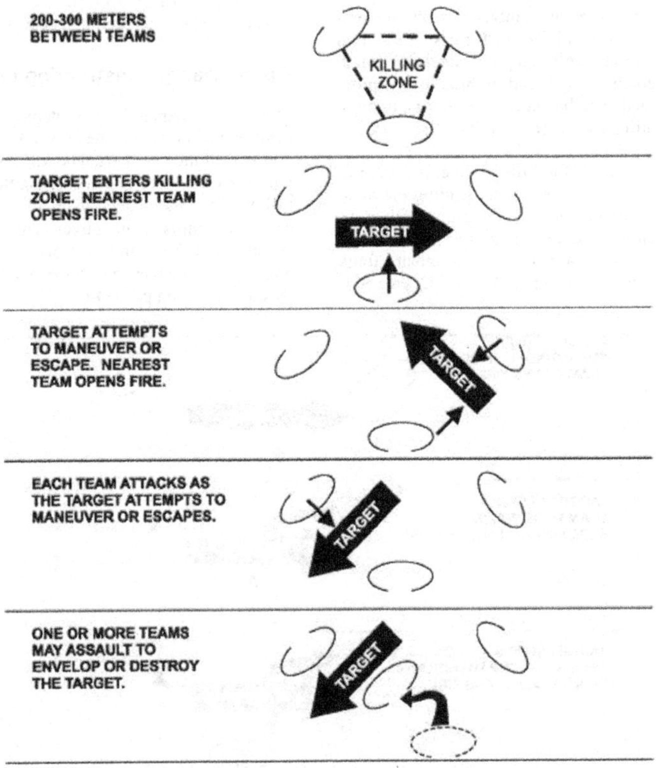

Figure D-13. Open Triangle Formation (Destruction Ambush).

This formation is suitable for platoon-size or larger ambush forces. A smaller force would be in too great a danger of being overrun. Another disadvantage is that control, in assaulting or maneuvering, is very difficult. Very close coordination and control are necessary to ensure that assaulting or maneuvering teams are not fired on by another team. The ambush site must be a fairly level open area that provides concealment around its border for the ambush force.

The Box Formation

The "box" formation is similar in purpose to the open triangle ambush. The assault element is deployed in four teams, positioned so that each team becomes a corner of a square or rectangle containing the killing zone. It can be used for a harassing or destruction ambush in the same manner as the variations of the open triangle formation. (See figs D-14 on page D-7 and D-15 on page D-8.)

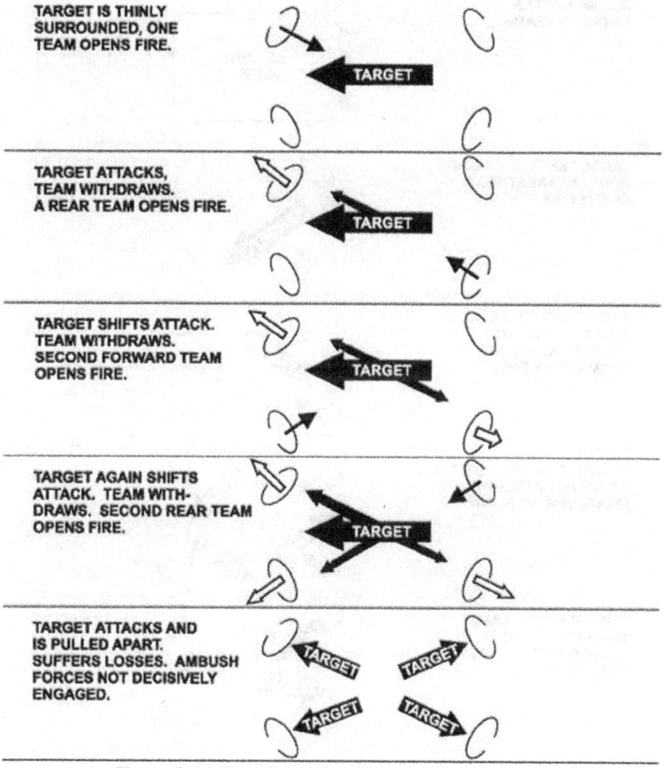

Figure D-14. Box Formation (Harrassing Ambush).

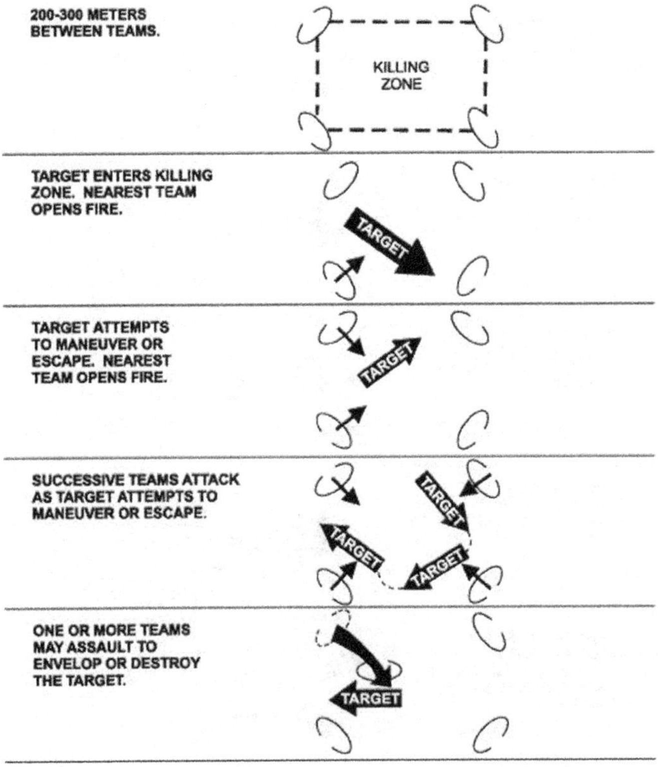

Figure D-15. Box Formation (Destruction Ambush).

APPENDIX E. ACRONYMS

A&S . . . . assault and security

CIT . . . . counterintelligence team

EOD . . . . explosive ordnance disposal

ITT . . . . interrogator-translator team

MAGTF . . . . Marine air-ground task force
MEDEVAC . . . . medical evacuation
METT-T . . . . mission, enemy, terrain and weather, troops and support available

NBC . . . . nuclear, biological, chemical
NCO . . . . noncommissioned officer

OTL . . . . overserver-target line

RCA . . . . riot control agents
ROE . . . . rules of engagement

SMAW . . . . shoulder-launched multipurpose assault weapon
SMEAC . . . . situation, mission, execution, administration and logistics, and command and signal
SOP . . . . standing operating procedures
SPOTREP . . . . spot report

TTP . . . . tactics, techniques, and procedures

APPENDIX F. REFERENCES

Joint Publication (JP)

1-02 DOD Dictionary of Military and Associated Terms

Fleet Marine Force Manual (FMFM)

6-5 Marine Rifle Squad (proposed MCWP 3-11.2)

Marine Corps Warfighting Publications (MCWP)

3-15.3 Scout Sniping (under development)

3-35.3 Military Operations on Urbanized Terrain (MOUT)

3-41.2 Raids (under development)

Marine Corps Reference Publications (MCRP)

2-15.3A Reconnaissance Patrol Leader's Planning Handbook (under development)

2-15.3B Reconnaissance Reports Guide

3-02C Water Survival Handbook (under development, currently exists as Fleet Marine Force Manual 0-13)

5-12A Operational Terms and Graphics

5-12C Marine Corps Supplement to the DOD Dictionary of Military and Associated Terms

Made in the USA
Monee, IL
05 November 2026

67811402R00266